T0099987

METHODS IN MOLECULAR BIOLOGY

Series Editor
John M. Walker
School of Life Sciences
University of Hertfordshire
Hatfield, Hertfordshire, AL10 9AB, UK

For further volumes:
http://www.springer.com/series/7651

Plant-Pathogen Interactions

Methods and Protocols

Second Edition

Edited by

Paul Birch

Dundee Effector Consortium, Cell and Molecular Sciences, The James Hutton Institute, Dundee, UK;
Division of Plant Sciences, College of Life Sciences, University of Dundee (at JHI), Dundee, UK

John T. Jones

Cell and Molecular Sciences, The James Hutton Institute, Dundee, UK

Jorunn I.B. Bos

Cell and Molecular Sciences, The James Hutton Institute, Dundee UK;
Division of Plant Sciences, College of Life Sciences, University of Dundee, Dundee UK

Humana Press

Editors
Paul Birch
Dundee Effector Consortium
Cell and Molecular Sciences
The James Hutton Institute
Dundee, UK

Division of Plant Sciences
College of Life Sciences
University of Dundee (at JHI)
Dundee, UK

Jorunn I.B. Bos
Cell and Molecular Sciences
The James Hutton Institute
Dundee, UK

Division of Plant Sciences
College of Life Sciences
University of Dundee
Dundee, UK

John T. Jones
Cell and Molecular Sciences
The James Hutton Institute
Dundee, UK

Additional material to this book can be downloaded from http://extras.springer.com

ISSN 1064-3745   ISSN 1940-6029 (electronic)
ISBN 978-1-62703-985-7   ISBN 978-1-62703-986-4 (eBook)
DOI 10.1007/978-1-62703-986-4
Springer New York Heidelberg Dordrecht London

Library of Congress Control Number: 2014933054

Cover illustration: TdTomato-tagged P. *infestans* infection surrounding a plant cell expressing a nuclear-located GFP-fusion. Numerous haustoria are evident as small projections from the hyphae. The plant plasma membrane separates the pathogen from the nucleus but is not visible as it is not fluorescently labelled. The image is a 3D shadow projection of a confocal Z series.
Photo credit for the front cover image goes to Petra Boevink.

Printed on acid-free paper

Humana Press is a brand of Springer
Springer is part of Springer Science+Business Media (www.springer.com)

Preface

Over the past decade a major paradigm shift has taken place, from studying disease resistance in plants to investigating the roles that plant pathogen effectors play in suppressing, triggering, or otherwise manipulating plant defenses. Effectors are secreted proteins or other molecules that can act either inside or outside plant cells. Many effectors are thought to be required for suppressing Pattern-Triggered Immunity (PTI), the front line of inducible plant defense. However, they can also be the targets for resistance proteins, leading to the activation of Effector-Triggered Immunity (ETI), making them central players in dictating the outcomes of plant–pathogen interactions. Effectors and their functions are being studied in their own right. However, they are emerging as major tools to dissect host defense pathways and as primary targets to develop new screens for host disease resistance genes. Effectors may also play other important roles in determining the success of a pathogen that are not necessarily related to suppression of host defenses. For example, effectors may be important for manipulation of the host metabolism to provide food to the pathogen. This is the case for the biotrophic plant-parasitic nematodes, which need to induce the formation of large and complex feeding structures in order to obtain nutrients, but is also likely to be true for other plant pathogens.

The past 3 years have seen an explosion in available plant pathogen genome sequences, revealing the blueprints for host interactions and the repertoires of effectors needed to overcome host immunity. Increased access to high-throughput sequencing platforms means that this pattern will continue and that generating genome and transcriptome sequences, and all the subsequent benefits of access to functional genomics approaches, will be achievable for most pathogens. Genomics has driven effector searches in eukaryotic pests and pathogens (such as fungi, oomycetes, nematodes, aphids). New bioinformatic methods for genome assembly, annotation, comparison and mining have emerged, indicating the phenomenal dynamics in genome evolution, and flooding labs with effector candidates to study interactions and functions. Chapter 1 (Cock and Pritchard) describes a Galaxy platform and workflows to identify candidate effectors from genome sequences, and Chapter 2 (Reid and Jones) describes approaches to identify effector candidates from expression data, applied to nematode pests. Chapter 3 (Saunders et al.) presents methods to analyze and visualize genome architecture, indicating gene-rich and -sparse regions. It allows researchers to portray patterns of gene expression, nucleotide polymorphisms, and the relative locations of effector candidates in the context of overall genome architecture. Chapter 4 (Pritchard and Broadhurst) takes a timely and cautionary look at the statistics of candidate effector prediction, and provides strategies to assist in improving design and evaluation of effector classifiers.

New cell biology approaches have been developed to image the molecular processes underlying plant–pathogen interactions. Chapter 5 (Beck et al.) presents a high-throughput method to visualize the earliest stages of PTI and, in particular, to quantify the dynamics of endocytic trafficking following activation of pattern recognition receptors. Cell biology is also an increasingly important tool to study effector delivery, subcellular localization, and interactions with host target proteins. In Chapter 6, Boevink et al. describe the use of bimolecular fluorescence complementation to study *in planta* interactions, or close proximity, between pathogen effector proteins and their "target" proteins in the host cell. Chapter 7 (Takemoto and Jones) describes a rapid procedure for particle bombardment-mediated transient expression of fluorescently tagged proteins in leaf epidermal cells. This procedure is applied to investigate subcellular localization of resistance proteins, and to identify associated targeting signals. The method also lends itself to detection of pathogen effector protease activities directed against target proteins in the plant cell and analysis of protease recognition sites within these target proteins. In Chapter 8 (Garnica and Rathjen) a method is presented for rapid purification of fungal haustoria, structures from which effectors are delivered inside plant cells. They describe a new technique which combines initial gradient purification of haustoria with flow-sorting based on labeling of haustoria with fluorescent Concanavalin A.

To study effector functions, methods have been developed to mutate them, or manipulate their expression, spawning techniques to study the effects of such changes on both host and pathogen performance. Chapter 9 (Elling and Jones) describes the use of RNA interference (RNAi) to knock down the expression of specific effector candidate genes in plant-parasitic cyst and root-knot nematodes in order to investigate their impact on host interactions. Plant-mediated RNAi is described in Chapter 10 (Coleman et al.) as an approach to knock down expression of candidate effectors in aphids, and Chapter 11 (Rodriguez et al.) presents a method to identify effectors that, when transiently expressed *in planta*, have an impact on aphid performance. Chapter 12 (Tomé et al.) provides a method to quantify colonization of plant material by an obligate biotrophic oomycete (*Hyaloperonospora arabidopsidis; Hpa*) pathogen, which lends itself to evaluation of the contributions of *Hpa* effectors to pathogenicity. Chapter 13 (Ayliffe et al.) describes a general approach to quantify fungal colonization (applied to the wheat pathogen *Puccinia graminis* f.sp. *tritici*), based upon the specific binding of the plant lectin wheat germ agglutinin to fungal chitin.

New methods are presented to study the functions of defense-associated proteins in plant hosts. Zhang and Thomma (Chapter 14) describe the methodology for *Tobacco rattle virus* (TRV)-based VIGS in *Nicotiana tabacum*. Following coexpression of the tomato immune receptor Ve1 and the corresponding *Verticillium* effector Ave1 they show how the VIGS approach can be used as a rapid system for assessing the requirement of candidate plant genes for Ve1-mediated immune signaling. Hong and van der Hoorn (Chapter 15) describe the use of "click-chemistry" to profile serine hydrolase activities in the apoplast of *Nicotiana benthamiana* challenged with *Pseudomonas syringae* p.v. *tomato* DC3000.

There is considerable interest in finding host targets of pathogen effectors as this helps to develop an understanding of how these proteins promote host susceptibility and disease. Steinbrenner et al. (Chapter 16) present a rapid co-immunoprecipitation protocol to identify effector–host protein complexes *in planta*. To explore the roles of effectors in suppressing the earliest events in PTI, Fraiture et al. (Chapter 17) describe a medium-throughput method to identify effectors that prevent activation of mitogen-activated protein kinases and upregulation of early marker genes in tomato mesophyll protoplasts.

As effector targets are identified, and effector functions are revealed, structural analysis of effectors in relation to function is an emerging area aimed at determining the detailed molecular basis of how these proteins manipulate host processes. Hughes and Banfield (Chapter 18) present a medium-throughput protocol for expression testing oomycete RXLR effectors in *Escherichia coli*, followed by methods to purify and crystallize soluble effector protein. The methods help investigators to fully assess *E. coli* as a host for soluble protein production before considering alternative hosts for heterologous protein expression.

Effector availability has spurred the development of new approaches to screen for durable disease resistance genes in host plants. This, in turn, has promoted the conception of techniques to rapidly accelerate *R* gene discovery. Du and Vleeshouwers (Chapter 19) draw on their extensive experience of "effectoromics" in recent years to share tips, do's, and don'ts of effector transient expression in host germplasm to seek responses indicative of ETI. Kanzaki et al. (Chapter 20) describe a rice protoplast cell death assay to identify candidate effectors based on their avirulence activities from *Magnaporthe oryzae*, and Upadhyaya et al. (Chapter 21) present the use of a bacterial type III secretion system to assay the functions of, and responses to, fungal effectors in cereals. Finally, Jupe et al. (Chapter 22) detail the use of "capture arrays" to annotate resistance genes in plant genomes, and to accelerate the discovery of resistance genes in combination with bulked segregant analysis.

In conclusion, this volume covers a breadth of new techniques (bioinformatics, cell biology, protein structural, biochemical, and functional assays) developed to identify and characterize effectors and to study their impacts on host immunity and their role in pathogen biology. It presents protocols to identify avirulence and resistance genes and new methods to investigate the roles of effector targets and other defense-associated proteins in plant immunity.

Dundee, UK

Paul Birch
John T. Jones
Jorunn I.B. Bos

Contents

Contributors

MICHAEL AYLIFFE • *CSIRO Plant Industry, Canberra, ACT, Australia*

MARK J. BANFIELD • *Department of Biological Chemistry, John Innes Centre, Norwich Research Park, Norwich, UK*

MARTINA BECK • *The Sainsbury Laboratory, Norwich, UK*

JIM L. BEYNON • *Warwick Systems Biology, School of Life Sciences, Warwick University, Coventry, UK*

PAUL BIRCH • *Dundee Effector Consortium, Cell and Molecular Sciences, The James Hutton Institute, Dundee, UK; Division of Plant Sciences, College of Life Sciences, University of Dundee (at JHI), Dundee, UK*

PETRA BOEVINK • *Dundee Effector Consortium, Cell and Molecular Sciences, James Hutton Institute, Dundee, UK*

JORUNN I.B. BOS • *Cell and Molecular Sciences, The James Hutton Institute, Dundee, UK; Division of Plant Sciences, College of Life Sciences, University of Dundee, Dundee, UK*

DAVID BROADHURST • *Department of Medicine, Katz Group Centre for Pharmacy & Health, University of Alberta, Edmonton, AB, Canada*

FRÉDÉRIC BRUNNER • *Department of Plant Biochemistry, Centre for Plant Molecular Biology, Eberhard Karls University, Tübingen, Germany*

TATYANA BUKHAROVA • *Dundee Effector Consortium, Cell and Molecular Sciences, James Hutton Institute, Dundee, UK; Division of Plant Sciences, College of Life Sciences, University of Dundee (at JHI), Dundee, UK*

XINWEI CHEN • *Dundee Effector Consortium, CMS, The James Hutton Institute, Dundee, UK*

PETER J.A. COCK • *Information and Computational Sciences, The James Hutton Institute, Dundee, UK*

ALEXANDER D. COLEMAN • *Department of Cell and Developmental Biology, The John Innes Centre, Norwich, UK*

PETER N. DODDS • *CSIRO Plant Industry, Canberra, ACT, Australia*

IAN DRY • *CSIRO Plant Industry, Glen Osmond, SA, Australia*

JUAN DU • *Wageningen UR Plant Breeding, Wageningen University & Research Centre, Wageningen, The Netherlands; Key Laboratory of Horticultural Plant Biology (Huazhong Agricultural University), Ministry of Education, Wuhan, Hubei, People's Republic of China; National Centre for Vegetable Improvement (Central China), Huazhong Agricultural University, Wuhan, Hubei, People's Republic of China; Potato Engineering and Technology Research Centre of Hubei Province, Huazhong Agricultural University, Wuhan, Hubei, People's Republic of China*

MATTHEW ELDRIDGE • *School of Life Sciences, Warwick University, Coventry, UK*

AXEL A. ELLING • *Department of Plant Pathology, Washington State University, Pullman, WA, USA*

JEFFERY G. ELLIS • *CSIRO Plant Industry, Canberra, ACT, Australia*

STEFAN ENGELHARDT • *Dundee Effector Consortium, Cell and Molecular Sciences, James Hutton Institute, Dundee, UK; Division of Plant Sciences, College of Life Sciences, University of Dundee (at JHI), Dundee, UK*

CHRISTINE FAULKNER • *The Sainsbury Laboratory, Norwich, UK; Faculty of Health and Life Sciences, Oxford Brookes University, Oxford, UK*

ANGELA FEECHAN • *CSIRO Plant Industry, Glen Osmond, SA, Australia*

MALOU FRAITURE • *Department of Plant Biochemistry, Centre for Plant Molecular Biology, Eberhard Karls University, Tübingen, Germany*

DIANA P. GARNICA • *The Australian National University, Research School of Biology, Linnaeus Way, ACT, Australia*

INGO HEIN • *Dundee Effector Consortium, CMS, The James Hutton Institute, Dundee, UK*

SASKIA A. HOGENHOUT • *Department of Cell and Developmental Biology, The John Innes Centre, Norwich, UK*

TRAM NGOC HONG • *The Plant Chemetics Lab, Max Planck Institute for Plant Breeding Research, Cologne, Germany*

RICHARD K. HUGHES • *Department of Biological Chemistry, John Innes Centre, Norwich Research Park, Norwich, UK*

JONATHAN D.G. JONES • *The Sainsbury Laboratory, Norwich, UK*

JOHN T. JONES • *Cell and Molecular Sciences, The James Hutton Institute, Dundee, UK*

DAVID A. JONES • *Plant Science Division, The Australian National University, Research School of Biology, Canberra, ACT, Australia*

FLORIAN JUPE • *The Sainsbury Laboratory, Norwich, UK*

SOPHIEN KAMOUN • *The Sainsbury Laboratory, Norwich, UK*

HIROYUKI KANZAKI • *Iwate Biotechnology Research Center, Kitakami, Iwate, Japan*

EVANS LAGUDAH • *CSIRO Plant Industry, Canberra, ACT, Australia*

HAZEL MCLELLAN • *Dundee Effector Consortium, Cell and Molecular Sciences, James Hutton Institute, Dundee, UK; Division of Plant Sciences, College of Life Sciences, University of Dundee (at JHI), Dundee, UK*

SAMBASIVAM K. PERIYANNAN • *CSIRO Plant Industry, Canberra, ACT, Australia*

MARCO PITINO • *Department of Cell and Developmental Biology, The John Innes Centre, Norwich, UK*

LEIGHTON PRITCHARD • *Information and Computational Sciences, The James Hutton Institute, Dundee, UK*

ANTHONY PRYOR • *CSIRO Plant Industry, Canberra, ACT, Australia*

SYLVAIN RAFFAELE • *Laboratoire des Interactions Plantes-Microorganismes (LIPM), UMR441 INRA–UMR2594 CNRS, Castanet-Tolosan, France*

JOHN P. RATHJEN • *The Australian National University, Research School of Biology, Linnaeus Way, ACT, Australia*

ADAM J. REID • *Parasite Genomics, Wellcome Trust Sanger Institute, Cambridge, UK*

SILKE ROBATZEK • *The Sainsbury Laboratory, Norwich, UK*

PATRICIA A. RODRIGUEZ • *Cell and Molecular Sciences Group, James Hutton Institute, Dundee, UK*

HIROMASA SAITOH • *Iwate Biotechnology Research Center, Kitakami, Iwate, Japan*

DIANE G.O. SAUNDERS • *The Sainsbury Laboratory, Norwich, UK*

ULRIKE SCHUMANN • *CSIRO Plant Industry, Canberra, ACT, Australia*

JENS STEINBRENNER • *School of Life Sciences, Warwick University, Coventry, UK*

DAIGO TAKEMOTO • *Plant Pathology Laboratory, Graduate School of Bioagricultural Sciences, Nagoya University, Nagoya, Japan*

MULUNEH TAMIRU • *Iwate Biotechnology Research Center, Kitakami, Iwate, Japan*

RYOHEI TERAUCHI • *Iwate Biotechnology Research Center, Kitakami, Iwate, Japan*

BART P.H.J. THOMMA • *Laboratory of Phytopathology, Wageningen University, Wageningen, The Netherlands*

DANIEL F.A. TOMÉ • *School of Life Sciences, Warwick University, Coventry, UK*

NARAYANA M. UPADHYAYA • *CSIRO Plant Industry, Canberra, ACT, Australia*

RENIER A.L. VAN DER HOORN • *The Plant Chemetics Laboratory, Department of Plant Sciences, University of Oxford, Oxford, UK*

WALTER VERWEIJ • *The Genome Analysis Centre, Norwich, UK*

VIVIANNE G.A.A. VLEESHOUWERS • *Wageningen UR Plant Breeding, Wageningen University & Research Centre, Wageningen, The Netherlands*

JOE WIN • *The Sainsbury Laboratory, Norwich, UK*

KAMEL WITEK • *The Sainsbury Laboratory, Norwich, UK*

KENTARO YOSHIDA • *Iwate Biotechnology Research Center, Kitakami, Iwate, Japan*

ZHAO ZHANG • *Laboratory of Phytopathology, Wageningen University, Wageningen, The Netherlands*

XIANGZI ZHENG • *Department of Plant Biochemistry, Centre for Plant Molecular Biology, Eberhard Karls University, Tübingen, Germany*

JI ZHOU • *The Sainsbury Laboratory, Norwich, UK*

Part I

Bioinformatics

Chapter 1

Galaxy as a Platform for Identifying Candidate Pathogen Effectors

Peter J.A. Cock and Leighton Pritchard

Abstract

The Galaxy web platform provides an integrated system for its users to run multiple computational tools, linking their output in order to perform sophisticated analysis without requiring any programming or installation of software beyond a modern web-browser. Analyses can be saved as reusable workflows, and shared with other Galaxy users, allowing them to easily perform the same analysis or protocol on their own data.

We describe example Galaxy workflows for the identification of candidate pathogen effector proteins. Our main example focuses on nematode plant pathogens where signal peptide and transmembrane prediction tools are used to identify predicted secreted proteins.

Key words Effectors, Workflow, Pipeline, Galaxy, Classification, Bioinformatics, Sequence analysis, Genomics, High-throughput screening

1 Introduction

1.1 Reproducible Bioinformatics Pipelines

Computational biology enables higher fidelity sharing of reusable protocols than a human readable written description of the method in a scientific paper. A computer program or script *in itself* can define the method completely and is easily and unambiguously shared.

Much exploratory bioinformatics analysis is done on an ad hoc basis, using a combination of single-use scripts, command line tools, web-services, or applications, often with some simple script programming. Good note taking, in the form of a computational biologist's electronic lab book or otherwise, is essential when trying to publish the results of such an exploratory process. Inadequate note taking is a possible contributory factor in the common complaint that many published Methods Sections do not cover the bioinformatics in enough depth for the work to be reproducible.

To reach the goal of a fully reproducible computational method, authors should minimally provide: their software, with clear licensing terms, and ideally as open source to encourage collaborative contributions; clear installation instructions, including a

Paul Birch et al. (eds.), *Plant-Pathogen Interactions: Methods and Protocols*, Methods in Molecular Biology, vol. 1127, DOI 10.1007/978-1-62703-986-4_1, © Springer Science+Business Media New York 2014

complete list of dependencies with version information; usage documentation; and sample data and tests, so that users may verify the installation process and that it works as expected. It is the authors' experience that only a minority of published scientific software reaches this level, possibly for the very pragmatic reasons that this takes additional time and work, and is not typically required by journals or referees.

1.2 Galaxy

Sharing of reproducible workflows and tools for computational biology is one of the key goals of the Galaxy Project [1, 2]. However, this is sometimes pragmatically seen as secondary to the practical benefits of a Galaxy server: making a range of tools available to non-bioinformaticians via a web-browser, without the need for them to install and run tools, or learn any scripting or programming language to chain tools together.

Basic installations of Galaxy provide wrappers for many tools that reflect the interests of the target user base of the public Galaxy Instance (http://usegalaxy.org). These tools are predominantly associated with high throughput sequencing analysis, such as read mapping. However, Galaxy is an extensible platform into which new tools and data-type definitions (i.e., additional file formats) can be added. The Galaxy Tool Shed (http://toolshed.g2.bx.psu.edu/) acts as a community hub for sharing these add-ons, allowing research groups or institutes to set up their own local Galaxy installation, and customize it by the addition of further tools (*see* **Note 1**). Galaxy may be installed on local physical hardware or on rented machines using cloud computing [3].

Galaxy also allows for the provision of reference datasets, such as genome sequences and their annotations, and quantitative data, in a central location. Users may be anonymous, or can be identified individually by a login process, and allocated membership of one or more "groups." Access to sensitive data can be restricted only to the appropriate groups or individuals.

Since 2011 we have been running an internal Galaxy installation at The James Hutton Institute, connected to our existing local computer cluster. This setup has increased the use of centrally funded resources and helped justify continued investment. However, the primary goal is user empowerment by making relevant tools and pipelines available in an easy to use way. MacLean and Kamoun [4] describe a similar situation at The Sainsbury Laboratory (Norwich, UK).

Our local Galaxy setup has initially been targeted at plant pathogen researchers, and our tool selection reflects this. We have focused on gene and protein analysis, with the applied goal of identifying candidate pathogen effectors. This has required considerable investment in developing Galaxy wrappers for existing tools of relevance, and enabling the workflows we discuss here [5].

1.3 Candidate Pathogen Effectors

Although the term "effector" is widely used in host-pathogen interaction studies, precise definitions vary. Here we use the term to mean any protein synthesized by a pathogen that is exported to a potential host, which has the effect of making the host environment more beneficial to the pathogen. This includes biochemical effector functions such as DNA binding or protein–protein interaction to modify host gene expression; for example, to modify host metabolism or suppress host defenses.

When working with a newly sequenced pathogen, functional annotation of candidate effector proteins is often a priority. The most appropriate prediction methods may differ between organism, and effector class, but they generally use a combination of: predicted localization (e.g., secreted proteins); biochemical function (e.g., protein domains); evidence of differential expression; and/or similarity to known effectors. This is a classification problem, and all methods produce a list of positive predictions (or, more generally, a score for each candidate). The output of any model will include false positives and omit false negatives, and should be interpreted with care (*see* Chapter 4 in this volume).

In this chapter we describe simple effector protein prediction pipelines as examples of reusable workflows within the Galaxy platform.

1.4 Secreted Proteins

The potato cyst nematode (*Globodera pallida*) effector finding protocol in Jones et al. [6] was one of the first workflows we wanted to reconstruct within Galaxy. This plant-parasitic nematode possesses glands that secrete proteins that are injected into the plant host through a hypodermic needle-like stylet. There was no known sequence or motif reliably associated with effector function for this pathogen, so in order to identify candidate effector proteins we instead looked for sequences characteristic of secreted proteins, of which effectors are a subset. The published method takes the organism's predicted protein complement (as a multiple entry FASTA file), and queries it to identify proteins that contain predicted signal peptides but do not contain predicted transmembrane domains, using the command line tools SignalP v3.0 [7] and TMHMM v2.0 [8].

One of the documented cases of false positive transmembrane domain predictions by TMHMM v2.0 is wrongly identifying a signal peptide as a transmembrane domain [8]. We initially therefore provided TMHMM with input that comprised mature peptides with the predicted signal peptide removed [6]. Galaxy contains a number of tools for manipulating sequences using genomic interval files which can, in principle, be generalized to work on arbitrary regions of gene or protein sequences. We hope to work on this with the Galaxy team, but currently our workflow provides TMHMM with full length proteins as input, and any false-positive transmembrane predictions result in a reduced set of predicted candidate effectors. In practice this is not a significant problem, as there are typically more candidates than it is feasible to investigate experimentally.

After wrapping the required tools (SignalP and TMHMM), and implementing a sequence-filtering tool, this slightly simplified version of the protocol was straightforward to implement as a pipeline within Galaxy. This workflow was used in the genome project for *Bursaphelenchus xylophilus* [9], and is currently being used for other nematode genome projects and in as yet unpublished work on other plant pathogens, including bacteria, aphids, oomycetes, and fungi.

Reid and Jones (*see* Chapter 2, this volume) describe a contrasting approach to implementation of this pipeline using Unix/Linux command line tools such as `grep`. This is an extremely powerful way to explore data, but unlikely to appeal to the majority of non-bioinformaticians, whereas running a pre-existing Galaxy Workflow is less intimidating.

2 Materials

To follow the methods described, you will need a local Galaxy server where our additional Galaxy tools and workflows and the underlying command line tools have been installed. It is not possible for an ordinary Galaxy user to do this, so tool installation must be handled by the local Galaxy administrator and/or Linux system administrator.

Our Galaxy wrappers for SignalP and TMHMM, and sequence-filtering tool are all open source licensed, and are available for automated installation from the main Galaxy Tool Shed in the following URLs:

- http://toolshed.g2.bx.psu.edu/view/peterjc/secreted_protein_workflow

- http://toolshed.g2.bx.psu.edu/view/peterjc/tmhmm_and_signalp

- http://toolshed.g2.bx.psu.edu/view/peterjc/seq_filter_by_id

These Galaxy add-ons include documentation and unit tests to allow verification that they are installed and working as intended.

Regrettably, the SignalP v3.0 and TMHMM v2.0 tools were released under a fee-based license, or a free non-exclusive license to use the software for internal research purposes only:

- http://www.cbs.dtu.dk/services/SignalP-3.0/

- http://www.cbs.dtu.dk/services/TMHMM/

This prevents anyone else from packaging these tools to offer automated installation. The tools must therefore be licensed, downloaded, and installed by a local administrator such that the binaries `signalp` and `tmhmm` are on the system $PATH.

3   Methods

Here we describe how to perform this workflow within Galaxy manually (step by step), using the shared workflow (automated), and how to recreate the saved workflow. We will close with some related effector finding approaches within Galaxy.

3.1   Performing the Secreted Protein Pipeline Manually Within Galaxy

This simple secreted protein workflow can be performed manually within Galaxy with eight steps as follows:

1. Upload or import a protein FASTA file of interest (Fig. 1).

2. Run SignalP 3.0 on the FASTA file from **step 1**, using the appropriate organism type, and a length cutoff of 60 amino acids (Fig. 2). This produces a tabular file of predictions.

3. Filter the output from **step 2** for positive results using the condition c14=='Y' or c15=='S', meaning the neural network model D-score has 'Y' for yes in column 14, and/or the HMM has 'S' for secreted in column 14.

4. Filter the original FASTA file from **step 1** with the "Filter sequences by ID" tool, selecting column c1 to use the identifiers from **step 3**. Select the "Just positive matches" option in the drop down list, to give a FASTA file of proteins that contain a predicted signal peptide.

Fig. 1 Galaxy with a protein FASTA file loaded as the first dataset (*green box* on *right*), displayed in the *central panel* (by clicking on the "eye" icon on the dataset). The *left-hand tool column* has been scrolled to show the workflows entry (Color figure online; *see* **Note 2**)

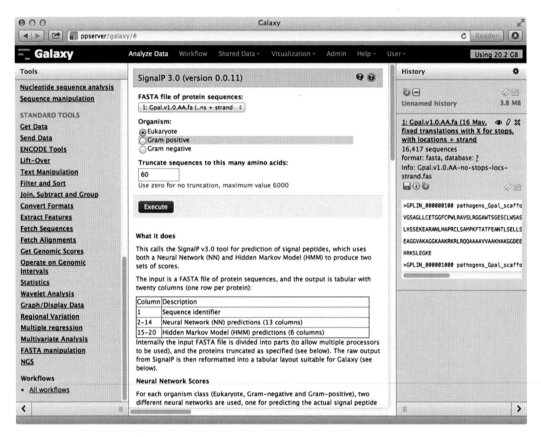

Fig. 2 Galaxy showing the SignalP 3.0 tool (in *central panel*), accessed from the list of tools in the *left-hand column*. This tool has three input parameters, a FASTA file of proteins (defaulting to the most recent FASTA file in the history, shown in the *right-hand column*), the organism type (selecting from three options), and a sequence truncation length (defaulting to 70 amino acids). This tool would be run on the selected data using the *blue* "execute" button (Color figure online)

5. Run TMHMM 2.0 on the FASTA file from **step 4**, selecting the appropriate organism type (e.g., eukaryote).

6. Filter the output from **step 5** for negative results using the condition c5==0 (zero transmembrane helices in column five).

7. Filter the FASTA file from **step 4** with the "Filter sequences by ID" tool, again with the "Just positive matches" option, and selecting column c1 to use the identifiers from **step 6**, giving a FASTA file of proteins with a predicted signal peptide but no predicted transmembrane domain.

8. (Optionally) delete the no longer required intermediate files from **steps 2** to **6**, leaving just the input and candidate secreted protein FASTA files (Fig. 3).

The most complex step here is deciding how to interpret the SignalP output. This produces two independent sets of predictions using a neural network and hidden Markov model (HMM), with

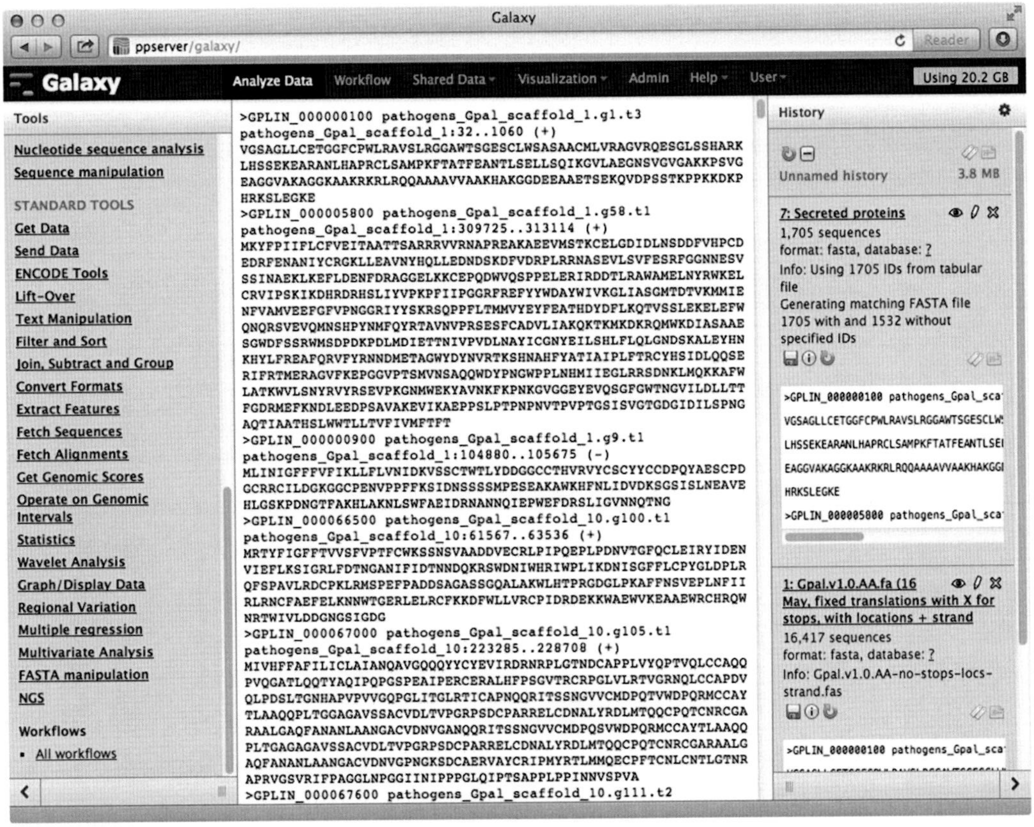

Fig. 3 Galaxy showing the results of the completed workflow. Here the input protein FASTA file was 16,417 predicted proteins from the nematode *Globodera pallida* (history entry 1), and the output was a protein FASTA file of 1,705 candidate secreted proteins (history entry 7)

both a yes/no categorization (which we use here) and an associated score or probability (which could be used for a more inclusive filter).

Having performed a task like this once in Galaxy, it can be turned into a workflow, a process discussed in more detail below.

3.2 Running the Secreted Protein Workflow Within Galaxy

Once created, a workflow can be shared within the local Galaxy instance (for all users or with specific users), or downloaded for transfer to a separate Galaxy instance, or inclusion on the Galaxy Tool Shed. Reusing this workflow within Galaxy is then extremely straightforward:

1. Upload or import protein FASTA file of interest (Fig. 1).

2. Select this workflow from the workflows list, accessed from the bottom of the left-hand tools pane (Fig. 4).

3. Confirm the FASTA file from **step 1** as the input dataset, pick the organism type for SignalP, and click on "Run workflow," then wait for the results (Fig. 3).

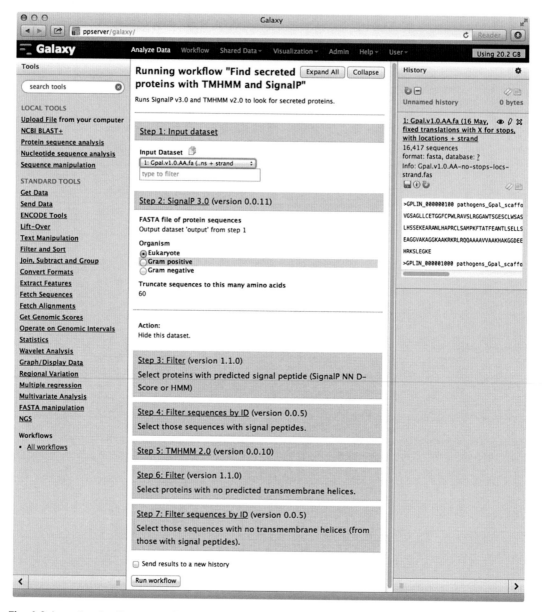

Fig. 4 Galaxy showing the secreted protein workflow (in the *central panel*), accessed via the "All workflows" link at the bottom of the tools list (*left column*). The input dataset and SignalP organism type can be altered, and the workflow executed using the "run workflow" button

Galaxy will then automatically run SignalP, filter the output, run TMHMM, filter the output, and delete the intermediate files without further user interactions. This example is only moderately complicated, but it is clear that rerunning the workflow requires less expertise or manual intervention, removing two potential sources for error, than performing it step by step, even with detailed instructions.

3.3 Recreating the Secreted Protein Workflow Within Galaxy

This workflow has been provided on the Galaxy Tool Shed so that it can be installed and reused elsewhere, but we will now use it as an example of how to create your own workflow within Galaxy.

The simplest way to create a new workflow within Galaxy is to start from a new empty history (workspace), import sample input data, and perform the necessary steps one by one by hand (as described above). Then, select "Extract Workflow" from the history menu. The starting data files will automatically be marked as workflow inputs, and by default, all the current tool parameters will be saved as is.

In this case we choose to mark the SignalP organism type as a workflow parameter (rather than always using the same option). This means when the workflow is run, the user must select both their input protein FASTA file and the appropriate organism type (eukaryotes, Gram positive or Gram negative prokaryotes), making the workflow more general than if it only applied to eukaryotes.

The workflow can then be edited within Galaxy (Fig. 5), which allows refinements such as assigning names to the data files produced. The workflow editor can also be used to add additional tool steps, or alternatively to build a workflow up piece by piece.

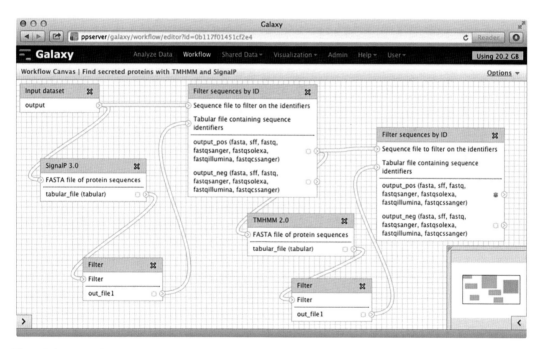

Fig. 5 Galaxy workflow editor showing the overview of the secreted protein workflow. Due to space limitations, only the *central panel* is shown, which displays each tool as a box with data file inputs on the *left* and output on the *right*, with the connecting *curved lines* indicating where the output of one tool becomes the input of another. Using the mouse the visual layout can be adjusted by dragging the tools, and the connections themselves edited. The (collapsed) *left-hand column* would show the available Galaxy tools that can be added to the workflow to perform additional steps. The (collapsed) *right-hand column* allows the existing workflow steps to be edited, for example, to add text annotation (*see* Fig. 4, for examples)

3.4 Identifying Functional Domains: The RxLR Translocation Motif

Many of the tools we have wrapped for use within Galaxy predict effector translocation, which can be combined into pipelines along the lines above.

One specific example is a tool implementing three different protocols for the identification of RXLR motifs found in oomycetes [10–12]. The tool takes a protein FASTA file as input and gives a tabular output file with one line per input sequence. A simple workflow would be to generate a FASTA file of the sequences with positive matches, using the same filtering tools as the previous example. Performed manually:

1. Upload or import the protein FASTA file of interest.
2. Run the "RXLR Motifs" tool on the FASTA file from **step 1**, selecting an appropriate classifier model.
3. Filter the output for positive results using the expression c2=='Y', indicating a positive prediction as described in the RXLR tool's documentation shown in **step 2**.
4. Filter the original FASTA file from **step 1** using the identifiers from **step 3**, giving a FASTA file of proteins with a predicted RXLR motif.
5. (Optionally) delete the no longer required intermediate files from **steps 2** and **3**.

When saving this manual analysis as a workflow, we might opt to mark the choice of RXLR model as a workflow parameter, meaning it must be chosen when applying the workflow to another dataset.

A more complicated variation could run all three RXLR models to compare their output with a Venn Diagram, demonstrated with the workflow http://toolshed.g2.bx.psu.edu/view/peterjc/rxlr_venn_workflow on the Galaxy Tool Shed. This shows visually that the Win et al. and Whisson et al. models give different subsets of the more inclusive Bhattacharjee et al. model (Fig. 6) [10–12].

3.5 Effector Finding Through Sequence Similarity

The NCBI BLAST tools are rightly one of the most commonly used tools in biology and are employed for a range of tasks. As early adopters of the Galaxy platform, out of our own needs we implemented the original BLAST+ [13] wrappers available on the Galaxy Tool Shed:

http://toolshed.g2.bx.psu.edu/view/devteam/ncbi_blast_plus

One extremely general idea for identifying novel effector proteins in a newly sequenced species or strain is to look for sequence similarity to known effectors in related organisms. Consider a project that produces a fresh transcriptome assembly (nucleotide), or predicted gene set from a draft genome (nucleotide or protein), and a separate set of known effectors from a sister organism (nucleotide or protein). The known effectors will be a FASTA file used as the BLAST queries, while the FASTA file of novel genes will be turned into a BLAST database. One of the standard BLAST tools (e.g.,

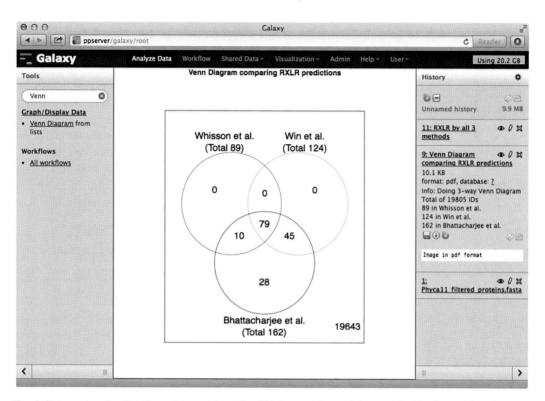

Fig. 6 Galaxy showing the Venn diagram from the RXLR example workflow applied to the predicted protein complement of *Phytophthora capsici*, showing visually that the Whisson et al. [9] and Win et al. [10] definitions are refinements of the original more inclusive Bhattacharjee et al. [11] definition

BLASTN or TBLASTX if both files are nucleotides) can then be used to look for matches, and these results can then be filtered by Galaxy according to criteria such as alignment length or bit-score.

Similarly, one might wish to identify potential orthologues by comparing the gene or protein set of two related organisms. With the exception of complex gene families, reciprocal best hits (RBH) using BLAST are a useful technique to filter sequence sets, and identify these relationships (subject to limitations discussed in ref. 14), and we are working on facilitating this analysis within Galaxy.

3.6 Sequence Selection with External Tools

Galaxy is useful when employed in combination with other software. A common example is where microarray analysis has been performed using a package like GeneSpring (Agilent Technologies, Inc.), yielding a list of candidate genes for further investigation. A common next step is to prepare a FASTA file of just those sequences, which can be done in Galaxy as follows:

1. Upload or import full gene set FASTA file.

2. Upload the tabular file of candidate gene identifiers.

3. Filter the original FASTA file from **step 1** using the identifiers from **step 2**, giving a FASTA file of the genes of interest.

4   Notes

1. A local Galaxy installation is worth considering as a way of deploying both existing third party tools and new programs developed in-house. If you have developed a novel tool for effector finding, wrapping it for use within Galaxy and sharing this on the Galaxy Tool Shed should prove an excellent way to make the method available to other researchers by including the URL in your manuscript. The same applies to new workflows built using existing Galaxy Tools; the workflow definition itself can be shared either as a supplementary file, or via the Galaxy Tool Shed with a unique URL.

   Our hope is that some of the simple effector finding protocols described here, and the associated Galaxy workflows, will be useful in themselves, but also as starting points or inspiration for more ambitious pipelines (using the Galaxy tools discussed here, or others).

2. The visual styling of Galaxy has changed over time, and so the details in the screenshots will date. Also, the contents of the left-hand column of tools will differ between Galaxy setups.

Acknowledgments

The James Hutton Institute receives funding from the Scottish Government.

References

1. Goecks J, Nekrutenko A, Taylor J, Galaxy Team (2010) Galaxy: a comprehensive approach for supporting accessible, reproducible, and transparent computational research in the life sciences. Genome Biol 11(8):R86

2. Blankenberg D, Kuster von G, Coraor N, Ananda G, Lazarus R, Mangan M, Nekrutenko A, Taylor J (2010) Galaxy: a web-based genome analysis tool for experimentalists. Curr Protoc Mol Biol Chapter 19:Unit 19.10.1–21

3. Afgan E, Baker D, Coraor N, Chapman B, Nekrutenko A, Taylor J (2010) Galaxy CloudMan: delivering cloud compute clusters. BMC Bioinforma 11(suppl 12):S4

4. MacLean D, Kamoun S (2012) Big data in small places. Nat Biotechnol 30(1):33–34

5. Cock PJ, Grüning BA, Paszkiewicz K, Pritchard L (2013) Galaxy tools and workflows for sequence analysis with applications in molecular plant pathology. PeerJ 1:e167 http://dx.doi.org/10.7717/peerj.167

6. Jones JT, Kumar A, Pylypenko LA, Thirugnanasambandam A, Castelli L, Chapman S, Cock PJA, Grenier E, Lilley CJ, Phillips MS, Blok VC (2009) Identification and functional characterization of effectors in expressed sequence tags from various life cycle stages of the potato cyst nematode *Globodera pallida*. Mol Plant Pathol 10(6):815–828

7. Bendtsen JD, Nielsen H, von Heijne G, Brunak S (2004) Improved prediction of signal peptides: SignalP 3.0. J Mol Biol 340(4): 783–795

8. Krogh A, Larsson B, von Heijne G, Sonnhammer E (2001) Predicting transmembrane protein topology with a hidden markov model: application to complete genomes. J Mol Biol 305:567–580

9. Kikuchi T, Cotton JA, Dalzell JJ, Hasegawa K, Kanzaki N, Mcveigh P, Takanashi T, Tsai IJ, Assefa SA, Cock PJA, Otto TD, Hunt M, Reid AJ, Sanchez-Flores A, Tsuchihara K, Yokoi T,

Larsson MC, Miwa J, Maule AG, Sahashi N, Jones JT, Berriman M (2011) Genomic insights into the origin of parasitism in the emerging plant pathogen *Bursaphelenchus xylophilus*. PLoS Pathog 7(9):e1002219

10. Whisson SC, Boevink PC, Moleleki L, Avrova AO, Morales JG, Gilroy EM, Armstrong MR, Grouffaud S, Van West P, Chapman S, Hein I, Toth IK, Pritchard L, Birch PRJ (2007) A translocation signal for delivery of oomycete effector proteins into host plant cells. Nature 450(7166):115–118

11. Win J, Morgan W, Bos J, Krasileva KV, Cano LM, Chaparro-Garcia A, Ammar R, Staskawicz BJ, Kamoun S (2007) Adaptive evolution has targeted the C-terminal domain of the RXLR effectors of plant pathogenic oomycetes. Plant Cell 19(8):2349–2369

12. Bhattacharjee S, Hiller NL, Liolios K, Win J, Kanneganti T-D, Young C, Kamoun S, Haldar K (2006) The malarial host-targeting signal is conserved in the Irish potato famine pathogen. PLoS Pathog 2(5):e50

13. Camacho C, Coulouris G, Avagyan V, Ma N, Papadopoulos J, Bealer K, Madden TL (2009) BLAST+: architecture and applications. BMC Bioinforma 10:421

14. Moreno-Hagelsieb G, Latimer K (2008) Choosing BLAST options for better detection of orthologs as reciprocal best hits. Bioinformatics 24(3):319–324

Chapter 2

Bioinformatic Analysis of Expression Data to Identify Effector Candidates

Adam J. Reid and John T. Jones

Abstract

Pathogens produce effectors that manipulate the host to the benefit of the pathogen. These effectors are often secreted proteins that are upregulated during the early phases of infection. These properties can be used to identify candidate effectors from genomes and transcriptomes of pathogens. Here we describe commonly used bioinformatic approaches that (1) allow identification of genes encoding predicted secreted proteins within a genome and (2) allow the identification of genes encoding predicted secreted proteins that are upregulated at important stages of the life cycle. Other approaches for bioinformatic identification of effector candidates, including OrthoMCL analysis to identify expanded gene families, are also described.

Key words Transcriptomics, OrthoMCL, Effector, Signal peptide

1 Introduction

Many studies on a wide variety of phylogenetically unrelated plant pathogens have shown that effectors are upregulated at specific life stages of the pathogen. For example, many *Phytophthora infestans* RXLRs are specifically upregulated during the biotrophic phase of infection [1]. In plant parasitic nematodes, several large-scale studies have shown that different pools of effectors are upregulated at different life stages, e.g., [2, 3]. This type of analysis can provide information about potential functional roles of nematode effectors. Those important in invasion, migration, and induction of the biotrophic feeding structure peak in expression soon after the invasive stage nematode emerges from dormancy. Other effectors have a role in suppression of host defenses and maintenance of the feeding structure and peak in expression during the established parasitic stages. These observations underpin the strategy of using bioinformatic analysis of expression data to identify effector candidates.

Paul Birch et al. (eds.), *Plant-Pathogen Interactions: Methods and Protocols*, Methods in Molecular Biology, vol. 1127, DOI 10.1007/978-1-62703-986-4_2, © Springer Science+Business Media New York 2014

In principle, bioinformatics approaches for effector candidate identification are relatively straightforward and based on effector candidates being defined as secreted proteins lacking a transmembrane domain that are upregulated at key life stages. However, it is important to note that applying an approach based on these two criteria will identify many predicted secreted proteins that are clearly not effectors. For example, when we apply this approach to an RNA-seq dataset from the potato cyst nematode *Globodera pallida*, these pipelines identify digestive proteinases and proteinaceous cuticle components that are produced in abundance after the nematode settles to feed and begins the molting cycle (P. Thorpe & J. Jones, unpublished). The researcher may therefore wish to add a BLAST search of all candidates that are identified in order to predict their functions based on sequence similarity and generate a priority list of candidates of interest. Upon applying a bioinformatics approach to identify effector candidates one can check whether previously identified effectors from the species being studied, or orthologues of effectors from related species, are detected to validate the results.

1.1 The Pros and Cons of RNA-seq Versus Microarrays

There are several good reasons for using microarrays, although as the technology matures RNA-seq is becoming more feasible for a larger range of applications. Microarray analysis is relatively cheap once a microarray platform has been established allowing many life stages and replicates to be analyzed. However, the costs associated with RNA-seq are decreasing as Illumina sequencing machines such as the HiSeq achieve greater yields. In addition, it has become feasible to run multiple samples in each lane by multiplexing, which can further reduce costs. While smaller quantities of RNA may be used for microarray analysis, improvements in RNA-seq library preparation may allow a reduction in the quantity of samples required in future. RNA-seq has several inherent advantages over microarray analysis. A microarray analysis will only ever analyze expression profiles of sequences that are present on the array. This may be an issue where only a limited cDNA dataset or a poorly annotated genome is available. By contrast, RNA-seq will identify all the expressed regions of a genome and is therefore not reliant on a cDNA dataset or detailed annotation. RNA-seq data can drastically improve annotation of unannotated or poorly annotated genomes as it clearly outlines the intron-exon structure of genes, even those expressed at low levels. A downside of RNA-seq is that it is less well established than microarray technology and produces much more data and thus requires a higher overhead in terms of informatics as well as hardware and data storage capacity.

In this chapter we provide examples of protocols used for enriching a genome scale dataset for effectors and describe how RNA-seq data have been used to further enrich this dataset.

2 Materials

Many standard protocols are available for purification of total RNA and mRNA, and these are not covered here. The integrity and purity of the RNA to be used for RNA-seq are of critical importance and need to be checked carefully before proceeding with this expensive technique. We have previously used a Bioanalyzer (Agilent) for this purpose.

Replication is essential for RNA-seq analysis. Ideally, three biological replicates are required as a minimum for each sample; several studies have shown that increasing the number of biological replicates will improve the accuracy of the analysis. It is worth emphasizing that biological replicates are required, rather than technical replicates. Biological replicates are independent collections of biological materials from separate runs of an experiment. These replicates need not be split up unless technical replicates are desired and should, ideally, not be pooled. Pooling can be used in cases where there is insufficient material for individual replicates; however, this may result in an unwarranted increase in power to detect differential expression due to an artificial reduction in biological variability.

There are two fundamentally different ways to begin analyzing RNA-seq data, one of which requires a reasonably well-annotated genome sequence and one which can be done without any reference to the genome. Here we describe the first, in which RNA-seq reads are mapped against a reference genome and the reads mapped to each known gene are counted. The second approach, using de novo transcript assembly from the reads, is described in ref. 4.

Requirements:

1. We assume here that gene models have been generated for the genome under consideration. Specifically, there are protein sequences in fasta format (for initial screening) and a GFF-formatted annotation of the genome sequence for subsequent expression analyses by RNA-seq.

2. Several stages of the life cycle need to have been interrogated by RNA-seq, including those where effectors are expected to be expressed and one or more where effectors are not thought to be expressed. For example, in the case of plant parasitic nematodes suitable timepoints would be the infectious J2 and parasitic stages (where effectors are likely to be expressed) and unhatched J2 (where effectors should not be expressed). An example of these types of comparisons can be found in the microarray analysis of *Heterodera glycines* [2].

3. Almost all bioinformatic analysis requires basic programming ability in Perl or a similar language. In particular, data will usually need to be reformatted. Furthermore, working with large datasets like RNA-seq data and performing BLAST searches against

large databases often requires the use of large computational resources in order to process in a reasonable time.

The example we use here is that of the potato cyst nematode, *Globodera pallida*, which was sequenced by the Wellcome Trust Sanger Institute in collaboration with the James Hutton Institute, The University of Leeds and Rothamsted Research [5]. To follow the example you will need the genome sequence, genome annotation, and predicted amino acid sequences for this genome. These are available from http://www.sanger.ac.uk/resources/downloads/helminths/globodera-pallida.html. You will also need the associated RNA-seq data, which is available from ArrayExpress. Alternatively we provide a subset of this data for pertinent life stages reduced to ten million reads per library to speed up mapping and subsequent analysis. This data can be downloaded from http://extras.springer.com/. A variety of software tools are used here and will need to be installed. We have provided information about where they can be downloaded from but not instructions or tips for installation. We assume that you are working in a Unix/Linux environment. Commands are shown in courier font and should be entered on the command line in a terminal.

3   Methods

The methods described here demonstrate first how to identify a subset of the protein sequences predicted from a genome (initial screening) based on features of effector candidates. RNA-seq data is then used to identify effector candidates upregulated at stages of the life cycle associated with parasitism.

3.1   Initial Identification of Candidate Genes

There are a variety of approaches that can be used for initial identification of candidates and these can be mixed and matched as appropriate. While many genes in the initial list will not be effectors and some genuine effectors may have been excluded, the aim is to use the full genome to predict a list of genes enriched for effector candidates for further downstream analyses.

3.1.1   Identification of the Pathogen Secretome: SignalP and TMHMM

The most commonly used enrichment approach is to determine the subset of genes whose protein sequences are predicted to contain an N-terminal signal peptide but that lack transmembrane domains. These are hallmarks of proteins that are trafficked via the Golgi-dependent pathway but that are not embedded in the cell membrane and are thus presumed to be secreted [6]. In some cases effectors may be exported through alternative pathways but these are not sufficiently well understood that we can identify their export signatures in the genome [6]. The researcher needs to be aware that using this strategy may, in some cases, be restrictive. In addition, predicting the correct start site for a protein can be computationally difficult. Genes may therefore be missing signal peptides erroneously due to incorrectly predicted start codons.

The most widely used tools for this analysis are SignalP ([7]; http://www.cbs.dtu.dk/services/SignalP/) and TMHMM ([8]; http://www.cbs.dtu.dk/services/TMHMM/). These tools identify signal peptides (signifying that the protein is exported) and transmembrane domains (signifying that the protein is membrane bound), respectively. These tools are available through webpages but the upper limit for submissions means that it is necessary to run a local installations for genome scale analysis. In practice, it is far better if all programs are installed locally and run, where possible, over multiple computers in order to reduce the analysis time. This also allows the generation of a bespoke pipeline which, once established, enshrines your protocol and can be easily run many times with different parameters and on new datasets to discover effectors in other species.

Several iterations of SignalP are available. SignalP 4.0 is recommended as this update is specifically designed to differentiate between signal peptides and N-terminal transmembrane domains. Note that SignalP imposes an upper limit of 10,000 sequences that can be searched at once. This can be adjusted in the script or you can split your sequences into batches. To run the program locally, use the following command:

```
signalp -t euk Gpal.v1.0.cds.fa > Gpal.v1.0.cds.sp
```

(Where the full list of sequences is in a file called Gpal. v1.0.cds. fa)

The output of this program will include an indication as to whether each protein is secreted. The list of sequences that are predicted to contain signal peptides can be obtained using this command:

```
grep Y Gpal.v1.0.cds.sp | grep -v "#" | awk -F
" " '{print $1}' > Gpal.v1.sp.ids
```

This shows that of 16,417 predicted genes in the *G. pallida* genome 1,897 are predicted to have a signal peptide.

In order to identify *G. pallida* proteins with a transmembrane domain TMHMM is run using this command:

```
tmhmm Gpal.v1.0.cds.fa > Gpal.v1.0.cds.tm
```

The list of proteins that are predicted to contain transmembrane domains can be obtained using the command:

```
grep TMhelix Gpal.v1.0.cds.tm | cut -f1 | sort
-u > Gpal.v1.tm.ids
```

This analysis shows that 3,541 proteins have transmembrane domains. Proteins that have signal peptides but that do not have transmembrane domains can be found from the two lists using the following command:

```
comm -23 Gpal.v1.sp.ids Gpal.v1.tm.ids > effec-
tor_candidates.ids
```

In this case almost all the genes with signal peptides lack transmembrane domains and we have identified 1,812 effector candidates.

3.1.2   Identification of Known Effectors Using a Bespoke BLAST Database

Where your organism has close relatives that have been studied in detail, a complementary approach may be to identify sequences in your genome that are similar to effectors from related species using BLAST. This approach can be used to supplement the output obtained in the approaches described in Subheading 3.1.1. For *G. pallida*, a set of effectors from *Meloidogyne hapla* and *H. glycines* [9, 10] was used to search the *G. pallida* genome yielding a total of 390 sequences [5].

3.1.3   Identification of Effector Gene Families Using OrthoMCL (http://www.orthomcl.org)

When sequencing the genome of a nematode or other pathogen, large families of similar proteins are often found that are specific to that species or genus. These are frequently associated with host–parasite interactions due to the rapid evolution of effector genes, which are under strong selection pressure to evade recognition by the host [11]. A large family of distinct genes in your genome of interest that is not present in closely related species is therefore likely to include candidate effectors, assuming a signal peptide is present. Furthermore, if these families are only present as "hypothetical proteins" in your genome and lack any detailed annotation (as is almost always the case for novel genes) then they will not share any annotation features and are unlikely to be recognized as significant using other analyses. A simple approach to identify such gene families is to use an orthologue clustering tool such as OrthoMCL [12]. One can either simply cluster the genes in the genome of interest and determine the top large gene families, or combine your genome of interest with one or more related genomes and look for species-specific families, e.g., those with no orthologue in the related species.

Here we identify large gene families in *G. pallida* using orthoMCL v1.4. This is computationally demanding, and ideally the BLAST stage should be run separately over multiple machines. The command that is used for this analysis is:

```
orthomcl.pl --mode 1 --fa_files ../Gpal.v1.cds.fa
```

As a result of this analysis 2,142 clusters or multigene families are identified from the *G. pallida* genome. The largest contains 398 genes and a simple BLAST search on the Uniprot webserver identifies them as SPRYSECs, a key family of *Globodera* effectors [13]. The second family contains 295 members and is similar to a dorsal gland protein from *Heterodera avenae* and to a similar sequence

(Hgg20) in *H. glycines*. This candidate effector family may therefore be specific to cyst nematodes. The third family contains 176 members and is similar to protein kinases from other nematodes. The fourth family includes 158 genes with BTB/POZ domains. These latter two families are unlikely to be effectors, but the two largest families are excellent effector candidates and demonstrate the utility of this approach for the identification of sequences involved in the host–parasite interaction in any organism.

3.2 Bioinformatic Tools for Analyzing RNA-seq Data to Identify Differentially Expressed Genes

There are several preparatory stages when analyzing RNA-seq datasets: QC, mapping, and read counting. These stages are performed independently for each replicate of each timepoint/condition. It is then possible to determine which genes are differently expressed between different timepoints. Depending on the quality of RNA-seq data, the size and quality of the genome assembly, formatting of genome annotation, and complexity of the transcriptome, each stage can require significant informatics overhead in terms of scripting skills and compute time. Where possible we present relatively straightforward examples with some discussion of potential complexities.

The analysis described here operates on the assumption that Illumina sequencing has been used, that the RNA-seq libraries are not strand specific, that the library preparation has worked well and that the libraries adequately represent the transcriptome of the target stages. The QC steps associated with each of these stages are described in ref. 14. For this analysis we describe the tuxedo suite pipeline comprising bowtie [15], tophat [16], and cufflinks [17], which is easy to use and produces good results with a reasonable number of biological replicates.

3.2.1   Mapping

The program tophat can be used to map transcriptome reads to a genome sequence. It is aware of splice sites and will split reads across introns. First, you will need to download and install bowtie2 (http://bowtie-bio.sourceforge.net/index.shtml) and tophat2 (http://tophat.cbcb.umd.edu/). You then need to index your genome sequence using bowtie2-build before running tophat to map your reads. If you are using paired-end reads you will need to specify "-r" the inner mate distance which is equal to the mean fragment size used in your sequencing library minus two times the read length. The maximum intron length default is set for mammals and for nematodes a more appropriate value would be 10,000. Here we have taken ten million pairs of reads from each library to reduce the mapping time. Even so the mapping may take 12 h and around 5 Gb of RAM per library. Your command will thus look something like this (assuming a paired-end library with a fragment size of 400 bp and a read length of 100 bp):

```
bowtie2-build Gpal.v1.0.fas Gpal.v1.0.fas
tophat -o egg1 -I 10000 -r 200 Gpal.v1.0.fas
egg1_1_10M.fastq egg1_2_10M.fastq
tophat -o egg2 -I 10000 -r 200 Gpal.v1.0.fas
egg2_1_10M.fastq egg2_2_10M.fastq
tophat -o 7dpi1 -I 10000 -r 200 Gpal.v1.0.fas
7dpi1_1_10M.fastq 7dpi1_2_10M.fastq
tophat -o 7dpi2 -I 10000 -r 200 Gpal.v1.0.fas
7dpi2_1_10M.fastq 7dpi2_2_10M.fastq
tophat -o J21 -I 10000 -r 200 Gpal.v1.0.fas
J21_1_10M.fastq J21_2_10M.fastq
tophat -o J22 -I 10000 -r 200 Gpal.v1.0.fas
J22_1_10M.fastq J22_2_10M.fastq
```

3.2.2 Analysis of Differential Expression

The most powerful tools for determining differential expression are accessed as libraries in the statistical package R and take read counts for each gene as input. This requires the independent determination of read counts. However, the tuxedo suite offers a tool which calculates differential expression directly from BAM files without the requirement for independently enumerating read counts. This tool is called cuffdiff. It requires that your genome annotation (e.g., gene models) must be in the appropriate GFF/GTF format. A description of the required format can be found at http://cufflinks.cbcb.umd.edu/gff.html. You may need to write a script in order to convert your particular type of GFF to the format required. Here we use cuffdiff to identify differentially expressed genes between all pairs of timepoints in the *G. pallida* dataset. This will take around 7 h to run and require around 4 Gb of memory:

```
cuffdiff Gpal.v1.0.gtf egg1/accepted_hits.
bam,egg2/ accepted_hits.bam J21/accepted_hits.
bam,J22/accepted_hits.bam 7dpi1/accepted_hits.
bam,7dpi2/accepted_hits.bam
```

In this case the life stages being examined are egg, J2, 7, 14, 21, 28, and 35 days post infection, parasitic and adult male. Each life stage has two replicates.

You will now need to extract the results, filtering for an appropriate *p*-value cutoff, direction of differential expression, and fold change. Subsequently you can cross-reference these with your dataset describing likely effectors identified informatically. It is also necessary to incorporate some functional information about your genes such as informative gene names or protein product descriptions. This will help you to interpret your results and identify known genes and novel effector candidates.

Candidate effectors involved in invasion, migration, and induction of biotrophic feeding structure are likely to be upregulated between egg and J2 or between J2 and 7dpi parasitic nematodes. We found that 859 genes are differentially expressed between egg and J2, of which 753 are upregulated in J2. These are identified with the command below and are exported into a file named "early effector.exp."

```
grep yes gene_exp.diff | grep q1 | grep q2 | cut
-f1,10 | perl -ne 'chomp;@a=split/\t/;print
"$a[0]\n" if $a[1] > 0' > early_effector.exp
```

Similarly, 1,466 genes are upregulated between J2 and 7dpi parasitic nematodes and are identified with the following command:

```
grep yes gene_exp.diff | grep q2 | grep q3 | cut
-f1,10 | perl -ne 'chomp;@a=split/\t/;print
"$a[0]\n" if $a[1] > 0' > late_effector.exp
```

The lists of genes upregulated at the key life stages and the list of genes encoding predicted secreted proteins can now be compared. This shows that of the 753 early expressed genes, 276 encode proteins with predicted signal peptides and no transmembrane domain. Twenty of these are previously characterized effectors. Of the 1,466 later upregulated genes, 264 encode predicted secreted proteins and 23 of these are known effectors. This analysis demonstrates that secreted proteins are highly enriched in the dataset of genes upregulated at key stages of parasitism and provides a list of candidate effectors that can be further analyzed.

3.2.3 Clustering RNA-seq Data

The protocol described above is useful in determining which genes are differentially expressed between two life stages. However, where you have an RNA-seq timecourse of multiple stages it may be useful to identify genes that are commonly regulated or that are regulated in multiple specific life stages. This can allow a more specific group of genes to be identified and, if appropriate life stages are selected, can further enrich for likely effectors. This analysis is more complicated than that presented above, requiring the writing of bespoke scripts. So while we outline the general approach we do not specify the steps involved.

The most commonly used method for clustering RNA-seq data is MBCluster.seq. This program is implemented in R and is well described in the accompanying manual (http://cran.r-project.org/web/packages/MBCluster.Seq/index.html). The main difficulty is deciding how many clusters to use. This is an unresolved problem in cluster analysis and we suggest starting with 50–75 clusters. If there are many noisy, unresolved clusters then it may be necessary to increase the number. If too many are chosen, it may be necessary to combine clusters that show similar expression profiles. The appropriate number also depends on the number of timepoints being examined, with more clusters likely to be required to resolve expression profiles for a larger number of timepoints.

A two-stage procedure can be used to identify novel effector candidates by clustering. This procedure requires some prior

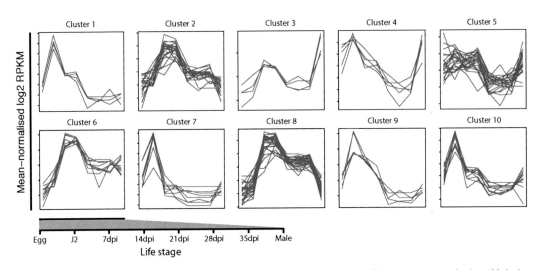

Fig. 1 Gene expression profile clusters for 125 known effectors in *G. pallida*. The genes were clustered into ten clusters using MBCluster.seq. The gene expression profiles across eight life stages are plotted for genes in each cluster as mean-normalized log$_2$ of the RPKM

knowledge of effectors in the species of interest. In the first stage a cluster analysis of expression profiles for known effectors is performed. For the *G. pallida* data this analysis suggests that they tend to be highly expressed in J2, upregulated in J2 and males, or upregulated in early parasitic stages (Fig. 1). Repeating this analysis with the full genome allows new genes that have the same expression profiles to be identified.

Acknowledgments

The James Hutton Institute receives funding from the Scottish Government. Adam James Reid was funded by the Wellcome Trust. We would like to thank Hayley Bennett for critical reading of the manuscript.

References

1. Whisson SC, Boevink PC, Moleleki L, Avrova AO, Morales JG, Gilroy EM, Armstrong MR, Grouffaud S, van West P, Chapman S et al (2007) A translocation signal for delivery of oomycete effector proteins into host plant cells. Nature 450:115–118

2. Elling AA, Mitreva M, Gai X, Martin J, Recknor J, Davis EL, Hussey RS, Nettleton D, McCarter JP, Baum TJ (2009) Sequence mining and transcript profiling to explore cyst nematode parasitism. BMC Genomics 10:58

3. Palomares-Rius JE, Hedley PE, Cock PJ, Morris JA, Jones JT, Vovlas N, Blok V (2012) Comparison of transcript profiles in different life stages of the nematode *Globodera pallida* under different host potato genotypes. Mol Plant Pathol 13:1120–1134

4. Martin JA, Wang Z (2011) Next-generation transcriptome assembly. Nat Rev Genet 12: 671–682

5. Cotton JA, Lilley CJ, Jones LM, Kikuchi T, Reid AJ, Thorpe P, Tsai IJ, Beasley H, Blok V,

Cock PJA, van den Akker SE, Holroyd N, Hunt M, Mantelin S, Naghra H, Pain A, Palomares-Rius JE, Zarowiecki M, Berriman M, Jones JT, Urwin PE (2014) The genome and life-stage specific transcriptomes of *Globodera pallida* elucidate key aspects of plant parasitism by a cyst nematode. Genome Biology in press

6. Dubreuil G, Magliano M, Deleury E, Abad P, Rosso MN (2007) Transcriptome analysis of root-knot nematode functions induced in the early stages of parasitism. New Phytol 176:426–436

7. Petersen TN, Brunak S, von Heijne G, Nielsen H (2011) SignalP 4.0: discriminating signal peptides from transmembrane regions. Nat Methods 8:785–786

8. Sonnhammer EL, von Heijne G, Krogh A (1998) A hidden Markov model for predicting transmembrane helices in protein sequences. Proc Int Conf Intell Syst Mol Biol 6:175–182

9. Huang G, Gao B, Maier T, Allen R, Davis EL, Baum TJ, Hussey RS (2003) A profile of putative parasitism genes expressed in the esophageal gland cells of the root-knot nematode *Meloidogyne incognita*. Mol Plant Microbe Interact 16:376–381

10. Gao B, Allen R, Maier T, Davis EL, Baum TJ, Hussey RS (2003) The parasitome of the phytonematode *Heterodera glycines*. Mol Plant Microbe Interact 16:720–726

11. Jones JD, Dangl JL (2006) The plant immune system. Nature 444:323–329

12. Li L, Stoeckert CJ Jr, Roos DS (2003) OrthoMCL: identification of ortholog groups for eukaryotic genomes. Genome Res 13:2178–2189

13. Jones JT, Kumar A, Pylypenko LA, Thirugnanasambandam A, Castelli L, Chapman S, Cock PJ, Grenier E, Lilley CJ, Phillips MS et al (2009) Identification and functional characterization of effectors in expressed sequence tags from various life cycle stages of the potato cyst nematode *Globodera pallida*. Mol Plant Pathol 10:815–828

14. DeLuca DS, Levin JZ, Sivachenko A, Fennell T, Nazaire MD, Williams C, Reich M, Winckler W, Getz G (2012) RNA-SeQC: RNA-seq metrics for quality control and process optimization. Bioinformatics 28:1530–1532

15. Langmead B, Trapnell C, Pop M, Salzberg SL (2009) Ultrafast and memory-efficient alignment of short DNA sequences to the human genome. Genome Biol 10:R25

16. Trapnell C, Pachter L, Salzberg SL (2009) TopHat: discovering splice junctions with RNA-Seq. Bioinformatics 25:1105–1111

17. Trapnell C, Williams BA, Pertea G, Mortazavi A, Kwan G, van Baren MJ, Salzberg SL, Wold BJ, Pachter L (2010) Transcript assembly and quantification by RNA-Seq reveals unannotated transcripts and isoform switching during cell differentiation. Nat Biotechnol 28:511–515

Chapter 3

Two-Dimensional Data Binning for the Analysis of Genome Architecture in Filamentous Plant Pathogens and Other Eukaryotes

Diane G.O. Saunders, Joe Win, Sophien Kamoun, and Sylvain Raffaele

Abstract

Genome architecture often reflects an organism's lifestyle and can therefore provide insights into gene function, regulation, and adaptation. In several lineages of plant pathogenic fungi and oomycetes, characteristic repeat-rich and gene-sparse regions harbor pathogenicity-related genes such as effectors. In these pathogens, analysis of genome architecture has assisted the mining for novel candidate effector genes and investigations into patterns of gene regulation and evolution at the whole genome level. Here we describe a two-dimensional data binning method in R with a heatmap-style graphical output to facilitate analysis and visualization of whole genome architecture. The method is flexible, combining whole genome architecture heatmaps with scatter plots of the genomic environment of selected gene sets. This enables analysis of specific values associated with genes such as gene expression and sequence polymorphisms, according to genome architecture. This method enables the investigation of whole genome architecture and reveals local properties of genomic neighborhoods in a clear and concise manner.

Key words Genome architecture, Data binning, Intergenic, R, Visualization, Heatmap, Effectors, Filamentous plant pathogen

1 Introduction

Comparative genomic analyses have revealed that evolutionary constraints often disparately affect coding, regulatory, and non-coding sequences. In addition, the distribution of these constraints largely depends on the relative position of genetic elements in the genome, referred to as the "genome architecture" [1]. In eukaryotic genomes, the organization of genes and genomic neighborhoods has evolved to ensure accurate regulation of gene expression and splicing. However, the plasticity of their structure can exceed that of their sequence, leading to structural re-arrangements over short evolutionary scales whereas genetic sequences, which are under strong constraints, diverge only minimally. The evolution of genome architecture is largely driven by genetic drift, but is also

Paul Birch et al. (eds.), *Plant-Pathogen Interactions: Methods and Protocols*, Methods in Molecular Biology, vol. 1127, DOI 10.1007/978-1-62703-986-4_3, © Springer Science+Business Media New York 2014

constrained by lifestyles and specific biological contexts [1]. Therefore, studies of genome architecture and its evolution can provide significant insight into gene regulation and function, and the underlying molecular bases of adaptation.

The recent generation of genomic data for plant pathogenic fungi and oomycetes has revealed an expansion in genome size in several unrelated lineages [2–4]. For example, in the late blight oomycete pathogen *Phytophthora infestans*, a repeat-driven expansion of the genome has created repeat-rich, gene-sparse regions that are distinct from the gene-dense conserved regions [2]. For every gene, the distance to its closest gene neighbors in either direction, designated as its flanking intergenic regions (FIRs), can be used to determine whether a gene resides in a gene-dense or gene-sparse environment. A dramatic enrichment in genes associated with virulence is observed in *P. infestans* repeat-rich, gene-sparse regions, [2, 5]. Accordingly, most *P. infestans* effector genes have longer FIRs than the genome average [6, 7]. Although less striking than *P. infestans*, repeat-rich genomic niches harboring pathogenicity-related genes are also found in several other filamentous plant pathogens, including *Magnaporthe oryzae* telomeric regions, *Leptosphaeria maculans* AT-rich isochores, and *Fusarium* spp. conditionally dispensable chromosomes [8]. Genes encoding *Vir* antigens in the malaria parasite *Plasmodium falciparum* reside in repeat-rich regions of the genome suggesting that some animal eukaryotic parasites have also evolved repeat-rich regions that host pathogenicity-associated genes [9].

Given that genes associated with pathogenicity tend to have long FIRs in some pathogen genomes, genome architecture can, in some cases, be used to identify new candidate pathogenicity genes. For example, the oomycete pathogen *Pythium ultimum* lacks a key family of oomycete effectors, known as "RXLR-type" effectors that contain a conserved Arginine-X-Leucine-Arginine N-terminal motif involved in translocation into host cells [10, 11]. Analysis of the architecture of the *P. ultimum* genome combined with Markov clustering resulted in the identification of a novel family of the candidate effector genes encoding small-secreted proteins with a conserved N-terminal domain. This domain contains a YXSL[RK] motif strongly enriched in secreted proteins and typically located between amino acid 60 and 80, similar to the RXLR motif [10, 11]. This example clearly illustrates how examining genome architecture can help to reveal novel effector candidates.

Analysis of genome architecture through the visualization of the length and distribution of intergenic regions is one way to gain insights into the biology and evolution of eukaryotes. Eukaryotic genomes are typically comprised of between 5,000 and 50,000 genes, each flanked by two intergenic regions (5′ and 3′ FIRs). To facilitate the mathematical analysis of such large datasets, "quantization," the process of mapping a large set of input values onto a smaller set, is often used. Data binning is a quantization method, replacing values that fall within a given interval (called a "bin") by a value representative of this interval, thereby reducing the number of values to analyze.

Here we report a method that was developed using two-dimensional data binning to visualize both the 5' and 3' intergenic regions on either side of each gene in a single representation. In this approach, genes are sorted into bins that are defined firstly by the length of the 5' intergenic regions (first dimension) and secondly by the length of 3' intergenic regions (second dimension). In the simplest implementation of the method, the representative value of a bin is the number of genes it contains. This is represented by a color code in a heatmap or as a third dimensional value in a surface plot, providing a view of gene-density distribution [2, 12]. Using this method the genome architecture for a whole genome can also be compared to a smaller dataset, such as the length of intergenic regions for a small subset of genes. Indeed, the method described here offers the opportunity to overlay a scatter plot of a particular subset of genes on the whole genome heatmap, thereby highlighting the position of selected genes in the genome architecture. This approach was used to compare the FIRs of effector genes to the architecture of whole genomes in *Phytophthora* spp., illustrating that effector genes preferentially reside in gene-sparse regions of these genomes [2, 5–7].

Data binning is also used as a pre-processing technique to accelerate analyses and reduce the bias caused by minor observation errors in metagenomics [13], mass spectrometry imaging [14], and modeling [15]. In comparative genomics, data binning has proved useful in revealing local properties of genomic neighborhoods. Values describing gene properties can be associated with every gene and processed per bin, providing a representative value for genes in the same bin that have FIRs of similar length. Using *in planta* expression data as an associated value, *P. infestans* gene-sparse regions were shown to be enriched in transcriptionally induced genes during plant infection [16]. Analyses of the frequency and the type of single nucleotide polymorphisms and gene copy number as associated values led to the concept of a "two-speed genome," with gene-sparse regions evolving faster than gene-dense regions in the *P. infestans* lineage [16].

We describe here a method to perform two-dimensional binning of genes based on the length of their FIRs, overlay a scatter plot over the resulting heatmap graph, and analyze the distribution of gene-associated values such as gene expression and sequence polymorphisms, according to genome architecture. The method, implemented in R, takes standard gff or gtf coordinate files as input and produces datasets and graphics that can be

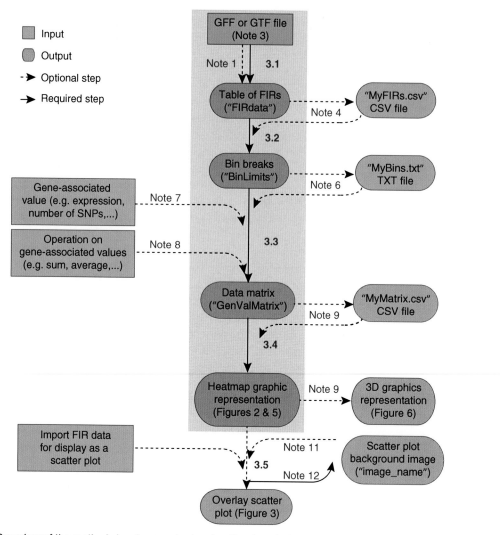

Fig. 1 Overview of the method showing required and optional analysis steps. Numbers refer to Subheadings 3 and 4. The core of the method is shown in a *gray box*, with optional inputs (*green*) and outputs (*red*) connected with *dotted arrows*

exported in any format handled by R. It is composed of five major sections (Fig. 1):

1. Calculation of FIRs length based on gff/gtf input.

2. Determination of optimal bins based on quantiles and exponential regression.

3. Data binning.

4. Representation of binned data as a heatmap graph.

5. Optional: Overlay of a scatter plot representing FIRs for a subset of genes.

Intermediary output files can be exported or imported at each major step of the analysis, as explained in the corresponding notes and outlined in Fig. 1.

The shape of genome architecture heatmaps is largely dependent on the choice of the bin size. If a bin size is too large local variations in genome environment are collapsed. Conversely, if a bin size is too small gene frequencies fluctuate greatly and patterns of genome architecture cannot be distinguished. In many eukaryotic genomes, the length of intergenic regions in a genome roughly follows a Gaussian distribution. Considering the form of Gaussian functions, we approximated that the length of intergenic regions varies proportionally to the logarithm of gene frequencies. To fit with this property, the proposed method determines bins of size increasing exponentially.

For global analysis of genome architecture, section five can be disregarded. The methods for data binning (Subheading 3.3) and **Notes 7** and **8** describe how to associate values with genes, providing the opportunity to analyze the distribution of these values according to genome architecture. In the following command lines, parameters to be set by the user are underlined. All commands should run if typed in or copied in the R console as shown, omitting the initial "> " prompt. Long command lines are split across lines for clarity reasons, and should run if typed in as a single line or copied in the R console as a whole (omitting the initial "> ").

2 Materials

2.1 Required R Packages

1. The following R base packages are required: "base," "graphics," "grDevices," "stats," "utils," installed by default with R.2.XX.X. We recommend running R 3.0.1 or above for better compatibility.

2. Three packages from the bioconductor suite are needed, and they can be downloaded and installed with the following instructions:

```
> source("http://bioconductor.org/biocLite.R")
> biocLite("GenomicRanges")
> biocLite("rtracklayer")
> biocLite("Rsamtools")
```

3. Four packages hosted by the CRAN mirror sites can be downloaded directly using the "install.packages" function as follows:

```
> install.packages("png")
> install.packages("gridExtra")
> install.packages("ggplot2")
```

2.2 Required Custom R Functions

1. The methods use the function "getFeat2" or "getFeat2b", modified versions of the "getFeat" function contributed by Thomas Girke. The codes of the "getFeat2" and "getFeat2b" functions can be downloaded from the Figshare repository using the links http://dx.doi.org/10.6084/m9.figshare.707325 and http://dx.doi.org/10.6084/m9.figshare.707326 respectively.

2. The methods use the function "filled.contour3" that can be downloaded using the link http://dx.doi.org/10.6084/m9.figshare.707327 to the Figshare repository. This is a modified version of the "filled.contour" function contributed by Ian Taylor, Carey McGilliard, and Bridget Ferris available at http://wiki.cbr.washington.edu/qerm/sites/qerm/images/1/16/Filled.contour3.R.

2.3 Optional R Packages

1. Alternative methods proposed in Subheading 4 use the "fields" and "rgl" packages hosted by the CRAN mirror sites and "EBImage" bioconductor package. They can be installed using the following instructions:

```
> install.packages("fields")
> install.packages("rgl")
> source("http://bioconductor.org/biocLite.R")
> biocLite("EBImage")
```

2.4 Sample Datasets

The description of methods in Subheadings 3 and 4 was performed on *P. infestans* datasets described in [2]. Pre-processed files, as used in the following Subheading 3, can be downloaded from the Figshare repository using the link http://dx.doi.org/10.6084/m9.figshare.707329. This archive includes:

1. The "Mygtf.gtf" file containing the final transcript calls for *P. infestans* genome version 2.4. The original file is available at http://www.broadinstitute.org/annotation/genome/phytophthora_infestans/MultiDownloads.html. The file "Mygtf.gtf" imported in **step 3** of Subheading 3.1 is a modified version in which lines were sorted in ascending order on the seqname and start position columns according to **Note 3**.

2. The "RXLR_FIRs.csv" file containing intergenic regions length for RXLR effector genes only, imported in **step 10** of Subheading 3.5.

3. The "Tom_expression.csv" file containing gene induction values at 2 days post inoculation on tomato, imported in **Note 7**.

3   Methods

The following methods describe the procedure to draw genome architecture heatmaps and overlay them with scatter plots in R. An alternative method for calculating intergenic region length (Subheading 3.1) in perl is proposed in **Note 1**.

3.1 Calculation of Flanking Intergenic Regions

1. Copy the file "getFeat2.R" into your R working directory. Specify the use of this file as the source for the "getFeat2" function as follows:

   ```
   > source("getFeat2.R")
   ```

   You may want to ignore genes at the ends of contigs in the analysis. In this case the "getFeat2b" function should be used instead of "getFeat2". *See* **Note 2** for details on how to use it.

2. Load libraries required for this section.

   ```
   > library(rtracklayer)
   > library(GenomicRanges)
   > library(Rsamtools)
   ```

3. Place a copy of the .gff file containing feature coordinates for the genome to analyze in your R working directory. Files in the .gff or .gtf format are accepted as input. They must contain either "gene," "exon" or "start codon," and "stop codon" features for the "get-Feat2" function to work (*see* **Note 3**). Import the gff file using the "import.gff" function of the GenomicRanges package, replacing "Mygtf.gtf" by the name and extension of the file to analyze.

   ```
   > gff<- import.gff("Mygtf.gtf", asRangedData
   =FALSE)
   ```

4. Generate a Range object called "gffgene" containing gene features using the "getFeat2" function. Specify the type of the file imported ("gff" or "gtf") after the "format=" parameter.

   ```
   >    gffgene<-getFeat2(x=gff,    format="gtf",
   range_types=c("gene"))
   ```

5. Restore the gene strand information lost after processing using the "getFeat2" function.

   ```
   > strand(gffgene)<-mcols(gffgene)$score
   ```

6. Clean up the temporary strand information column.

   ```
   > mcols(gffgene)$score<-NULL
   ```

7. Generate a Range object called "gffintg" containing intergenic region features using the "getFeat2" function. Specify the type of the file imported ("gff" or "gtf") after the "format=" parameter. This operation can take some time to complete.

   ```
   > gffintg<-getFeat2(x=gff, format="gtf",
   range_types=c("intergenic"))
   ```

8. Create a data frame object called "length_intg" containing one column with intergenic regions indexes and another column with intergenic regions length.

   ```
   > length_intg<-
   as.data.frame(cbind(seq(1:length(ranges(gffi
   ntg))), as.numeric(mcols(gffintg)$length)))
   ```

9. Rename columns of the "length_intg" data frame as "index" and "length" for convenience in future references.

```
> colnames(length_intg)<-c("index", "length")
```

10. Create a list called "three_intg_index" of indexes for intergenic regions following every gene (located at the 3′ end of each gene), using the "precede" function of the GenomicRanges package.

```
> three_intg_index<-precede(gffgene, gffintg)
```

11. Create a list called "five_intg_index" of indexes for intergenic regions preceding every gene (located at the 5′ end of each gene), using the "follow" function of the GenomicRanges package.

```
> five_intg_index<-follow(gffgene, gffintg)
```

12. Create a data frame object called "gene_data" containing columns with gene identifiers, gene strand, five prime, and three prime intergenic region indexes.

```
> gene_data<-
  as.data.frame(cbind(as.character(mcols(gffg
  ene)$group), as.character(strand(gffgene)),
  as.numeric(five_intg_index),     as.numeric
  (three_intg_index)))
```

13. Rename columns of the "gene_data" data frame as "geneid," "strand," "FivePrime_index," and "ThreePrime_index" for convenience in future references.

```
> colnames(gene_data)<-c("geneid", "strand",
"FivePrime_index", "ThreePrime_index")
```

14. Merge "gene_data" and "length_intg" data frames into a "tempdata" data frame using indexes for five prime intergenic regions as a common data column between the two data frames. The "all.x=TRUE" parameter allows lines of the "gene_data" data frame to be kept with no five prime intergenic region index.

```
> tempdata<-merge(x=gene_data, y=length_intg,
      by.x="FivePrime_index", by.y="index",
  all.x=TRUE)
```

15. Rename columns of the "temp_data" data frame as "delete1", "geneid", "strand", "ThreePrime_index," and "fiveprime" to avoid conflicts with the following steps of the process.

```
> colnames(tempdata)<-c("delete1", "geneid",
"strand", "ThreePrime_index", "fiveprime")
```

16. Merge "tempdata" and "length_intg" data frames into a "FIRdata" data frame using indexes for three prime intergenic regions as a common data column between the two data frames. The "all.x=TRUE" parameter allows lines of the "gene_data" data frame to be kept with no three prime intergenic region index.

```
> FIRdata<-merge(x=tempdata, y=length_intg,
    by.x="ThreePrime_index",    by.y="index",
  all.x=TRUE)
```

17. Clean up unnecessary columns in the "FIRdata" data frame object.

```
> FIRdata$ThreePrime_index<-NULL
> FIRdata$delete1<-NULL
```

18. Rename columns of the "FIRdata" data frame as "geneid," "strand," "fiveprime," and "threeprime" for convenience in future references. The FIRdata table can be exported for external use (*see* **Note 4**).

```
> colnames(FIRdata)<-
    c("geneid", "strand", "fiveprime", "threeprime")
```

3.2 Bin Breaks Setup

1. Set the number of bins to use by replacing "40" in the code line below. This number is stored under the variable name "NumBins." We recommend values between 10 and 60 for an appropriate representation of most eukaryotic genomes.

```
> NumBins=40
```

2. The size and breaks of bins will be calculated based either on five prime or three prime intergenic regions length, depending on which list contains the longest intergenic region of the genome. The corresponding list is stored under the variable name "FIR2Bin"

```
> if ((max(FIRdata$fiveprime, na.rm=TRUE)>max
(FIRdata$threeprime, na.rm=TRUE)) == TRUE)
{
FIR2Bin<-FIRdata$fiveprime
} else {
FIR2Bin<-FIRdata$threeprime
}
```

3. Remove intergenic regions of length 0 or with no length assigned from the "FIR2Bin" list.

```
> FIR2Bin=FIR2Bin[which(FIR2Bin!=0)]
> FIR2Bin<-na.omit(FIR2Bin)
```

4. Determine the number of entries "BinSteps" per bin to separate all entries in the "FIR2Bin" list into a number of quantiles equal to "Numbins-1"

```
> BinSteps<-round(length(FIR2Bin)/(Num Bins-1),
  digits=0)
```

5. Sort the length of intergenic regions stored in the "FIR2Bin" list in ascending order; store the ordered list under the variable name "FIR2BinOrd".

```
> FIR2BinOrd<-sort(FIR2Bin)
```

6. Determine the temporary bin breaks stored based on intergenic region lengths, the position of which delimits quantiles of length "BinSteps." The list of temporary bin breaks is stored under the list named "TempBinLimits."

```
> TempBinLimits<-FIR2BinOrd[seq(FIR2BinOrd[2*
  BinSteps],
    length(FIR2BinOrd),BinSteps)]
```

7. To allow a more complete coverage of the range of intergenic region length present in the genome analyze, the maximum value in the list "FIR2Bin" is added as the last break in the list "TempBinLimit".

```
> TempBinLimits[length(TempBinLimits)+1]<-
    max(FIR2Bin, na.rm=TRUE)
```

8. Fit an exponential distribution to the values of temporary bin breaks in "TemBinLimits" list using the non-linear regression function "nls." A prevalent weight is given to the extreme values in the "TempBinLimits" list during the regression to ensure a better representation of genes with extreme intergenic region length (*see* **Note 5**).

```
> x<-seq(length(TempBinLimits))
> fit<-nls(log(TempBinLimits) ~ a*x + b, start
= c(a=0, b=0),
    algorithm='port',weights=((x-0.5
 * NumBins)^2))
```

9. Calculates the predicted bin breaks based on the exponential regression and store values in the "BinLimits" list.

```
>pred=predict(fit, x)
>BinLimits=c(1, round(exp(pred),0), max(FIR2Bin))
```

Comparisons of the architecture of several genomes, using the same set of bin breaks might be required. **Note 6** describes how to export the list of bin breaks calculated here and how to import an external list of bin breaks.

3.3 Data Binning

1. Assign five prime and three prime intergenic regions of every gene to bins delimited by breaks in the "BinLimits" list. Assigned bins are stored under variables called "xbin" and "ybin" for five prime and three prime intergenic regions, respectively.

```
>xbin=cut(FIRdata$fiveprime,breaks=c(BinLimits))
>ybin=cut(FIRdata$threeprime, breaks= c(BinLimits))
```

2. Concatenate the list of assigned bins to the "FIRdata" data frame. This step also concatenates a list of gene-associated values called "genevalue" that is set to "1" for every gene. A custom list of gene-associated values can be provided instead (*see* **Note 7**).

```
> FIRdata<-cbind(FIRdata, xbin, ybin,
    genevalue=rep(1,length(FIRdata$fiveprime)))
```

3. Create the matrix "GenValMatrix" containing the number of genes in each bin. The number of genes is calculated by summing up gene-associated values of 1 for each gene, using the function "sum." Other operations can be performed on gene-associated values using different built-in or custom functions (*see* **Note 8**). **Note 9** describes how to export the matrix calculated here or how to import an external matrix of values.

```
> GenValMatrix<-with(FIRdata, tapply(genevalue,
list(xbin, ybin), sum))
```

3.4 Heatmap Drawing

1. Set units of the graph axes in the three dimensions using the number of columns in "GenValMatrix", the number of rows in "GenValMatrix," and the maximum value in "GenValMatrix" for *x*, *y*, and *z* dimensions, respectively.

```
> x<-1:ncol(GenValMatrix)
> y<-1:nrow(GenValMatrix)
> zlim = range(as.numeric (unlist(GenVal
Matrix)) , finite=TRUE)
```

2. Set the color palette "mypalette" using the "colorRampPalette" function. The number and identifiers of colors can be freely chosen to modify the aspect of the heatmap, by changing the list proposed below.

```
> mypalette<-colorRampPalette(c( "white",
"darkblue", "forestgreen", "goldenrod1",
"orangered", "red3", "darkred"), space="rgb")
```

3. Create a list of colors "mycol" based on the "mypalette" color palette and the number of subdivisions "mynlevels."

```
> mycol=mypalette(2*max(GenValMatrix, na.rm=TRUE))
```

4. Create a list of labels "mylabels" for axes by concatenating the length of the shortest and longest intergenic regions in each bin (stored in the "BinLimits" list) separated by a hyphen.

```
> mylabels<-paste(BinLimits[1:length(BinLim
its)-1], BinLimits[2:length(BinLimits)], sep="
- ", collapse=NULL)
```

5. Perform heatmap graph rendering using the "filled.contour" function (Fig. 2). 3D rendering of the "GenValMatrix" matrix is also possible (*see* **Note 10**).

```
> filled.contour(x, y, z=GenValMatrix,
    plot.title = title(main ="Phytophthora infestans
genome",
    xlab = "five prime intergenic regions",
    ylab = "three prime intergenic regions",
    cex.main=0.8, cex.lab=0.5),
    key.title = title(main ="Number of
genes", cex.main=0.5,
             line=1),
    col=mycol,
    levels = pretty(zlim, 2*max(GenValMatrix,
na.rm=TRUE)),
    plot.axes={axis(1,at=x, labels=mylabels, las=2,
             cex.axis=0.5);
         axis(2,at=y, labels=mylabels,
    cex.axis=0.5)})
```

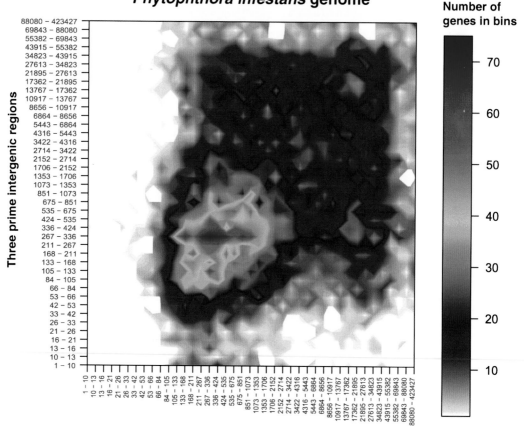

Fig. 2 The heatmap plot output of **step 5** of Subheading 3.4, representing gene counts in *P. infestans* genome. The initial gtf file was obtained at http://www.broadinstitute.org/annotation/genome/phytophthora_infestans/ MultiDownloads.html. Minor edits were performed in Adobe Illustrator to include italic fonts in the main title and improve the legibility of the color key and axis labels

3.5 Overlaying a Scatter Plot Over a Genome Architecture Heatmap

1. Copy the code of the "filled.contour3" function given in Subheading 2 into a text file named "filled.contour3.R" and place this file in your R working directory. Specify the use of this file as the source for the "filled.contour3" function as follows:

```
> source('filled.contour3.R')
```

2. Load libraries required for this section.

```
> library(png)
> library(gridExtra)
> library(ggplot2)
```

3. *See* **Note 11** if you want to re-use a scatter plot background image generated from a previous analysis. Automatically set the name of the heatmap image file to be used as a background for the scatter plot. The name is composed of the system date (year, month, and day) and time (hours, minutes, seconds)

followed by "_graph" and stored in the variable "image_name" for future reference.

```
> image_name<-paste(as.character(format(Sys.time(),
      "%Y%m%d%H%M%S")), "_graph", sep="")
```

4. Create a .png image file in your R working directory named after the "image_name" variable. For alternative image file formats *see* **Note 12**.

```
> png(filename = paste(image_name, ".png", sep=""))
```

5. Remove margins of the graph so that it fully covers the background of the scatter plot to be overlaid.

```
> par(mar=c(0,0,0,0))
```

6. Perform heatmap graph rendering using the "filled.contour3" function. This modified version of the "filled.contour" function does not draw the color key scale allowing the heatmap to fully cover the graphic device area.

```
> filled.contour3(x, y, z=GenValMatrix,
      col=mycol,
      levels = pretty(zlim, 2*max(GenValMatrix,
  na.rm=TRUE)),
      frame.plot = FALSE,
      axes = FALSE)
```

7. Save current graph to the file specified in step 4 and close the graphic device.

```
> dev.off()
> quartz.save("heatmap.pdf", type="pdf")
```

8. Load the heatmap background image created previously in a variable called "img" using the "readPNG" function of the "png" package.

```
> img <- readPNG(paste(image_name, ".png", sep=""))
```

9. Convert the background image into a grid graphical object using the "rasterGrob" function of the "gridExtra" package. This object is stored under the name "g."

```
> g <- rasterGrob(img, interpolate=TRUE)
```

10. Import data to be shown as a scatter plot. In the example below, a comma delimited table containing five prime and three prime intergenic regions for RXLR effector genes (called "RXLR_FIRs. csv") is imported using the "as.data.frame" and "read.csv" functions. Note that the file should be located in the R working directory to allow the omission of the filepath when calling the file.

```
> rxlrData<-as.data.frame(read.csv('RXLR FIRs.csv',
header=TRUE))
```

11. Plot the "rxlrData" data as a scatter plot over the grid object "g" corresponding to the background heatmap image using the "ggplot" function of the "ggplot2" package. After "aes(",

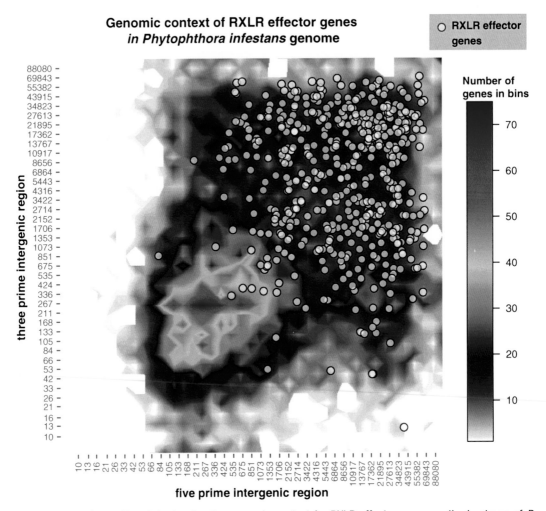

Fig. 3 Overlay of a scatter plot showing the genomic context for RXLR effector genes on the heatmap of *P. infestans* genome architecture. The figure shows the output of **step 10** of Subheading 3.5 with edits performed in Adobe Illustrator to display italic fonts in the main title and include a color key and scatter plot legend

the "x=" and "y=" arguments specify data to use as 5′ and 3′ FIRs, respectively. This should be provided as the name of the variable referring to the dataset, followed by "$" and the name of the column containing either 5′ or 3′ FIR length data ("rxlrData$rxlr_five" and "rxlrData$rxlr_three, respectively in the example below). The following options are used to format the graph: "annotation_custom" specifies the object "g" as a background image and sets its dimension to the whole plot area; ""coord_fixed" locks the x and y ratio to 1 so that the background always covers the whole plot area; "geom._point" specifies the style of dot to use for the scatter plot; "scale_y_log10" and "scale_x_log10" set axes in logarithm scale and define their limits; "theme(axis" formats axes labels and titles. The resulting graph is shown in Fig. 3. A warning message may inform you of data points outside the scatter plot range that are not shown.

```
> ggplot(data=rxlrData,
    aes(x=rxlrData$rxlr_five, y=rxlrData$rxlr_three,
        geom="blank")) +
    annotation_custom(g, xmin=-Inf, xmax=Inf, ymin=
  -Inf,
        ymax=Inf) +
    coord_fixed(ratio=1) +
    geom_point(shape=21,fill="white",colour="black",
        size=4, alpha=0.7, na.rm=FALSE) +
    scale_y_log10(breaks = BinLimits[2:length(BinLimits)],
        limits = c(BinLimits[2], BinLimits
        [NumBins +1])) +
    scale_x_log10(breaks=          BinLimits
  [2:length(BinLimits)],
        limits=c(BinLimits[2],
        BinLimits[NumBins +1])) +
    theme(axis.text.y=element_text(size = 10,
  vjust=0.5))    +    theme(axis.text.x=element_
  text(size=10, vjust=0.5,
        angle=90)) +
    theme(axis.title.x = element_text(face=
  "bold",size=12)) +
    xlab("five prime intergenic region") +
    theme(axis.title.y    =    element_
  text(face="bold",size=12)) + ylab("three
  prime intergenic region")
```

---

4   Notes

1. For the calculation of intergenic region length in perl (Subheading 3.1), the "Calculate_FIR_length.pl" script available from the Figshare repository using the link http://dx.doi.org/10.6084/m9.figshare.707328. This script takes either .gff3 or .gtf files as input. It sets the values of FIRs at the border of contigs to "NA" similar to the "getFeat2b" R function described in **Note 2**. To run it, use:

```
> perl Calculate_FIR_length.pl
```

Several prompts will invite the user to select appropriate options. First, type in "gff3" or "gtf" to specify whether the input file is in .gff3 or .gtf format:

Is the file in gff3 or gtf format? (gff3/gtf)

Second, type in the name and extension of the input file (e.g., "Myfile.gff3")

Please enter the name of the gff or gtf file:

Third, type in "gene", "mRNA," or "exon" to specify the type of feature included in the input file to use as gene models:

Which feature/type to process? (gene/mRNA/exon):

Finally, enter the name of the output file, including the ".csv" extension (e.g., "MyFIRs.csv")

Enter name of output file:

This will produce a comma delimited file containing FIR lengths (stored in the folder where the script was executed). To import this file in R and continue with steps under Subheading 3.2 of the analysis, use:

```
> FIRdata<-read.csv(file="MyFIRs.csv", sep=",")
```

2. By default, the "getFeat2" function calculates distance to the end of contigs as the FIRs for genes residing at the edge of contigs. Because contig edges are often difficult to assemble, these FIRs are likely underestimated. As an alternative, you can use the "getFeat2b" function to set FIRs at the edge of contigs to "NA," and therefore ignore them in subsequent steps of the analysis. To do so, copy the file "getFeat2b.R" into your R working directory. Specify to use this file as the source for the "getFeat2b" function as follows:

```
> source("getFeat2b.R")
```

Then at **steps.4** and 7 of Subheading 3.1 call the "getFeat2b" function instead of "getFeat2" as follows:

```
> gffgene<-getFeat2b(x=gff,format="gff", range_
types=c("gene"))
> gffintg<-getFeat2b(x=gff, format="gff",
        range_types=c("intergenic")
        )
```

3. The methods described here take .gff and .gtf files as input. Gene and intergenic regions' features are generated using the custom "getFeat2" function, a modified version of the "get-Feat" function written by T. Girke available at http://faculty.ucr.edu/~tgirke/Documents/R_BioCond/My_R_Scripts/gffMod.R. The file type "gff" should be selected when the annotation file contains gene features, and "gtf" file type should be selected otherwise. If the wrong file type is selected, the following error message might show:

```
Error in .Call2("solve_user_SEW0", start, end,
width, PACKAGE = "IRanges") : solving row 1: range
cannot be determined from the supplied arguments
(too many NAs)
```

To avoid errors while running "getFeat2", the annotation file should not contain any blank lines and must be sorted by

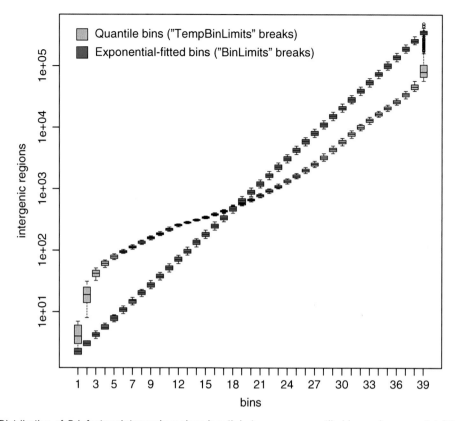

Fig. 4 Distribution of *P. infestans* intergenic regions length in temporary quantile bins and exponential-fitted bins

ascending order on the seqname (first column) first and then start position (third column).

4. To export the "FIRdata" table as a comma delimited table (.csv file, exported to R working directory by default), the following instruction can be used:

```
> write.table(FIRdata,file="MyFIRs.csv", sep=",",
row.names=FALSE)
```

To import a .csv file to start the analysis directly at Subheading 3.2 use:

```
> FIRdata<-read.csv(file="MyFIRs.csv",sep=",")
```

5. A comparison of temporary bins based on quantiles (with breaks defined by "TempBinLimits") and bins derived from exponential regression (with breaks defined by "BinLimits") allows to control whether the distribution of intergenic regions length in bins has been appropriately optimized. The following code creates a boxplot representation of data in each bin, for temporary (grey) and final bins (red), as shown in Fig. 4. It

required installation of the "fields" package to run. On *P. infestans* intergenic region data, a significant reduction in the variance in bins is obtained with the proposed exponential regression. Alternative regression models may be considered in step 8 of Subheading 3.2 for some datasets.

```
> x<-seq(length(TempBinLimits))
> library(fields)
> bplot(FIR2Bin, as.numeric(cut(FIR2Bin,
     breaks=TempBinLimits)),
     breaks=TempBinLimits, xlim=c(1,length(T
     empBinLimits)), col="grey", log="y",
     xlab="bins", ylab="intergenic regions")
> par(new=TRUE)
> bplot(FIR2Bin, as.numeric(cut(FIR2Bin, breaks
=BinLimits)),
     breaks=BinLimits, xlim=c(1,length(BinLimits)),
     col="red", log="y", axes=FALSE, ann=FALSE)
```

6. To export the "BinLimits" list as list in a text file (.txt file, exported to R working directory by default), the following instruction can be used:

```
> write.table(BinLimits,file="MyBins.txt")
```

To import an external set of bin breaks saved as a .txt file, use:

```
> BinLimits<-
as.numeric(unlist(read.table(file="MyBins.txt",
            header=TRUE, row.names=1)))
```

To start the analysis directly from Subheading 3.3, both "FIRdata" and "BinLimits" must be imported as described in **Notes 4** and **6,** respectively.

7. Typical gene-associated values are gene expression values or number of SNPs per genes. The following example illustrates how to replace the gene-associated value column in the "FIRdata" table by imported gene expression data using the "merge" function. Datasets used in this example are available in the Figshare repository using the link http://dx.doi. org/10.6084/m9.figshare.707329.

```
> GeneExpression<-
as.data.frame(read.csv('Tom_expression.csv',
            header=TRUE)) #import gene expression data
> colnames(GeneExpression)<-c("geneid", "genevalue")
            #rename columns in gene expression
     data frame
> FIRdata$genevalue<-NULL
            #delete former gene associated value
> tempdata<-as.data.frame(FIRdata)
     # convert to data frame to perform merge
     function
> FIRdata<-merge(x=tempdata, y=GeneExpression,
            by.x="geneid", by.y="geneid", all.
     x=TRUE)
```

8. Virtually any function can be applied on gene-associated values, the function "mean" being among the more useful when analyzing gene expression data. The following creates a "GenValMatrix" containing average gene expression in bins using data imported in **Note 7**.

```
> GenValMatrix<-with(FIRdata, tapply(genevalue,
  list(xbin, ybin), mean))
```

Functions such as "mean" introduce significant noise in the heatmap due to bins containing few genes. It is therefore advisable to ignore bins with too few elements. The following lines create a "CountMatrix" containing the number of genes in bins that is used to replace the gene expression values of bins with less than four genes by "0" in the GenValMatrix.

```
> CountMatrix<-with(FIRdata, tapply(genevalue,
                list(xbin, ybin), length))
> GenValMatrix[CountMatrix < 4]<-0
```

The color scale may need to be modified to fit with the gene-associated values. In particular, gene expression data contain negative and positive values and the color scale needs to be adjusted so that "0" values are shown with a neutral color (here "white"). The following lines generate the gene expression heatmap shown in Fig. 5.

```
> mypalette<-colorRampPalette(c("black", "black",
"darkblue",
    "blue",      "dodgerblue3",      "dodgerblue",
    "white",      "goldenrod1",      "orangered",
    "red3", "red4", "red4"), space="rgb")
>mycol=mypalette(7+50*(max(GenValMatrix,na.rm=TRUE)-
    min(GenValMatrix, na.rm=TRUE)))
> filled.contour(x, y, z=GenValMatrix,
    plot.title = title(main ="Phytophthora infestans
gene
        induction at 2 days post inoculation on
    tomato",
    xlab = "five prime intergenic regions",
    ylab = "three prime intergenic regions", cex.main=0.8,
        cex.lab=0.5),
    key.title   =   title(main   ="Average   gene
    induction",
        cex.main=0.5, line=1),
    col=mycol,
    levels=pretty(zlim, 50*(max(GenValMatrix,
na.rm=TRUE) -
            min(GenValMatrix, na.rm=TRUE))),
    plot.axes={axis(1,at=x, labels=mylabels, las=2,
            cex.axis=0.5),
            axis(2,at=y, labels=mylabels, cex.
        axis=0.5)})
```

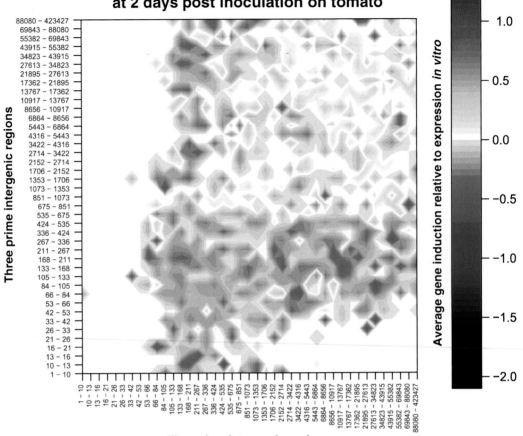

Fig. 5 Heatmap representing the average induction fold at 2 days postinoculation on tomato for *P. infestans* genes, relative to gene expression in vitro. Gene induction fold was averaged per bin and plotted as described in **Note 8**. Minor edits were performed in Adobe Illustrator to include italic fonts in the main title and improve the legibility of the color key and axis labels

9. To export the "GenValMatrix" table as a comma delimited table (.csv file, exported to R working directory by default), the following instruction can be used:

```
> write.table(GenValMatrix, file="MyMatrix.
csv", sep=",", row.names=FALSE)
```

To import an external data matrix saved as a .csv file, use:

```
>GenValMatrix<-as.matrix(read.csv(file="MyMatrix.
csv", sep=","))
```

To start the analysis directly from Subheading 3.4, both "BinLimits" and "GenValMatrix" must be imported as described in **Notes 6** and **9** respectively.

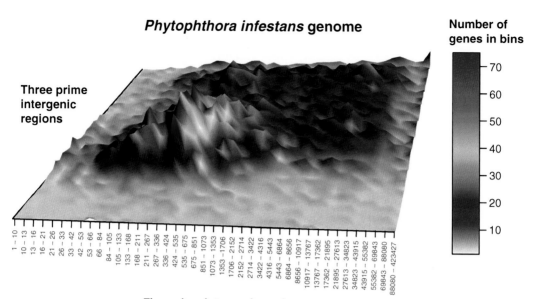

Fig. 6 A 3D view of gene counts in *P. infestans* genome created using RGL package. Legends and axis labels were added in Adobe Illustrator

10. Data in the "GenValMatrix" matrix can be visualized in 3D using the "rgl" package. The following describes how, from **step 6** of the Subheading 3.4 "Heatmap drawing", a 3D real-time visualization of the genome architecture can be obtained (Fig. 6). The "rgl" package must be installed for this method.

```
> library(rgl)                    #load the rgl library
> clear3d("all")                  #clear the graphic device
> rgl.light(theta = 45, phi = 45, viewpoint.
rel=TRUE)
#light parameters
> z<-2*GenValMatrix              #Exaggerate the relief
> z[is.na(z)]<-0    #Replace <NA> by '0'
> x<-20*(1:nrow(z))
> y<-20*(1:ncol(z))    #set row and column spac-
ing to 20
> zlim<-range(z, na.rm=TRUE) #determine maximum z
values
> zlen<-zlim[2]-zlim[1] +1
> col <- mycol[z-zlim[1]+1] #determine          the
matrix of colors
> rgl.surface(x, y, z, color=col, alpha=1,
back="fill")
```

11. To display several gene subsets from the same genome as a scatter plot, a scatter plot background image generated from a previous analysis can be re-used. For this, skip steps 3–7 of

Subheading 3.5 and replace the variable "image_name" by the name of the image file to import in step 8 of Subheading 3.5.

12. File formats JPEG and TIFF are supported for the scatter plot background image. To write the background image file in the chosen format, replace step 4 of Subheading 3.5 with one of the following:

```
> jpeg(filename = paste(image_name, ".jpeg", sep=""))
> tiff(filename = paste(image_name, ".tiff", sep=""))
```

Loading JPEG or TIFF files in R will require the "EBImage" package. Replace **step 8** of Subheading 3.5 by the following, using either ".jpeg" or ".tiff" as the file extension:

```
> library("EBImage")
> img <- readImage(files = paste(image_name, ".jpeg", sep=""))
```

Acknowledgments

We are grateful to Adam Taranto for comments on drafts of this chapter.

References

1. Koonin EV, Wolf YI (2010) Constraints and plasticity in genome and molecular-phenome evolution. Nat Rev Genet 11:487–498

2. Haas BJ et al (2009) Genome sequence and analysis of the Irish potato famine pathogen *Phytophthora infestans*. Nature 461:393–398

3. Spanu PD, Abbott JC, Amselem J, Burgis TA, Soanes DM, Stüber K, Loren van Themaat EV, Brown JKM, Butcher SA, Gurr SJ (2010) Genome Expansion and Gene Loss in Powdery Mildew Fungi Reveal Tradeoffs in Extreme Parasitism. Science 330:1543

4. Duplessis S, Cuomo CA, Lin YC, Aerts A, Tisserant E, Veneault-Fourrey C, Joly DL, Hacquard S, Amselem J, Cantarel BL (2011) Obligate biotrophy features unraveled by the genomic analysis of rust fungi. Proc Natl Acad Sci 108:9166

5. Raffaele S, Win J, Cano L, Kamoun S (2010) Analyses of genome architecture and gene expression reveal novel candidate virulence factors in the secretome of *Phytophthora infestans*. BMC Genomics 11:637

6. van Damme M, Cano L, Oliva R, Schornack S, Segretin ME, Kamoun S, Raffaele, S (2012) Evolutionary and Functional Dynamics of Oomycete Effector Genes. In: Martin F, Kamoun S (eds) Effectors in plant-microbe interactions. Wiley, pp 103–122

7. Vleeshouwers VGAA, Raffaele S, Vossen JH, Champouret N, Oliva R, Segretin ME, Rietman H, Cano LM, Lokossou A, Kessel G (2011) Understanding and Exploiting Late Blight Resistance in the Age of Effectors. Annu Rev Phytopathol 49:507–531

8. Raffaele S, Kamoun S (2012) Genome evolution in filamentous plant pathogens: why bigger can be better. Nat Rev Microbiol 10:417–430

9. Pain A et al (2008) The genome of the simian and human malaria parasite *Plasmodium knowlesi*. Nature 455:799–803

10. Lévesque CA et al (2010) Genome sequence of the necrotrophic plant pathogen *Pythium ultimum* reveals original pathogenicity mechanisms and effector repertoire. Genome Biol 11:R73

11. Win J, Morgan W, Bos J, Krasileva KV, Cano LM, Chaparro-Garcia A, Ammar R, Staskawicz BJ, Kamoun S (2007) Adaptive Evolution Has Targeted the C-Terminal Domain of the RXLR Effectors of Plant Pathogenic Oomycetes. Plant Cell 19:2349–2369

12. Kemen E, Gardiner A, Schultz-Larsen T, Kemen AC, Balmuth AL, Robert-Seilaniantz A,

Bailey K, Holub E, Studholme DJ, MacLean D (2011) Gene Gain and Loss during Evolution of Obligate Parasitism in the White Rust Pathogen of *Arabidopsis thaliana*. PLoS Biol 9:e1001094

13. Sharma VK, Kumar N, Prakash T, Taylor TD (2012) Fast and Accurate Taxonomic Assignments of Metagenomic Sequences Using MetaBin. PLoS One 7:e34030

14. Xiong X, Xu W, Eberlin LS, Wiseman JM, Fang X, Jiang Y, Huang Z, Zhang Y, Cooks RG, Ouyang Z (2012) Data Processing for 3D Mass Spectrometry Imaging. J Am Soc Mass Spectrom 23:1147–1156

15. Mendhurwar KA, Devabhaktuni VK, Raut R (2008) Binning algorithm for accurate computer aided device modeling. In: International symposium on circuits and systems. pp 2773–2776

16. Raffaele S, Farrer RA, Cano LM, Studholme DJ, MacLean D, Thines M, Jiang RHY, Zody MC, Kunjeti SG, Donofrio NM (2010) Genome Evolution Following Host Jumps in the Irish Potato Famine Pathogen Lineage. Science 330:1540

Chapter 4

On the Statistics of Identifying Candidate Pathogen Effectors

Leighton Pritchard and David Broadhurst

Abstract

High-throughput sequencing is an increasingly accessible tool for cataloging gene complements of plant pathogens and their hosts. It has had great impact in plant pathology, enabling rapid acquisition of data for a wide range of pathogens and hosts, leading to the selection of novel candidate effector proteins, and/or associated host targets (Bart et al., Proc Nat Acad Sci U S A doi:10.1073/pnas.1208003109, 2012; Agbor and McCormick, Cell Microbiol 13:1858–1869, 2011; Fabro et al., PLoS Pathog 7:e1002348, 2011; Kim et al., Mol Plant Pathol 2:715–730, 2011; Kimbrel et al., Mol Plant Pathol 12:580–594, 2011; O'Brien et al., Curr Opin Microbiol 14:24–30, 2011; Vleeshouwers et al., Annu Rev Phytopathol 49:507–531, 2011; Sarris et al., Mol Plant Pathol 11:795–804, 2010; Boch and Bonas, Annu Rev Phytopathol 48:419–436, 2010; Mcdermott et al., Infect Immun 79:23–32, 2011).

Identification of candidate effectors from genome data is not different from classification in any other high-content or high-throughput experiment. The primary aim is to discover a set of qualitative or quantitative sequence characteristics that discriminate, with a defined level of certainty, between proteins that have previously been identified as being either "effector" (positive) or "not effector" (negative). Combination of these characteristics in a mathematical model, or *classifier*, enables prediction of whether a protein is or is not an effector, with a defined level of certainty. High-throughput screening of the gene complement is then performed to identify candidate effectors; this may seem straightforward, but it is unfortunately very easy to identify seemingly persuasive candidate effectors that are, in fact, entirely spurious.

The main sources of danger in this area of statistical modeling are not entirely independent of each other, and include: inappropriate choice of classifier model; poor selection of reference sequences (known positive and negative examples); poor definition of classes (what is, and what is not, an effector); inadequate training sample size; poor model validation; and lack of adequate model performance metrics (Xia et al., Metabolomics doi:10.1007/s11306-012-0482-9, 2012). Many studies fail to take these issues into account, and thereby fail to discover anything of true significance or, worse, report spurious findings that are impossible to validate. Here we summarize the impact of these issues and present strategies to assist in improving design and evaluation of effector classifiers, enabling robust scientific conclusions to be drawn from the available data.

Key words Effectors, Statistical modeling, Classification, Bioinformatics, Sequence analysis, Genomics, High-throughput screening

Paul Birch et al. (eds.), *Plant-Pathogen Interactions: Methods and Protocols*, Methods in Molecular Biology, vol. 1127, DOI 10.1007/978-1-62703-986-4_4, © Springer Science+Business Media New York 2014

1   Introduction: Domain-Specific Assumptions and Applicability

Remember that all models are wrong; the practical question is how wrong do they have to be to not be useful. George E.P. Box, "Empirical Model Building"

The assignment of putative effector function to a gene's protein product is a binary classification problem that can be characterized as declaring that a candidate sequence either has, or does not have, "effector function." The *classifier* is presented with each protein in turn and asked to decide whether it is an effector. This can be done in many ways, but all serious attempts must be based on clearly defined a priori knowledge of characteristics specific to the problem domain (here the biological function of effectors), such as the facts that effectors must leave the pathogen, and enter the host (e.g., via type III secretion [12], or using the RxLR translocation signal [13]).

In the absence of such a priori knowledge, classification is no more than a guess and could, we might imagine, be achieved by tossing a fair coin and calling the tested protein "effector" if the coin lands face-up. However, even this naïve "model" is incorrect, as it assumes (all assumptions are a priori knowledge) that there is equal chance of the outcome being positive or negative. We might at least include information (again a priori) about how often effectors occur in a typical genome: their *prevalence*. For a "coin-toss" model, this would mean flipping a heavily weighted coin that is much more likely to fall tails than heads. This effect of prevalence, or *base rate*, is discussed in detail in Subheading 3.1 and, fortunately, simple Bayesian statistics allows us to accommodate its influence.

Ultimately, the performance of a classifier that uses protein sequences as input is only as good as the understanding that goes into abstracting key characteristics of effector sequences for use in the classifier, and the quality of the reference data set used to optimize the parameters of the model. Typically there are relatively few verified effector proteins for any chosen pathogen, and their precise functions are usually not well characterized. As a result, naïve (but theoretically unbiased) use of raw sequence data as input to off-the-shelf "black box" machine learning algorithms (e.g., Support Vector Machines, Artificial Neural Networks, etc. [10, 14–17]) is to be discouraged in this context; these techniques are extremely effective under certain conditions, but here the number of variables v far exceeds the number of training examples n—the so-called *curse of dimensionality* [18]. Under these circumstances, the resulting models are typically highly prone to uncovering random multivariate correlations that are impossible to validate without extremely large reference data sets, or that are not *generalizable*, i.e., unstable when presented with data that falls outside the tight constraints of the training examples.

In addition to constraining the structural characteristics of a classifier, it is also wise to constrain the scope of its utility: a classifier trained on sequences that derive from one organism should be used only with caution on sequences that originate from a different organism, with the degree of caution tracking the degree of organism divergence. Pathogen effectors are a highly diverse set of proteins [2, 10, 19–22], and it is likely that tailoring a classifier to the family of interest will be more useful than naïve application of a one-size-fits-all tool.

2   Constructing a Classifier of Effectors

Our goal in constructing a classifier of effector candidates is to create a *model* that, when presented with a protein sequence in silico, returns a result, or *prediction*, that in vivo the protein does, or does not, function as an effector [11]. For this we must first decide the appropriate domain-specific a priori knowledge that will be used to form the basis for the classifier. A non-exhaustive and unordered list of *model characteristics* that have previously been considered includes:

1. Presence of a generic export signal, such as a predicted signal peptide [13, 22–24].

2. Presence of a pathogenicity-specific export signal, such as a predicted type III secretion signal [25–28].

3. Presence of a sequence motif associated with translocation into the host cell, such as a predicted RxLR motif [13, 22–24].

4. The absence of a motif associated with some function that could sequester the protein, such as a predicted transmembrane domain [13, 22].

5. Amino acid composition and/or k-mer frequency [16, 29, 30].

6. Protein size (as sequence length, or molecular mass) [14].

7. Predicted structural features [15].

8. Collocation with a specific promoter, or coregulation [25, 26].

9. Other predicted or measured biochemical function [31–33].

10. Sequence similarity to proteins of known effector function; MCL/OrthoMCL and other clustering methods are essentially applications of the determination of distance by sequence similarity [8, 15, 26, 28, 32].

Characteristics may be used singly, but it is more usual to combine them into a *model structure*: a logical combination of many input characteristics producing a single output variable.

Many measures of the presence of a sequence feature are not "yes/no" binary values, but rather are continuous or probabilistic (e.g., output from SignalP may be either a Y/N prediction, or a probability score; sequence similarity assessed using BLAST may be determined as an E-value, bit score, or a threshold on one or more of these values). The choice of discriminating value for declaring "presence" or "absence" of a characteristic is a *model parameter*. It is important to note that if, for example, one of the characteristics of the model involves a sequence alignment step to identify a common sequence region, then the way in which the alignment was done, including the software used and its settings, are also *parameters* of one element of your model.

It should now be clear that construction of a predictive model of effector function requires optimization of three fundamental elements:

(a) The set of defining sequence *characteristics*.

(b) The *parameters* associated with determining presence or absence of those characteristics.

(c) The model *structure* linking those characteristics together.

Ultimately, the end product of the model construction process will be a model in which each candidate protein has its sequence reduced down to an *optimal* derived characteristic profile, an *optimal* model structure, and an *optimal* set of parameter values (*see* Fig. 1).

2.1 Optimization and Evaluation of the Classifier

In this section we discuss techniques used to compare alternative models on their predictive capacity. This enables determination of whether a particular *characteristic* gives any additional useful diagnostic information, and what model *structure* and *parameter* values give the best predictive performance. While classifier construction typically does, and should, begin with biologically plausible reasons to include some characteristics and exclude others, their interactions via the model *structure* and *parameters* may still be optimized. For example, an initial choice of *characteristics* may be redundant if one characteristic can wholly be explained by the presence or absence of another.

Having selected a suitable set of *characteristics* and a complementary set of candidate model *structures*, the next step is to optimize the model's *parameter* values. This is not always straightforward: parameter values can influence each other and should not be optimized in isolation but rather as a whole. Unfortunately this can quickly become computationally difficult; a full discussion is beyond the scope of this article, and interested readers are directed toward these excellent reviews [34, 35]. The complexity of the problem can be reduced by constraining the "legal" set of parameter values so that the number of combinations is small enough that models can be run with all combinations of parameters in an *exhaustive search*, and then ranked in order of fitness.

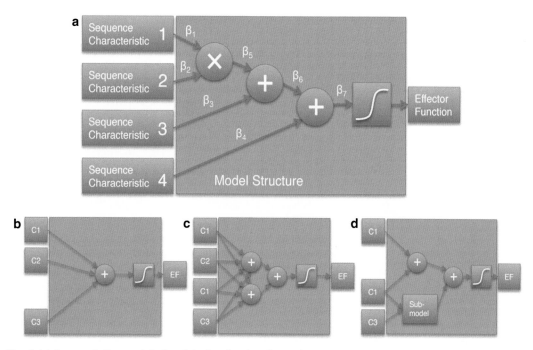

Fig. 1 (**a**) Any predictive model of protein function has three elements: the input *characteristics* of a sequence to be classified (four arbitrary characteristics C1–C4 are indicated here, representing concepts such as, e.g., presence or absence of a signal peptide, protein length, etc.); the model *structure*, i.e., the ways in which the characteristics are combined to produce the single output variable (represented as *addition* and *multiplication*); and the *parameters* of the model (β_1–β_7) that modify interactions between characteristics. The sigmoidal curve represents integration of the model characteristics, structure, and parameters to produce a single output response (EF: "effector function"—this may be simply binary, i.e., effector/not effector, or express some level of confidence in the prediction of effector function). Alternative models may differ in terms of *parameters*, the number and choice of *characteristics* (e.g., three in (**b**), four in (**c**)), the internal model *structure*, or all three elements. In the model optimization step all three factors can, and should be, optimized. The four different structures in the figure represent (**a**) mixed logic model (**b**) linear sum (logistic regression) model (**c**) projection model, and (**d**) embedded sub-model—all of which could potentially compete against each other through CV and ROC analysis

However the set of parameter values is chosen, each candidate classifier must be robustly evaluated, and this evaluation should be realistic in representing real-world performance of the classifier. The accuracy of this performance can be expected to be dependent on two things: quality of the reference data set, and the metric used to evaluate classifier performance.

To evaluate a binary (yes/no) classifier, we require two sequence datasets: *positive* examples of known effectors, and *negative* examples of proteins that are not effectors. To be most useful, positive and negative sequences should be matched as closely as possible, e.g., in terms of length, sequence composition, and other features not thought to bear directly upon effector function. The basic principle is that we measure the classifier's ability correctly to classify *positive* sequence examples as being effectors, and *negative* sequence examples as not being effectors. It is important to note that the number of proteins in the reference dataset is inextricably linked to the confidence we can have in a classifier's predictive ability: the larger the dataset, the smaller the uncertainty we have in the performance metric (by the Central Limit Theorem). However, for a fixed number of reference sequences, the better the classifier is, the smaller the uncertainty we have in the performance metric and, consequently, the choice of sample size is to some extent subjective.

A standard approach to evaluating model performance is to use cross-validation (CV) [13, 14, 18, 36–38]. The simplest approach to CV is to partition the sequence data into two parts: ideally about two thirds of the sequence data should be placed in a *training set*, selected randomly from both *positive* and *negative* sequence sets. The remaining sequences are termed the *holdout set*. Care must be taken that the *holdout* set should be a representative subsample of the complete sequence data set. Optimization of the model—the selection of *parameters, characteristics,* and *structure*— should take place on the *training* set only. The predictive ability of the model should be assessed only on the *holdout* set. This is done to minimize *overfitting*, in which the classifier is unwittingly trained specifically to perform well on the sample data set, but loses the ability to *generalize* (i.e., to classify well sequences that were not part of the reference set).

A single round of data partitioning is not typically sufficient to produce a realistic estimate of classifier performance. Multiple rounds of CV, partitioning the same data into several distinct *training* and *holdout* sets should be performed. The classifier's performance with a specific *model* (i.e., a single set of characteristics and *parameters*) is taken to be its average over all these rounds of partitioning.

Several modes of partitioning (e.g., bootstrap, *k*-fold, and leave-one-out (LOO) cross-validation) are common, and some such as LOO are particularly useful for effector prediction problems that involve small sample sizes of known effectors, or when only positive examples are known, for which a one third-two thirds split is not reasonable.

The results from CV should be considered as the frequencies with which the model produces four categories of result:

1. TP: count of known effector proteins correctly classified as "effector".

2. TN: count of known non-effector proteins correctly classified as "non-effector".

3. FP: count of known non-effector proteins incorrectly classified as "effector".

4. FN: count of known effector proteins incorrectly classified as "non-effector".

These may be combined to give useful summary estimates of performance:

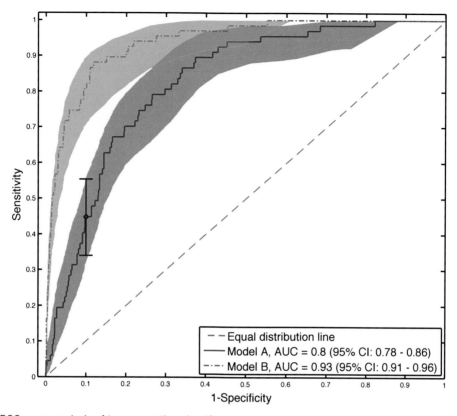

Fig. 2 ROC curve analysis of two competing classifier models. Model A has an estimated Area Under the ROC curve (AUC) of 0.8, and Model B has an AUC of 0.93. The *shaded* regions of both curves are the 95 % confidence interval (CI) calculated using bootstrap resampling (in this case $n=100$), from which the 95 % CI can be calculated for both the overall AUC, and for the Sensitivity corresponding to a given fixed Specificity. In this figure, for Model A, the 95 % CI for Sensitivity (0.35–0.59) is estimated for a fixed Specificity of 0.9. Note that, even though the competing ROC curves are significantly different (i.e., the AUC 95 % CI do not overlap), the overlapping 95 % CI *shaded* regions implies that, for the given reference data set, the performance is not significantly different at a Specificity <0.75

1. Sensitivity (Sn): $TP/(TP+FN)$; interpreted as the probability that an effector protein will be correctly classified.

2. Specificity (Sp): $TN/(TN+FP)$; interpreted as the probability that a non-effector will be correctly classified.

   Taken in isolation, sensitivity and specificity may be misleading of overall performance (especially when *prevalence* is low—*see* Subheading 3), and derived measures are useful to avoid this problem, such as:

1. Positive Predictive Value (PPV, precision): $TP/(TP+FP)$; interpreted as the probability that a positive classification result is correct.

2. *F*-score: $(2 \times PPV \times Sn)/(PPV+Sn)$; this is a weighted average of PPV and Sn, and ranges between 0 and 1, where an *F*-score closer to 1 is better.

One of the simpler and better ways to visualize change in classifier performance as model parameters are modified is the ROC (Receiver Operating Characteristic) curve [10, 11, 14, 15] (Fig. 2). ROC curves show how sensitivity and specificity vary as the model changes, and the area under the curve (AUC) may be interpreted as the probability that the classifier would rank a randomly chosen effector more highly than a randomly chosen non-effector. The ROC curve is widely considered to be an objective and statistically valid method for model performance evaluation [39–41].

3   Interpretation of Classifier Results

Having chosen the optimal model using the CV protocol outlined above, it should be tested, and its parameters fine-tuned, using *all* the available reference set data. Typically, effector prediction is applied on the basis of a small set of reference proteins (total number of positive and negative examples: $n < 100$) compared to the total number of gene products that will be tested for any individual (e.g., $\approx$14,000 for *P. infestans*) or a population. Any performance measure is thus only an approximation, based on a sample, to the (unmeasurable) performance of the model applied to the target population on which it may be used, which potentially includes all sequenced isolates of the target organism, or even all sequenced bacteria and archaea. Therefore, just as one should always quote a standard error when calculating sample means, one should always provide confidence intervals for the performance metrics described above, and this is typically done at the 95 % level. In the context of binary classifier performance, the 95 % CI for Sensitivity with a fixed Specificity may be determined, which allows us to state, for example, that "we calculated with 95 % confidence that the true Sensitivity of our classifier, given a fixed Specificity of 0.8, is within the range 0.7–0.9". It is imperative that a final performance metric is calculated in this way. A straightforward and widely applicable non-parametric method for calculating CIs is bootstrap percentile resampling, and other methods are readily available in most modern statistics or epidemiology textbooks [36].

Having constructed and optimized a predictive model, producing estimates of its performance as outlined above, only then is the model ready to be used to identify candidate effectors from a predicted gene product complement. This application should produce a *positive* (predicted effector) or *negative* (predicted non-effector) classification for each sequence. This is less an act of "prediction", rather one of *hypothesis generation*, in which the "prediction" that a gene codes for an effector is actually a hypothesis that the gene product has effector function. This process may drastically reduce the overall sequence search space, but still requires to be followed by rigorous experimental testing of each individual candidate effector [10].

3.1 The Base Rate Fallacy and Bayesian Statistics

There is a common statistical fallacy known as the *Base Rate Fallacy* that, in the context of an effector classifier, can have the effect of significantly increasing the number of false positive predictions made, relative to what would be expected from a naïve interpretation of the performance statistics. Simply, the very large proportion of an organism's gene complement that does not code for an effector can still give a significant *number* of false positive predictions, even with an acceptably high specificity for the predictive model that gives a low *rate* of false positives [14, 15].

For example, consider a genome with 5,000 predicted gene products, of which approximately 50 are effectors. This is an effector *base rate* of 1 %. If our classifier has a sensitivity of 0.99 (only 1 % of effectors are misclassified) and a specificity of 0.99 (only 1 % of non-effectors are misclassified) from our estimates of performance above, we find that:

1. The classifier acts on 4,950 non-effectors, predicting that $(1-0.99) \times 4{,}950 \approx 49$ of them are effectors (these are *false positives*).

2. The classifier acts on 50 effectors, predicting that $0.99 \times 50 \approx 49$ of them are effectors (these are *true positives*).

We have 98 positive predictions, of which half are true positives, and half false positives, yet our classifier had 99 % sensitivity and 99 % specificity!

This is due to the effect of the low *base rate*, or *prevalence*, of effector occurrence. It is straightforward to account for this effect with a Bayesian representation of the test result as a conditional probability, where **E** means that the candidate sequence under test is truly an effector (~**E** means that it is not), and + means that the classifier called the candidate sequence as *positive* (i.e., predicted that it was an effector):

$$P(\mathbf{E} \mid +) = P(+ \mid \mathbf{E}) P(\mathbf{E}) / P(+) = P(+ \mid \mathbf{E}) P(\mathbf{E}) / \big(P(+ \mid \mathbf{E}) P(\mathbf{E}) + P(+ \mid \sim \mathbf{E}) P(\sim \mathbf{E}) \big).$$

Plugging in the numbers from our example:

$$P(\mathbf{E} \mid +) = 0.99 \times 0.01 / (0.99 \times 0.01 + 0.01 \times 0.99) = 0.5,$$

demonstrates that, for this classifier performance and this base rate of effector occurrence, the probability that a positively classified protein is truly an effector is only 0.5. We should clearly, in this case, be careful of overstating the validity of our predictions. In almost every circumstance we should cautiously note that, purely because of the low-base rate of occurrence of effector proteins in genomes, the expected performance of any classifier on a genome is likely to be poorer than the reported estimate on the reference sequences, if this effect is not explicitly taken into account.

4 Guideline Protocol for Constructing a Classifier of Effectors

1. Identify a set of m known (experimentally demonstrated) effector sequences that share common characteristics: the *positive* sequence set.

2. Identify a set of $\approx m$ known (experimentally demonstrated) non-effector sequences, similar to the known effectors identified above: the *negative* sequence set.

3. Generate k sets of *training* and *holdout* sequences from the positive and negative sequence sets (the reference dataset) for k-fold cross-validation (CV). This should be done by first randomly splitting the reference dataset into k groups. Then for each round of CV one of the k groups is used as the *holdout* set, and the other $k-1$ groups as the training set. Ideally, the training and *holdout* sets should contain approximately the same proportions of positive and negative sequence examples.

4. Apply each *model* (which consists of a specific set of *characteristics, parameters,* and *structure,* as described above) to each training set (e.g., to optimize parameters, or to generate a HMM profile for HMMer), and test the model's performance on the corresponding holdout set, collecting TP, TN, FP, and FN for each training/holdout set combination.

5. Construct a ROC curve based on the k pairs of Specificity/Sensitivity statistics, and calculate the area under the ROC curve (AUC). This is the estimated model performance.

6. If required, compare the relative performance of models and parameter sets on the basis of their estimated performance, and proceed by preferring the best-performing model.

7. Assess the best-performing model using *all* the dataset. Optionally, re-optimize parameter values for this *fixed structural model.* Calculate 95 % confidence interval for Sensitivity given a fixed, predicated Specificity.

8. Apply the chosen model(s) to the predicted gene complement, noting the positive and negative predictions. Use Bayes' equation to interpret the likelihood that an individual positive prediction actually corresponds to a true positive prediction, given the model's estimated performance.

5 Recommendations for Reporting an Effector Classifier

Record and report the positive and negative example sequences used to train and test the classifier.

Report the classification and modeling methods used, the validation steps performed, and the estimated predictive performance of the optimized model (including confidence intervals and the correction for base rate).

Report and make public any and all software code that was used in the generation and testing of the model.

Compare the performance of the classifier to previously published classifiers using appropriate statistical or quantitative methods.

References

1. Bart R, Cohn M, Kassen A, McCallum EJ, Shybut M et al (2012) High-throughput genomic sequencing of cassava bacterial blight strains identifies conserved effectors to target for durable resistance. Proc Natl Acad Sci U S A. doi:10.1073/pnas.1208003109

2. Agbor TA, McCormick BA (2011) *Salmonella* effectors: important players modulating host cell function during infection. Cell Microbiol 13:1858–1869.doi:10.1111/j.1462-5822.2011.01701.x

3. Fabro G, Steinbrenner J, Coates M, Ishaque N, Baxter L et al (2011) Multiple candidate effectors from the oomycete pathogen *Hyaloperonospora arabidopsidis* suppress host plant immunity. PLoS Pathog 7:e1002348. doi:10.1371/journal.ppat.1002348

4. Kim J-G, Taylor KW, Mudgett MB (2011) Comparative analysis of the XopD type III secretion (T3S) effector family in plant pathogenic bacteria. Mol Plant Pathol 12:715–730. doi:10.1111/j.1364-3703.2011.00706.x

5. Kimbrel JA, Givan SA, Temple TN, Johnson KB, Chang JH (2011) Genome sequencing and comparative analysis of the carrot bacterial blight pathogen, *Xanthomonas hortorum* pv. *carotae* M081, for insights into pathogenicity and applications in molecular diagnostics. Mol Plant Pathol 12:580–594. doi:10.1111/j.1364-3703.2010.00694.x

6. O'Brien HE, Desveaux D, Guttman DS (2011) Next-generation genomics of *Pseudomonas syringae*. Curr Opin Microbiol 14:24–30. doi:10.1016/j.mib.2010.12.007

7. Vleeshouwers VGAA, Raffaele S, Vossen JH, Champouret N, Oliva R et al (2011) Understanding and exploiting late blight resistance in the age of effectors. Annu Rev Phytopathol 49:507–531. doi:10.1146/annurev-phyto-072910-095326

8. Sarris PF, Skandalis N, Kokkinidis M, Panopoulos NJ (2010) In silico analysis reveals multiple putative type VI secretion systems and effector proteins in *Pseudomonas syringae* pathovars. Mol Plant Pathol 11:795–804. doi:10.1111/j.1364-3703.2010.00644.x

9. Boch J, Bonas U (2010) *Xanthomonas* AvrBs3 family-type III effectors: discovery and function. Annu Rev Phytopathol 48:419–436. doi:10.1146/annurev-phyto-080508-081936

10. McDermott JE, Corrigan A, Peterson E, Oehmen C, Niemann G et al (2011) Computational prediction of type III and IV secreted effectors in gram-negative bacteria. Infect Immun 79:23–32. doi:10.1128/IAI.00537-10

11. Xia J, Broadhurst DI, Wilson M, Wishart DS (2012) Translational biomarker discovery in clinical metabolomics: an introductory tutorial. Metabolomics. doi:10.1007/s11306-012-0482-9

12. Cornelis GR (2006) The type III secretion injectisome. Nat Rev Microbiol 4:811–825. doi:10.1038/nrmicro1526

13. Whisson SC, Boevink PC, Moleleki L, Avrova AO, Morales JG et al (2007) A translocation signal for delivery of oomycete effector proteins into host plant cells. Nature 450:115–118. doi:10.1038/nature06203

14. Löwer M, Schneider G (2009) Prediction of type III secretion signals in genomes of gram-negative bacteria. PLoS ONE 4:e5917. doi:10.1371/journal.pone.0005917

15. Arnold R, Brandmaier S, Kleine F, Tischler P, Heinz E et al (2009) Sequence-based prediction of type III secreted proteins. PLoS Pathog 5:e1000376. doi:10.1371/journal.ppat.1000376

16. Sui T, Yang Y, Wang X (2013) Sequence-based feature extraction for type III effector prediction. Int J Biosci Biochem Bioinforma 3:246–251. doi:10.7763/IJBBB.2013.V3.206

17. Liu C, Che D, Liu X, Song Y (2013) Applications of machine learning in genomics and systems biology. Comput Math Methods Med 2013:587492. doi:10.1155/2013/587492

18. Broadhurst D, Kell DB (2006) Statistical strategies for avoiding false discoveries in metabolomics and related experiments. Metabolomics 2:171–196

19. O'Brien HE, Thakur S, Gong Y, Fung P, Zhang J et al (2012) Extensive remodeling of

the *Pseudomonas syringae* pv. *avellanae* type III secretome associated with two independent host shifts onto hazelnut. BMC Microbiol 12:141

20. McNally RR, Toth IK, Cock PJA, Pritchard L, Hedley PE et al (2012) Genetic characterization of the HrpL regulon of the fire blight pathogen *Erwinia amylovora* reveals novel virulence factors. Mol Plant Pathol 13:160–173. doi:10.1111/j.1364-3703.2011.00738.x

21. Arnold DL, Jackson RW (2011) Bacterial genomes: evolution of pathogenicity. Curr Opin Plant Biol 14:385–391. doi:10.1016/j.pbi.2011.03.001

22. Haas BJ, Kamoun S, Zody MC, Jiang RHY, Handsaker RE et al (2009) Genome sequence and analysis of the Irish potato famine pathogen *Phytophthora infestans*. Nature 461: 393–398. doi:10.1038/nature08358

23. Win J, Morgan W, Bos JIB, Krasileva KV, Cano LM et al (2007) Adaptive evolution has targeted the C-terminal domain of the RXLR effectors of plant pathogenic oomycetes. Plant Cell 19:2349–2369. doi:10.1105/tpc.107.051037

24. Bhattacharjee S, Hiller NL, Liolios K, Win J, Kanneganti T-D et al (2006) The malarial host-targeting signal is conserved in the Irish potato famine pathogen. PLoS Pathog 2:e50. doi:10.1371/journal.ppat.0020050

25. Petnicki-Ocwieja T, Schneider DJ, Tam VC, Chancey ST, Shan L et al (2002) Genomewide identification of proteins secreted by the Hrp type III protein secretion system of *Pseudomonas syringae* pv. tomato DC3000. Proc Natl Acad Sci U S A 99:7652–7657. doi:10.1073/pnas.112183899

26. Greenberg JT, Vinatzer B (2003) Identifying type III effectors of plant pathogens and analyzing their interaction with plant cells. Curr Opin Microbiol 6(1):20–28

27. Bogdanove AJ, Schornack S, Lahaye T (2010) TAL effectors: finding plant genes for disease and defense. Curr Opin Plant Biol 13: 394–401. doi:10.1016/j.pbi.2010.04.010

28. Boch J, Scholze H, Schornack S, Landgraf A, Hahn S et al (2009) Breaking the code of DNA-binding specificity of TAL-type III effectors. Science. doi:10.1126/science.1178811

29. Yang Y (2012) Identification of novel type III effectors using latent Dirichlet allocation. Comput Math Methods Med 2012:696190. doi:10.1155/2012/696190

30. Wang Y, Zhang Q, Sun M-A, Guo D (2011) High-accuracy prediction of bacterial type III secreted effectors based on position-specific amino acid composition profiles. Bioinformatics 27:777–784. doi:10.1093/bioinformatics/btr021

31. Macho AP, Ruiz-Albert J, Tornero P, Beuzón CR (2009) Identification of new type III effectors and analysis of the plant response by competitive index. Mol Plant Pathol 10:69–80. doi:10.1111/j.1364-3703.2008.00511.x

32. Xu S, Zhang C, Miao Y, Gao J, Xu D (2010) Effector prediction in host-pathogen interaction based on a Markov model of a ubiquitous EPIYA motif. BMC Genomics 11(Suppl 3):S1. doi:10.1186/1471-2164-11-S3-S1

33. Jehl M-A, Arnold R, Rattei T (2010) Effective – a database of predicted secreted bacterial proteins. Nucleic Acids Res. doi:10.1093/nar/gkq1154

34. Guyon I, Elisseeff A (2003) An introduction to variable and feature selection. J Mach Learn Res 3:1157–1182

35. Saeys Y, Inza I, Larrañaga P (2007) A review of feature selection techniques in bioinformatics. Bioinformatics 23:2507–2517. doi:10.1093/bioinformatics/btm344

36. Eriksson L, Johansson E, Kettaneh-Wold N, Wold S (2001) Multi- and megavariate data analysis: principles and applications. Umetrics AB, Umea

37. Brereton RG (2003) Chemometrics: data analysis for the laboratory and chemical plant. Wiley, Chichester UK

38. Efron B, Tibshirani R (1997) Improvements on cross-validation: the .632+ bootstrap method. J Am Stat Assoc 92:548–560. doi:10.1080/01621459.1997.10474007

39. Obuchowski NA, Lieber ML, Wians FH (2004) ROC curves in clinical chemistry: uses, misuses, and possible solutions. Clin Chem 50:1118–1125. doi:10.1373/clinchem.2004.031823

40. Zweig MH, Campbell G (1993) Receiver-operating characteristic (ROC) plots: a fundamental evaluation tool in clinical medicine. Clin Chem 39(4):561–577

41. Lasko TA, Bhagwat JG, Zou KH (2005) The use of receiver operating characteristic curves in biomedical informatics. J Biomed Inform 38(5):404–415

Part II

Cell Biology

Chapter 5

High-Throughput Imaging of Plant Immune Responses

Martina Beck, Ji Zhou, Christine Faulkner, and Silke Robatzek

Abstract

Fluorescence confocal microscopy has emerged in the past decade as an important method for studying the cellular changes associated with plant–microbe interactions. One such change is the internalization into endosomes of the cell surface receptor FLAGELLIN SENSING 2 (FLS2) upon activation by its ligand, bacterial flagellin (flg22). Quantification of endosomes containing FLS2 can thus be used as a direct readout of immune response activation at the cellular level. High-throughput imaging of cellular events is routinely applied in chemical screening for pharmaceutical drug discovery, and we have adapted this system for quantification of plant leaf cellular parameters. In this chapter we describe the instrument setup for high-throughput imaging of leaves, protocols for flg22-induced endocytosis, image acquisition for fluorescent-tagged FLS2 receptors and subcellular markers, automated image analysis of cellular parameters, and data outputs of FLS2 endocytosis.

Key words Plant, Microbe, Flagellin, flg22, FLS2, Immunity, Endocytosis, Receptor, Confocal microscopy, High-throughput imaging, Bioimage informatics

1 Introduction

Pattern recognition receptors located at the cell surface are key components of the molecular machinery used to detect the presence of potentially pathogenic microbes and initiate signaling pathways that regulate plant immunity [1]. The FLAGELLIN SENSING 2 (FLS2) receptor is known to mediate primary immune responses to bacterial infection through the perception of the conserved flagellin peptide flg22 [2, 3]. Pathogenic bacteria secrete effectors that suppress signaling triggered by FLS2 [4, 5], highlighting the importance of this receptor in the plant immunity. FLS2 is a leucine-rich repeat receptor kinase that interacts with the BRASSINOSTEROID INSENSITIVE 1 (BRI1)-ASSOCIATED KINASE 1 (BAK1) in a flg22-dependent manner to form an active signaling-receptor complex [6]. This induces internalization of the FLS2 receptor from the plasma membrane into endosomes [7]. The subcellular dynamics of this receptor can be visualized by confocal microscopy of plants stably expressing a functional fusion of

Paul Birch et al. (eds.), *Plant-Pathogen Interactions: Methods and Protocols*, Methods in Molecular Biology, vol. 1127, DOI 10.1007/978-1-62703-986-4_5, © Springer Science+Business Media New York 2014

FLS2 to green fluorescent protein (GFP) under the transcriptional control of its endogenous promoter [8].

There is an increasing demand for high-throughput imaging to overcome the labor- and time-intensive limitations that arise when studying plant–microbe interactions at a cellular scale. Flg22-induced endocytosis of FLS2-GFP depends on the activation status of the receptor complex and therefore provides a measurable readout of the activation of plant immune responses at the cellular level. We have developed a high-throughput confocal microscopy method that allows us to quantify receptor-mediated endocytosis in plant immunity and to dissect the endocytic pathway. This technology is used routinely in the pharmaceutical industry for drug discovery and we have modified protocols for imaging of the plant leaves. In combination with automated bioimaging analysis this approach allows extraction of meaningful data from defined leaf cellular and subcellular parameters [9]. This high-throughput imaging method can be used for monitoring flg22-induced FLS2 endocytosis over time, in response to treatments with hormones and chemical compounds such as membrane trafficking inhibitors as well as in backgrounds of different genetic mutations [7]. In addition, high-throughput imaging can be applied to quantify a range of subcellular compartments [7, 9] that are visualized by stable expression of fluorescent-tagged markers of, e.g., Golgi, *trans*-Golgi network (TGN), early endosomes (EE), late endosomes (LE), and multivesicular bodies (MVBs) (*see* published examples in Fig. 1). These markers can also be used for co-localization studies, for example together with flg22-induced FLS2 endosomes over time [7], thereby providing time-dependent resolution of the endocytic trafficking pathway.

In this chapter, we provide a protocol for triggering FLS2 endocytosis with flg22 suited for microscopic analysis. We outline methods for high-throughput confocal microscopy of plant leaves and detail the image acquisition of FLS2-GFP expressing and Red Fluorescent Protein (RFP)-tagged marker plants. To handle large image data sets generated from high-throughput microscopy, and to extract meaningful data from micrographs, we describe how to use our automated image analysis pipeline based on the Opera® Content Screening platform. Although we focus on flg22-induced endocytosis of FLS2 in this chapter this method can be easily adapted for quantification of other cellular and subcellular parameters, and additionally combined with cell segmentation [9–11].

2 Materials

2.1 Samples

1. *pFLS2*::FLS2-GFP Arabidopsis (*Arabidopsis thaliana* accession Col-0) transgenic plants [4, 7].

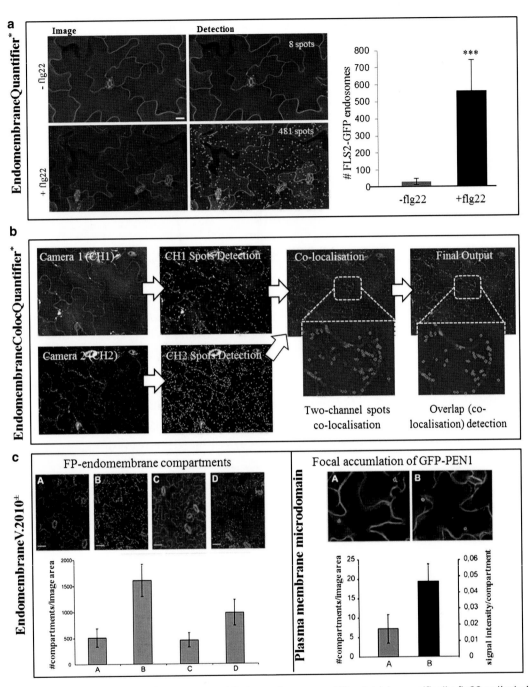

Fig. 1 Examples of quantitative imaging. (**a**)*Endomembrane quantifier* detects specifically flg22-activated FLS2-GFP endosomes. Spot number can be plotted against the time to follow kinetics—reprint from Beck et al. 2012. (**b**) *EndomembraneCoLocQuantifier* detects spots in two different channels and calculates the overlapped spot population between these channels. (**c**) *Right panel: Endomembrane V.2010* can be used for the specific detection of several endomembrane compartments A, GFP-2xFYVE. B, YFP-VTI12. C, YFP-Rab C1. D, YFP-RabF2b/ARA7. *Left panel: PlasmaMembrane microdomain* used to quantify GFP-PEN1 accumulation at fungal entry site 24 h post inoculation with *B. graminis* conidiospores—reprint from Salomon et al. (2010)

2. Dual color lines expressing *pFLS2::FLS2-GFP* and *pUBQ10::RFP-ARA7/*Rab F2b or *pUBQ10::ARA6/*Rab F1-RFP, respectively [7].

3. Accession numbers

AT5G46330 FLAGELLIN SENSING 2.
AT4g19640 ARA7/Rab F2b.
AT3g54840 ARA6/Rab F1.

2.2 Experiment/ Treatments

flg22 peptide: QRLSTGSRINSAKDDAAGLQIA (custom produced; EZBiolab/USA).

MW = 2,272.5 g/l.

2.3 Image Acquisition

2.3.1 Imaging/ Sample Device

Samples are mounted on a *custom-made 96 pin aluminum stamp* (400 g). Pins are 4 mm in diameter, 1.6 cm in length, and maximize contact of the cotyledon with the *optical glass bottom* of the 96-well plate (Greiner Sensoplate) (*see* **Note 2**). The tip of each pin contains a 4 × 4 mm-thick *cellular-rubber pin* (ethylene propylene diene rubber, JACOB NETTEKOVEN) coated with *a polyvinyl chloride film* (Mactac 8900) to minimize leaf tissue damage.

2.3.2 Opera® High Content Screening Microscope (PerkinElmer Cellular Technologies, Germany)

1. *High resolution* spinning disc confocal imaging plate reader.

2. *High speed* multicolor image acquisition.

3. *4 laser* based excitation sources 405 nm—violet, 488 nm—blue, 561 nm—green/yellow, 640 nm—red.

4. Xenon high-pressure lamp (optional) 360–410 nm.

5. *4 independent CCD detectors:* Maximum of 3 confocal cameras, 1 non-confocal camera (1.3 Megapixel, 12 bit resolution, up to 6 bandpass filters in front of each camera, customized filter sets).

6. 4 exchangeable objective lenses (maximum three *water immersion*, one air lens).

7. Automatic immersion water supply.

8. *Auto focus* system.

9. Minimum *exposure time of 40 ms.*

10. Compatible with the variety of plate types, including optical *glass bottom plate Sensoplate©Greiner* (Germany).

11. *Spinning disc:* Yokogawa Modified Nipkow CSU10, Concurrent spots: ~1,000, Pinhole diameter: 50 μm; Axial resolution: 1.3–10 μm; Lateral resolution to 0.25 μm.

These optical features allow the user to

12. Eliminate background fluorescence.

13. Significantly improve signal-to-noise performance.

14. Minimize photobleaching and phototoxicity for living cells.

15. Perform comparative imaging assays/studies.

16. Study subcellular features with sufficient resolution using a high numerical aperture (NA) water immersion lens.

17. Detect multiple parameters by simultaneous four-color imaging.

2.4 Image Processing

2.4.1 Acapella® High Content Imaging and Analysis Software System (PerkinElmer)

The Opera® high content screening system operates on the Acapella software platform. This software enables users to extract quantitative data from images for statistical analysis in batch-processing mode.

Acapella facilitates rapid online image data processing (embedded in the Columbus™ system) as well as flexible image analysis for high content cellular applications, including multiparametric multiplex assays.

System requirements

1. Acapella Windows Edition (v2.0) requires Microsoft® Windows® XP or Vista, v2.6 or above and can be operated on Windows 7.

2. *Minimum specification*: Pentium® III 1 GHz, 512 MB RAM, 1 GB free HD.

3. *Recommended specification*: Pentium® 4.3 GHz, 1 GB RAM,10 GB free HD.

4. *Acapella Linux Edition* requires OpenSuSE 10.3 64 bit.

5. *Minimum specification*: Intel® Core™ 2, 2 GHz, 2 GB RAM, 1 GB free HD.

6. *Recommended specification*: Intel® QuadCore 4.3 GHz, 4 GB RAM, 10 GB free HD.

Image formats

7. Opera image format .flex files.

8. Over 30 standard image formats such as TIFF, JPEG, GIF, PNG, etc.

Data outputs

9. Analysis results can be exported from the Acapella system in .xml, .csv, and .xls file formats.

2.5 Scripts

1. *EndomembraneQuantifier*and*EndomembraneCoLocQuantifier* can be freely obtained online: https://sourceforge.net/projects/bioimage/files/Endosome Detection/.

2. *MaxProjectionAndImageSlices* can be freely obtained online: https://sourceforge.net/projects/bioimage/files/OperaMIP/.

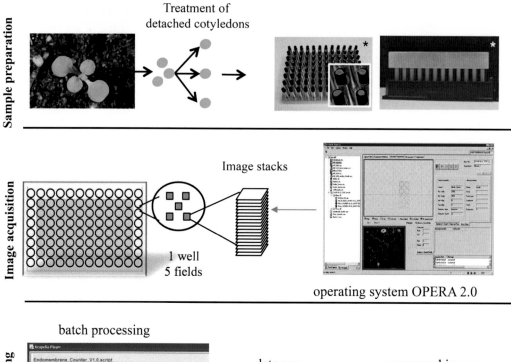

Fig. 2 Quantitative imaging—general work flow. For sample preparation, detached cotyledons are mounted on custom-made stamp which is placed in a 96-well glass-bottom plate. Sample treatments can be performed before the mounting step or within the 96-well plate. The *image acquisition* operates with OPERA V2.0, layout, field numbers per well, and slice number per field can be chosen as well as the microscopy setup. The generated .flex files are batch processed in Acapella V.2.0 with custom tailored script. The *data output* is subdivided in processed images and specific output parameters in table format

3 Methods

General workflow: *see* Fig. 2

Sample preparation including treatments (Subheadings 3.1 and 3.2), image acquisition of 60 samples in 96-well plate (Subheading 3.3), image processing and computational detection (Subheading 3.4).

3.1 Samples

For high-throughput confocal imaging, seedlings are grown for 2 weeks on soil under controlled environment conditions (12 h light and 60 % humidity). Imaging is performed on cotyledons or leaf discs.

**3.2 Experiment/
Treatment**

1. Detached cotyledons of 2-week old *Arabidopsis* plants are positioned with upper epidermis facing upward on stamp pins. The stamp is then fitted into optical plates containing appropriate aqueous solutions (*see* **Note 1**). Silicon grease can be used to help mount samples on the stamp pins.

2. For chemical treatment and/or cell staining (prior to mounting), detach cotyledons and place in treatment solution (*see* Table 1: Inhibitors and Staining). Vacuum infiltrate for 5 min and incubate for a further 55 min at room temperature (incubation time may vary for different treatments). Mount cotyledons on the imaging stamp and transfer to a 96-well-imaging plate containing relevant treatment solution (*see* **Note 2**).

3. Flg22 treatment can be carried out in the 96-well-imaging plate by directly applying 50 μl of 10 μM flg22 solution to each well of the optical plate. Carefully invert the mounted stamp into the wells of the plate. The timing of flg22-incubation prior imaging depends on individual experiments. To visualize the flg22-induced FLS2-GFP endosomes, we recommend 30 min incubation with flg22 prior to image acquisition. The kinetics of FLS2 incorporation into flg22-induced endosomes can be then measured 30–70 min after flg22-treatment.

3.3 Image Acquisition

(*See* **Note 3**).

3.3.1 Exposure Parameters

(a) Excitation of the samples is performed with the 488 nm laser line for GFP and the 561 nm laser line for RFP.

(b) Water immersion 40× objective (40× PlanApo NA = 0.9) is used for subcellular imaging of FP-labeled endosomes.

	GFP	RFP
(a)	Laser 488 Power: 7,000 μW	Laser 561 Power: 1,500 μW
(b)	Filter for Cam 1: 540/75	Filter for Cam 2: 600/40
(c)	Detection Dichro: 568	Detection Dichro: 510
(d)	Primary Dichro: 405/488/640	Primary Dichro: 405/561/640
(e)	Cam 1 exposure time: 120 ms	Cam 2 exposure time: 120 ms
(f)	Focus height: 10 μm	Focus height: 10 μm

3.3.2 Experiment Parameters

.lay: Plate layout consists of 60 wells, outer columns and rows remain free when using water immersion objectives.

.sly: Sublayout can contain six fields of interest per well for single exposure experiment or three fields per well for double exposure (*see* **Note 3(f)**).

Table 1
Inhibitors and stains used for dissection of endomembrane pathways

Name	Stock	in	Working	Target	Cell biological effect
flg22	10 mM	dH$_2$O	10 μM	FLS2 receptor	Triggers internalization in endosomes
Brefeldin A (BFA)	10 mM	EtOH	30 μM	GNOM-positive endosomes	Causes formation of BFA bodies
Wortmannin (Wm)	10 mM	DMSO	30 μM	Phosphoinositide (PI) 3-kinase	Enlarges LE/MVB and interferes with endocytic uptake from the plasma membrane
ConcanamycinA (ConcA)	10 mM	DMSO	10 μM	Vacuolar ATPase activity	Inhibits vacuolar transport and acidification of LE/MVB
Tyrphostin A23 (Tyr A23)	30 mM	DMSO	100 μM	Clathrin-mediated endocytosis via the YxxF motif	Inhibits receptor-mediated endocytosis
Endosidin 1 (ES1)	1.13 mM	DMSO	10 μM	TGN/EE and the actin cytoskeleton	Triggers formation of ES1 bodies and stabilizes actin
Latrunculin B (LatB)	10 mM	EtOH	30 μM	Actin cytoskeleton	Depolymerises actin
2,3-Butanedione monoxime (BDM)	Prepare fresh	dH$_2$O	500 mM/50 mM	Myosin ATPase activity	Inhibits intracellular myosin dependent trafficking
FM4-64	5 mg/ml	dH$_2$O	5 μg/ml	Lipophilic membranes	Stains plasma membranes, endosomes, and tonoplasts
Propidium iodide (PI)	5 mg/ml	dH$_2$O	5 μg/ml	Membrane impermeable and DNA intercalating	Stains cell walls and nuclei (nuclear staining only in dead cells)

.stc: Definition of stack by adjusting slice numbers and slice distance. For subcellular imaging we use a total of 21 slices (10 μm focus height ± 10) with 1 μm distance between each slice. This parameter must be adjusted for different sample types and different objectives (*see* **Note 3**).

3.3.3 Start Measurement

The experiment is saved with all parameters and then started as an automatic experiment.

Single camera measurement takes about 40 min, double camera measurements about 80 min.

The images are automatically saved and archived and can be used for image analysis.

3.4 Image Processing

(*See* **Note 4**)

Before conducting analysis on images captured by the Opera screening microscope, users should first ensure that the OperaDB (The Opera database management system) has transferred all captured images to an appropriate image directory (normally an internal image server).

1. Open the Acapella image analysis software system.

2. Drag and drop an image analysis software solution ("Acapella script") into the Acapella software interface. After loading the Acapella script, the interface of the "Acapella Player" will change (*see* Fig. 3). To conduct high-throughput image analysis of flg22-induced endocytosis, three Acapella scripts are normally used:

   MaxProjectionAndImageSlices: this script can perform maximum intensity projections (MIP) and generate z-plane images of input flex files.

   EndomembraneQuantifier [7]: this algorithm can subtract background signals (those derived from chloroplasts, deeper tissues, and guard cells), detect genuine endomembrane compartment signals with very high accuracy, and monitors flg22-induced FLS2 endocytosis over time.

   EndomembraneCoLocQuantifier [7]: in addition to functions provided by EndomembraneQuantifier, this script analyzes individual and overlapping endosomal numbers of GFP- and RFP-labeled endosomes. In addition, it allows the detection of spot-like objects labeled with two different fluorophores and determines the degree of overlap between objects in two different channels.

All three scripts can be freely obtained online (*see* Subheading 2.5 for URLs).

3. In the Data Selection section box (Fig. 3a), choose "Single Step" for analyzing one flex file and "Batch Mode" for analyzing

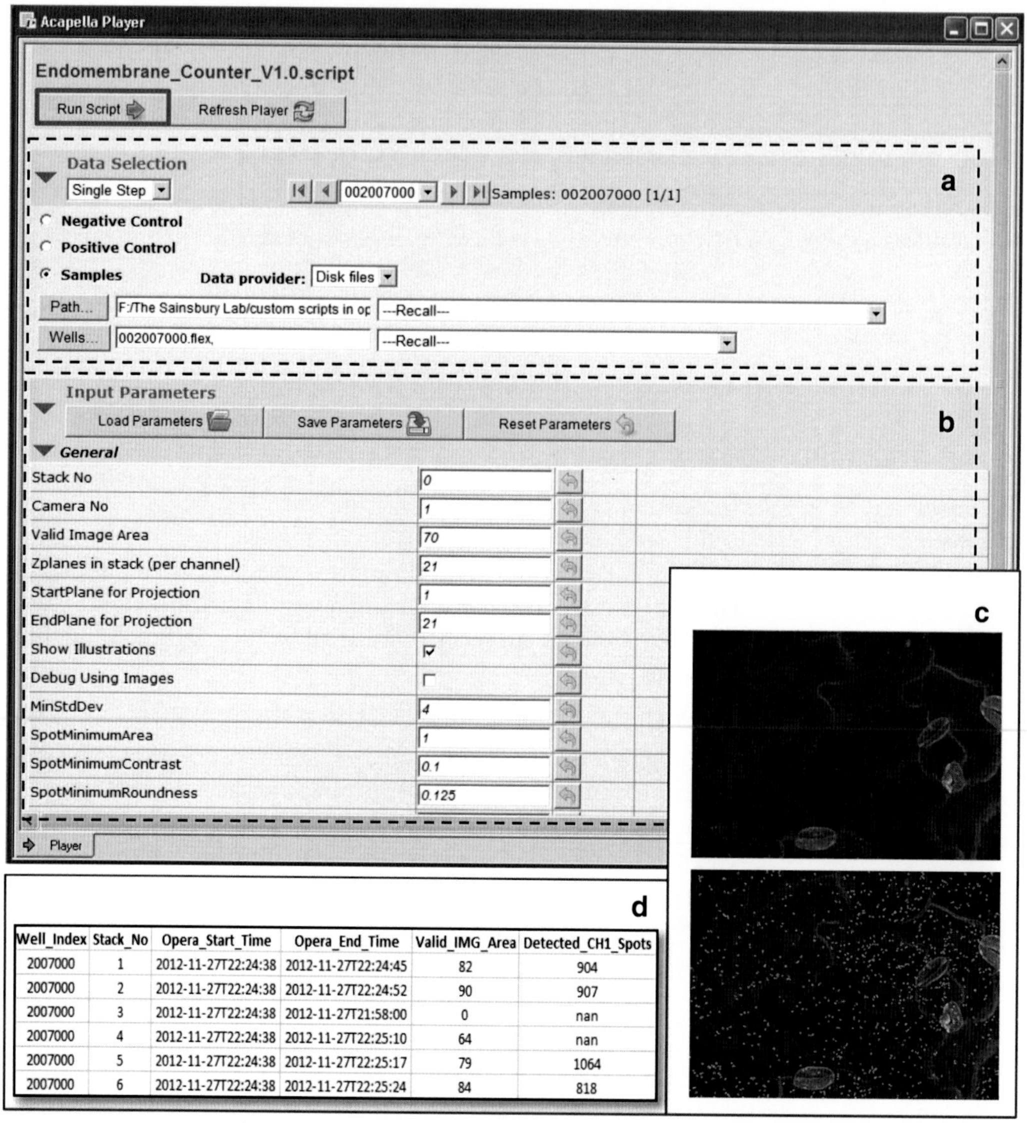

Well_Index	Stack_No	Opera_Start_Time	Opera_End_Time	Valid_IMG_Area	Detected_CH1_Spots
2007000	1	2012-11-27T22:24:38	2012-11-27T22:24:45	82	904
2007000	2	2012-11-27T22:24:38	2012-11-27T22:24:52	90	907
2007000	3	2012-11-27T22:24:38	2012-11-27T21:58:00	0	nan
2007000	4	2012-11-27T22:24:38	2012-11-27T22:25:10	64	nan
2007000	5	2012-11-27T22:24:38	2012-11-27T22:25:17	79	1064
2007000	6	2012-11-27T22:24:38	2012-11-27T22:25:24	84	818

Fig. 3 Acapella software interface. (**a**) In the *Data selection* section, users can choose the directory of their image files (either in OperaDB or in a file system). After selecting the images, they can choose to analyze these images one by one (*Single Step*) or batch processing them (*Batch Mode*). (**b**) In the *Input Parameters* section, users can change a set of input parameters according to their experiment/imaging settings (e.g., the number of z-planes captured in image acquisition). (**c**) Original Opera image and processed image with detected endosomes (highlighted in *green*). Both types are saved in PNG format. (**d**) Quantifiable features such as Opera imaging time, endosome number (per image), and valid image area are exported to a CSV file for further statistical analysis

a group of flex files. Click the "Path" button to select the image directory and "Well" to choose which flex file(s) will be included in the high-throughput image analysis.

4. In the Input Parameters section (Fig. 3b), users can change a set of input parameters:

Stack no: Which image area (the actual z-stack) in a flex file will be analyzed? Enter "0" to evaluate all z-stacks, "1" for analyzing only the first z-stack, "2" for the second stack and so on.

Camera no: how many cameras (up to four cameras can be used by the Opera) were used during image acquisition.

Valid image area: specify the threshold of the valid image area. For example, entering "50" means that if over 50 % of an image is too dark or contains invalid information (e.g., out of focus or over exposed) this image will be discarded from image analysis.

z-planes in stack (per channel): how many z-planes are contained in the stack.

StartPlane and EndPlane for Projection: from which z-plane (the start plane) to which z-plane (the end plane), users would like to perform their image analysis.

Show Illustrations: denote whether to illustrate processed images in the Acapella Player after the image analysis or not.

Debug using images: whether to show intermediary image analysis steps together with related images or not.

SpotMinimumArea,   *SpotMinimumRoundness*, *SpotMinimumContrast*,   *MinStdDev*,   and *SpotMinimumWidth2LengthRatio*: default setting for detecting endosomal signals. Changes to any of the values in these input fields will result in the quantification of compartments with different parameters.

5. After setting the input parameters the analysis workflow can be initiated by clicking the "Run Script" button on the Acapella Player interface.

6. Following the image processing a set of PNG images (maximum intensity projection images and processed images with recognized endosomes, colored green) are produced (Fig. 3c).

7. Features such as well index, stack number, Opera imaging time, valid image area, and detected endosome number are exported and saved in a CSV file (Fig. 3d). If users want to measure roundness, intensity, area, length, and width of endosomal signals, they can click "Generate Report" button to obtain a more detailed report for the endosome detection.

4   Notes

1. Samples

   (a) Spring-loaded scissors facilitate easy cutting and handling of cotyledons (micro scissors 120 mm/Mueller/Braun-Aesulap).

   (b) Small amount of Silicon gel or grease helps to mount leaves on stamp.

    (c) Differences in cuticle thickness of leaves of soil-grown and media-grown plants may require the user to adjust the focus height for different experiments.

    (d) For quantification of dual color marker lines, it is important to compare levels of marker gene expression and fluorescent protein production with parental lines. In our case we generated dual color lines by crossing relevant marker lines. Quantitative PCR and Western blot analysis were used to confirm that homozygous dual color progeny (F2) had comparable levels of marker protein to parental lines.

2. Experiment/treatment

    (a) Stamps are custom made and not commercially available.

    (b) Clean the stamps after use with dH_2O and EtOH, especially the rubber pins.

    (c) Do not place the stamp in vacuum—the rubber pins will be damaged.

    (d) Replace the pins after a significant period of use.

    (e) If possible, keep different stamps for different experimental treatments to prevent cross-contaminations between experiments (e.g., (1) biotic/pathogen; (2) chemical/inhibitors; (3) control/no treatments).

3. Image acquisition

    (a) DO NOT use Table in/out function with stamp on plate.

    (b) Check water supply and rinse water lenses before use.

    (c) If you are not able to find the focus, then check plate type (e.g., Adjustment plate versus company plates) and check whether the plate is uneven in the automatic table, clean the plate bottom with a lens tissue.

    (d) Adjust the laser power, camera exposure time, and band path filter settings for different FP marker lines prior to image acquisition.

    (e) For different tissues (cotyledons vs. primary leaves or leaf discs) the user may need to adjust the focus height and the z-stack step size.

    (f) Do not exceed the recommended amount of images per well (*see* **Note 4**).

    (g) It is possible to use the Opera as a normal microscope or for rapid FP-screening (genotyping) of plant lines by using the provided *MaxProjectionAndImageSlices* script.

    (h) The outer rows/columns of the 96-well plate are restricted when imaged with water immersion high NA objectives and therefore these rows/columns should not be used for these conditions.

4. Image processing

(a) In our experience images require a minimum of 70 % valid image area for correct analysis. Under certain circumstances it may be necessary to manually remove images that have a high level of noise but that fall within the detection parameters.

(b) Adjustment of threshold/sizes/intensities may be necessary for different FP marker lines.

(c) Bioinformatics or scripting experience is highly recommended.

(d) As there are internal memory constraint issues in Acapella (V2.0), we recommend that no more than 150 images per well (i.e., 150 z-planes) should be captured during image acquisition. This equates to around 80–120 MB per flex file. Software versions Acapella V2.6 or onward are not limited in internal memory space.

(e) In *Arabidopsis* cotyledons, EndomembraneQuantifier detects ARA7-positive endosomes, Golgi, and the TGN without changing the object detection parameters.

Acknowledgments

We like to thank the members of the Robatzek laboratory for fruitful discussions and help. Research in S.R.'s laboratory is supported by the Gatsby Charitable Foundation and a grant by the European Research Council (ERC). M.B. is funded by a fellowship from German Research Council (DFG).

References

1. Boller T, Felix G (2009) A renaissance of elicitors: perception of microbe-associated molecular patterns and danger signals by pattern-recognition receptors. Annu Rev Plant Biol 60:379–406

2. Chinchilla D, Bauer Z, Regenass M, Boller T, Felix G (2006) The Arabidopsis receptor kinase FLS2 binds flg22 and determines the specificity of flagellin perception. Plant Cell 18(2):465–476

3. Zipfel C et al (2004) Bacterial disease resistance in Arabidopsis through flagellin perception. Nature 428(6984):764–767

4. Göhre V et al (2008) Plant pattern-recognition receptor FLS2 is directed for degradation by the bacterial ubiquitin ligase AvrPtoB. Curr Biol 18(23):1824–1832

5. Xiang T et al (2011) BAK1 is not a target of the Pseudomonas syringae effector AvrPto. Mol Plant Microbe Interact 24(1):100–107

6. Chinchilla D et al (2007) A flagellin-induced complex of the receptor FLS2 and BAK1 initiates plant defence. Nature 448(7152):497–500

7. Beck M, Zhou J, Faulkner C, MacLean D, Robatzek S (2012) Spatio-temporal cellular dynamics of the Arabidopsis flagellin receptor reveal activation status-dependent endosomal sorting. Plant Cell 24(10):4205–4219

8. Robatzek S, Chinchilla D, Boller T (2006) Ligand-induced endocytosis of the pattern recognition receptor FLS2 in Arabidopsis. Genes Dev 20(5):537–542

9. Salomon S et al (2010) High-throughput confocal imaging of intact live tissue enables quantification of membrane trafficking in Arabidopsis. Plant Physiol 154(3):1096–1104

10. Fitzgibbon J et al (2013) A developmental framework for complex plasmodesmata formation revealed by large-scale imaging of the Arabidopsis leaf epidermis. Plant Cell 25(1):57–70

11. Zhou J, Spallek T, Faulkner C, Robatzek S (2012) CalloseMeasurer: a novel software solution to measure callose deposition and recognise spreading callose patterns. Plant Methods 8(1):49

Chapter 6

In Vivo Protein–Protein Interaction Studies with BiFC: Conditions, Cautions, and Caveats

Petra Boevink, Hazel McLellan, Tatyana Bukharova, Stefan Engelhardt, and Paul Birch

Abstract

Bimolecular fluorescence complementation (BiFC), performed with suitable controls and the right conditions, can be a straightforward and simple method to assess protein–protein interactions accessible to anyone with basic confocal microscopy skills. It is of course not without its own potential pitfalls and requires specific controls. Here we describe its use to study the interactions between pathogen effector proteins and host proteins inside plant cells.

Key words Split-YFP, BiFC, Agroinfiltration

1 Introduction

One of the most interesting questions that arise when novel pathogen effector proteins are identified is: what are their host targets? Screening for host targets has been commonly done using yeast-2-hybrid systems [1]. Putative interactions must be confirmed by at least one other method and in planta methods provide greater confidence that an interaction is genuine, since this is where the putative interaction will be occurring if genuine.

Most effector protein localization and interaction assays are performed with overexpressed proteins with peptide or protein tags. The reason for this is that natural protein levels of either the effectors or their interactors are generally very low. In addition, for cytoplasmic effectors, such as the oomycete RXLR class, the amount of effector translocated from the pathogen to the host may be only a small fraction of the protein expressed. Specific antibodies that may be used to study the localization of unmodified effectors or their targets are time-consuming and costly to generate and are therefore particularly unsuitable for the initial stages of effector

Paul Birch et al. (eds.), *Plant-Pathogen Interactions: Methods and Protocols*, Methods in Molecular Biology, vol. 1127, DOI 10.1007/978-1-62703-986-4_6, © Springer Science+Business Media New York 2014

investigation when it is generally uncertain whether a putative interaction is real or will be tractable to study.

Of the other commonly used protein–protein interaction assays, Co-immunoprecipitation (Co-IP) assays have the advantage that antibody-recognition tags, such as myc and HA, attached to the test proteins can be small, around 10–15 amino acids, and are thus less likely to interfere with interactions. Co-IP assays, however, are generally conducted using high levels of overexpression of the test proteins which can result in cell death, protein aggregation, mislocalization, or other artifacts. We have observed that membrane-associated proteins are particularly disruptive to cells when overexpressed. In Fig. 1 the localization of a GFP-tagged RXLR effector PITG_03192 is shown at low and high levels of transient expression in *Nicotiana benthamiana* cells and the effect of an HA-tagged *Phytophthora infestans* effector PITG_03192 at a high level of transient expression is shown in a plant containing GFP in the ER lumen. At high levels of expression, the ER membranes with which the effector associates are highly disrupted. Not surprisingly this effector eventually causes cell death. What effect such disruption of membranes would have on the results of Co-IP assays is unknown, but the cell biology suggests that overexpression effects should be considered and discussed when presenting Co-IP results.

Fluorescent protein tags are larger than the common antigenic tags but are extensively used as they provide localization information in addition to interaction data. Fluorescence resonance energy transfer (FRET) assays can be performed with full-length fluorescent protein fusions to measure protein–protein interactions but have not been as widely published as might be expected. This may be because in practice the assays require extensive training, very careful attention to standards and controls, and a high level of skill in microscopy. The most accurate measure of FRET is by fluorescence lifetime imaging microscopy (FLIM) which involves specific, expensive equipment [2]. FRET requires that the fluorophores are very close (1–10 nm) and correctly oriented for resonance transfer to occur. This needs a very particular alignment of the interacting fusion proteins to give a positive result, and a negative result, with FRET may not mean that the proteins do not interact.

Bimolecular fluorescence complementation (BiFC) involves a splitting the sequence of a fluorescent protein, commonly yellow fluorescent protein (YFP), and fusing each part to a putative interacting protein [3]. When the proteins interact in a suitable orientation and conformation the fluorescent protein is non-covalently reconstituted and fluorescence is generated. If the proteins do not interact, or do not do so in the appropriate orientation, there is no fluorescence. It can be a straightforward and simple method to assess interactions with only basic confocal training. The difficulties many researchers have with the technique are due to the stability of the re-formed fluorescent protein. If there are sufficient amounts of each fusion protein in the same cellular compartment, they

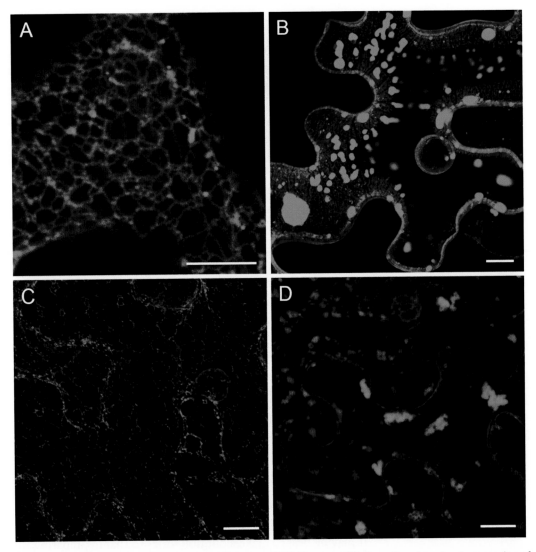

Fig. 1 High levels of overexpression of an ER-targeted RXLR causes ER disruption. Low-level expression of GFP-tagged RXLR PITG_03192 reveals labeling of the ER network (**a**). Higher level expression causes disruption of the ER network (**b**). Note that in image (**b**) the high-level overexpression of the GFP fusion also results in plasma membrane labeling. ER structure in a control leaf of a transgenic plant with GFP in the ER (**c**) and in a leaf expressing a high level of a HA-tagged PITG_03192 (**d**). Scale bars = 10 μm

encounter each other by random chance and the fluorophore is able to re-form to a level that produces a detectable signal. There is even the potential for an interaction to be forced by the association of the fluorophore sequences. A form of split venus YFP has been reported to have reduced self-association stability [4]. However, Lee et al. [5] reported that this variant gave poor fluorescent signal in plant cells even with known strong interactors. In a recent review Kodama and Hu [6] discuss the varied results with a number of different split YFP forms and indicated that no one variant may be ideal for all systems. We have obtained robust results from BiFC studies on effector-host protein interactions using the original split-EYFP. To achieve this we performed BiFC assays with agrobacterium-mediated transient expression using low concentrations of bacteria, observing cells at the earliest possible time points and using the most stringent controls available.

2   Materials

1. YN and YC fusion constructs of test proteins (*see* **Note 1**).
2. Plastic 1 ml syringes without needles.
3. Liquid cultures of agrobacteria transformed with constructs to be tested (separate cultures for each construct-containing strain; *see* **Note 2**).
4. Agroinfiltration medium: 0.01 M $MgCL_2$, 0.01 M MES (2-(*N*-Morpholino) ethanesulfonic acid), 15 µM acetosyringone (*see* **Note 3**).
5. Plants: 4–8-week-old *N. benthamiana*, grown under standard conditions, e.g., in a glasshouse at 22 °C (day temperature) and 18 °C (night temperature) with a minimum of 16 h light.

3   Methods

3.1   Agroinfiltration

1. Centrifuge 1–2 ml (or an appropriate volume) of each agrobacterial culture, remove supernatant, and resuspend the pellet in agroinfiltration medium.
2. Dilute the suspension 1:5–1:10 depending on the density of the bacteria and measure the absorbance at 600 nm (OD_{600}). From this calculate the dilution required for the final mixtures. It may be best to make a dilution series so that in preparing the final mixtures one is not adding very small volumes of highly concentrated suspensions to large volumes of buffer as this leads to inaccuracy. Mix aliquots of the suspensions of the different constructs and controls as required to achieve the final concentrations (*see* **Note 4**).
3. Infiltrate expanded leaves from the abaxial (lower) side of the leaf through the stomata if they are open using the minimum pressure required to achieve a seal around the mouth of the syringe. If the stomata are closed it is best to make a very small nick in the lower epidermis with a sharp blade or needle rather than causing greater damage by trying to force the solution in with greater pressure. The cut needs only to be of minimal depth, just breaking the lower epidermis, and just a pin-prick in size.
4. Leave the plants in suitable conditions in the glasshouse or growth chamber (*see* **Note 5**).

3.2 Imaging

1. Two days after agroinfiltration prepare sections of leaves for imaging (*see* **Note 6**).

2. Image with a confocal microscope using 514 nm excitation and collect emissions between about 520 and 560 nm for YFP. The collection window may be narrowed if expression levels are high enough or if required for multicolor imaging with other fluorophores. Collect multiple images of the most representative cells (*see* **Note 7**). With an epifluorescence microscope suitable bandpass filters are required for optimal imaging and are readily available from all major microscope manufacturers. Image test and control infiltrations using identical settings (*see* **Note 8**).

3.3 Quantification

1. If control protein combinations give some fluorescence it may indicate a degree of instability of the fusion proteins (*see* **Note 9**) or that some nonspecific interaction is possible. For these situations, in particular, it is important to obtain quantifiable measurements of the level of fluorescence. Quantification of the overall fluorescence intensity in images is possible with a number of different image analysis software packages, including those that are associated with confocal microscopes (*see* **Note 10**).

2. Measurements of fluorescence with a fluorimeter can be performed on leaf discs or on ground tissue (*see* **Note 11**). Leaf discs should be as flat as possible and the maximum size possible for the plate wells (*see* **Note 12**). Cutting leaves with a cork borer will give consistent disc sizes. Each well of the plate should be filled with exactly the same volume of water and the leaf discs floated on top, abaxial-side up. Numerous "top-reads", i.e., readings from the upper side such that one is measuring fluorescence from the surface cells of the leaf disc, should be collected from each disc.

3. To prepare ground material for fluorimeter measurements, equal amounts of tissue should be collected from infiltrated and control (non-infiltrated or infiltrated with empty vector and infiltrated with an intact YFP-fusion construct, ideally one giving a similar level and location of fluorescence as the test combination) leaves and ground in a simple buffer; for example, phosphate buffered saline, with a mixture of protease inhibitors added (*see* **Note 13**). We use the same protease inhibitor cocktail that we use for preparing western immunoblot samples. The ground suspension should then be centrifuged briefly, for example, 5 min in a microfuge at $16,000 \times g$ to remove the larger debris. The supernatant should be aliquoted into a 96-well plate such that there are several replicates of each sample in different locations on the plate. Total protein measurements, for example Bradford assays, should be performed on the ground material so that the fluorescence measurements may be normalized for the level of protein present.

4   Notes

1. There are a number of split YFP binary vectors available for Gateway or standard cloning methods. We generally use pBAT-TL-B-sYFP-N, pBAT-TL-B-sYFP-C [7] for fusions to the C-termini of test proteins and two home-made variants of these, pCL112 and 113 for fusions to the N-termini. Unless there is specific information about the test proteins, which indicates that protein fusions to one terminus will disrupt essential functions, modifications, or the interaction then it is necessary to make fusions in both orientations. The location and co-localization of the test proteins should also be assessed with fusions to full-length fluorescent proteins before commencing BiFC studies. This may reveal problems with particular orientations of the fusion proteins or changes in the location or stability of one of the test proteins in the presence of the other.

2. We have found that it is important for agroinfiltration experiments (not only for BiFC) to use cultures of agrobacteria that are actively growing; cultures that have been grown for 24 h or less at 28 °C in a shaking incubator and which are not opaque. It is generally best to setup cultures from fairly fresh plates, i.e., plates no more than 2 weeks old.

3. Other groups successfully use different agroinfiltration solutions (lacking MES for example) so the composition of this does not appear to be critical. Some *Agrobacterium* strains, such as those containing the virG gene [8], do not require acetosyringone. The infiltration solution is usually prepared fresh and generally does not need to be sterile.

4. We have found that using low concentrations of bacteria for agroinfiltration results in lower expression levels of fluorescent fusion proteins both generally across the leaf (which is in part due to fewer cells expressing) and in individual cells. The absolute concentration of the bacteria suitable for infiltration will depend on the promoter used for expression, the stability of the constructs, the sensitivity of the cells to overexpression of the test proteins, and the sensitivity of the plant species used for the analysis to agrobacteria. In *N. benthamiana* plants with 35S promoter-based constructs and the AGL1 strain of *Agrobacterium* we generally do not use concentrations higher than an OD_{600} of 0.01 for BiFC and if possible will use concentrations of around OD_{600} 0.001.

5. We leave our plants in a designated high containment glasshouse chamber in sealed trays and sterilize the watering run-off.

6. We usually conduct imaging no later than 2 days post-infiltration. Many fusion proteins are visible 1 day after agroinfiltration but the levels of expression are generally too low to

give images acceptable for publication. Minimal preparation of leaves is required for confocal microscopy. We generally infiltrate the leaf pieces with water to improve image quality; it also stops the leaf from drying too rapidly (the cellular disruption caused by drying is dramatic and leaf samples should never be allowed to dry and wilt; it is best to prepare just one or a small number of samples at a time). For imaging at low magnification or with water-dipping lenses no cover slip is required so the leaf pieces are simply stuck to slides with double-sided tape. The double-sided tape helps the leaf sit flat and prevents sample movement when the microscope stage is moved.

7. We do not image the brightest cells on a leaf but focus on those that have the most common level of fluorescence, i.e., the most representative cells. When conducting plant–pathogen interaction studies in particular one must be aware that plants grown in a glasshouse are not sterile and may be exposed to microbes. Individual cells on a leaf may be undergoing a MAMP-triggered response and this may affect the interaction, stability, or other behavior of test proteins. If the behavior of one or a few cells differs from the majority of cells on a leaf, they should be treated with caution.

8. A very important feature of any experiment is the use of suitable controls. For BiFC the ideal control is a fusion to a variant of one of the test proteins for which there is evidence that it does not interact. We were fortunate in our analyses of the *P. infestans* effector Avr3a that we had supporting evidence to indicate certain variants did not interact with our test proteins. Yeast-2-hybrid and phenotypic data indicated the interaction with CMPG1 was dependent on the terminal tyrosine residue and thus a tyrosine deletion form was used as a control for BiFC [9]. The resistance protein R3a strongly recognizes the KI form of Avr3a (with the residues K and I) but only weakly recognizes the form with residues E and M at those positions, thus the EM form was a suitable control in that experiment (Fig. 2 and [10]). In the latter case we interpreted the positive BiFC result to indicate that the R3a and Avr3a were in close proximity, as we have no supporting evidence from other experimental methods for a direct interaction. If there is insufficient information about a protein and its interaction or function then the ideal control may not be known. For effector studies we opt to use a closely related effector. This strategy may be applied to the putative interactor as well, or instead, if there are other family members available. These control proteins should also be tested for interactions with other methods such as yeast-2-hybrid. Kodama and Hu [6] recommend a competition assay with co-expression of unlabelled form of one of the test proteins along with the split-YFP forms. In plant tissues this would have to be carefully quantified because in any co-agroinfiltration not all

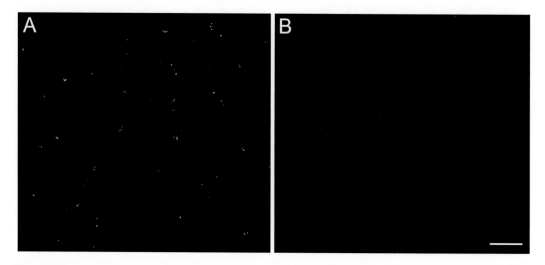

Fig. 2 BiFC results underpinned by strong controls. The resistance protein R3a recognizes the KI form of *P. infestans* Avr3a and relocates to endomosal compartments where the BiFC signal is detected (**a**). The EM form of Avr3a differs by only two amino acids but is very weakly recognized by R3a and thus acts as a specific control for BiFC analysis (**b**). Scale bar = 100 μm

cells will contain all constructs nor express them to the same levels. The component parts of the fluorescent protein expressed unfused in the cell, i.e., "free" YN or YC, as controls have been found to give high background levels of fluorescence, probably due to their small size meaning that they are not prevented from associating by chance with the other part, whether or not it is fused to a test protein.

9. The stability of all fusion proteins must be checked by western blotting. Only completely stable fusions have the potential to give clearly defined results. Partially cleaved proteins are likely to give a background level of no fluorescence.

10. Quantification using image analysis requires intensity measurements of a large number of representative confocal images of identical size (magnification and depth) taken with identical settings. The large numbers of images necessary to obtain statistically significant results can make it a time-consuming approach and it may be more effective to use a fluorimeter.

11. We have found that our SpectraMax M5 fluorimeter (Molecular Devices) is not sensitive enough to give reliable measurements from leaves infiltrated with the agrobacteria concentrations we use for confocal imaging, instead we have found it necessary to use OD_{600} of up to 0.1. More modern fluorimeters should be more sensitive.

12. Smaller discs, for example, for a 96-well plate are more time-consuming and difficult to handle; however, they will generally remain flatter than larger leaf discs. Leaf discs cut for a

24-well plate are quick to cut and handle but tend to curve with time, and the presence of large veins is difficult to avoid. Readings from protruding, large veins will be different to those from the leaf lamina. A greater number of reads should be collected from larger leaf discs. Uninfiltrated leaf material of the same age, and ideally from the same leaves as the infiltrated sample, should be present on every plate read so that background autofluorescence can be accounted for. If the fluorescence is highly localized within the cells, for example in the nucleolus, leaf disc measurements may contain reads in which the region of the cell or cells measured does not contain any fluorescent objects. These reads will be within the range of the background autofluorescence.

13. The actual amount of leaf material and the volume of buffer will need to be empirically determined for the sensitivity of the fluorimeter to be used, using material with similar levels of fluorescence as the test samples. As a guide only, we have successfully used two 10 mm leaf discs ground in 300 μl of buffer.

Final caveat: As an experimental scientist one must accept that any method to assess protein–protein interactions has its limitations. Proteins may only associate under specific physiological conditions or may require the action of other proteins to modify one or more of the protein partners. Some of these conditions may be testable. Even if proteins do interact at all times the orientation of the interaction and other constraints such as intolerance to the presence of tags at specific ends of the proteins may make it difficult or impossible to prove they interact with currently available tools.

Acknowledgment

This work was supported by BBSRC and the Scottish Government's Rural and Environment Science and Analytical Services (RESAS) Division.

References

1. Mukhtar MS, Carvunis A, Dreze M, Epple P, Steinbrenner J, Moore J, Tasan M, Galli M, Hao T, Nishimura MT, Pevzner SJ, Donovan SE, Ghamsari L, Santhanam B, Romero V, Poulin MM, Gebreab F, Gutierrez BJ, Tam S, Monachello D, Boxem M, Harbort CJ, McDonald N, Gai L, Chen H, He Y, European Union Effectoromics Consortium, Vandenhaute J, Roth FP, Hill DE, Ecker JR, Vidal M, Beynon J, Braun P, Dangl JL (2011) Independently evolved virulence effectors converge onto hubs in a plant immune system network. Science 333:596–601

2. Lleres D, Swift S, Lamond AI (2007) Detecting protein-protein interactions in vivo with FRET using multiphoton fluorescence lifetime imaging microscopy (FLIM). Curr Protoc Cytom Chapter 12:Unit12.10

3. Hu C-D, Chinenov Y, Kerppola TK (2002) Visualization of interactions among bZIP and Rel family proteins in living cells using bimolecular fluorescence complementation. Mol Cell 9:789–798

4. Kodama Y, Hu C-D (2010) An improved bimolecular fluorescence complementation assay with a high signal-to-noise ratio. Biotechniques 49:793–805

5. Lee L-Y, Wu F-H, Hsu C-T, Shen S-C, Yeh H-Y, Liao D-C, Fang M-J, Liu N-T, Yen Y-C, Dokládal L, Sýkorová E, Gelvin SB, Lin C-S (2012) Screening a cDNA library for protein–protein interactions directly in planta. Plant Cell 24:1746–1759

6. Kodama Y, Hu C-D (2012) Bimolecular fluorescence complementation (BiFC): a 5-year update and future perspectives. Biotechniques 53:285–298

7. Uhrig JF, Mutondo M, Zimmermann I, Deeks MJ, Machesky LM, Thomas P, Uhrig S, Rambke C, Hussey PJ, Hülskamp M (2007) The role of Arabidopsis SCAR genes in ARP2-ARP3-dependent cell morphogenesis. Development 134:967–977

8. Vain P, Harvey A, Worland B, Ross S, Snape JW, Lonsdale D (2004) The effect of additional virulence genes on transformation efficiency, transgene integration and expression in rice plants using the pGreen/pSoup dual binary vector system. Transgenic Res 13:593–603

9. Bos JIB, Armstrong MR, Gilroy EM, Boevink PC, Hein I, Taylor RM, Zhendong T, Engelhardt S, Vetukuri RR, Harrower B, Dixelius C, Bryan G, Sadanandom A, Whisson SC, Kamoun S, Birch PRJ (2010) *Phytophthora infestans* effector AVR3a is essential for virulence and manipulates plant immunity by stabilizing host E3 ligase CMPG1. Proc Natl Acad Sci USA 107:9909–9914

10. Engelhardt S, Boevink PC, Armstrong MR, Ramos MB, Hein I, Birch PRJ (2012) Re-localisation of late blight resistance protein R3a to endosomal compartments is associated with effector recognition and required for the immune response. Plant Cell 24:5142–5158

Chapter 7

Particle Bombardment-Mediated Transient Expression to Identify Localization Signals in Plant Disease Resistance Proteins and Target Sites for the Proteolytic Activity of Pathogen Effectors

Daigo Takemoto and David A. Jones

Abstract

Plant pathogens, including fungi, oomycetes, bacteria, aphids, and nematodes, produce a variety of effector proteins to counter plant disease resistance mechanisms. After delivery into the cytosol of the plant cell, effectors may target proteins localized to different compartments within the plant cell. Plants, in turn, have evolved disease resistance (R) proteins to recognize the action of effectors. Elucidation of the subcellular localization of pathogen effectors, the plant proteins they target, and plant disease resistance proteins is essential to fully understand their interactions during pathogen challenge. In recent years, expression of fluorescent protein fusions has been widely used to determine the subcellular localization of plant proteins and pathogen effectors. Use of fluorescent proteins enables researchers to monitor the dynamic behavior of proteins in living cells. Among various methods available for the introduction of genes into plant cells, particle bombardment-mediated transient expression is the most rapid method suitable for both the identification of localization signals in proteins of interest and their dissection via amino acid substitutions generated using site-directed mutagenesis. This chapter describes a rapid procedure for particle bombardment-mediated transient expression in leaf epidermal cells. This method is also applicable to detection of pathogen effector protease activities directed against target proteins in the plant cell and analysis of protease recognition sites within these target proteins.

Key words Particle bombardment, Subcellular localization, Disease resistance proteins, Effectors

1 Introduction

The construction, expression, and visualization of fluorescent protein fusions to help determine the subcellular localization of plant proteins has become a very powerful technique in plant molecular cell biology. It allows the localization of proteins in intact living plant cells and thereby complements other techniques such as immunolocalization (based on fixed, nonliving cells) and subcellular fractionation (based on cellular disruption). The four main variables to consider in setting up a subcellular localization

Paul Birch et al. (eds.), *Plant-Pathogen Interactions: Methods and Protocols*, Methods in Molecular Biology, vol. 1127, DOI 10.1007/978-1-62703-986-4_7, © Springer Science+Business Media New York 2014

experiment using transgenes encoding plant proteins tagged with fluorescent proteins are (1) the choice of fluorescent protein, (2) the location of the fluorescent protein within the fusion protein, (3) the level of fusion-protein expression, and (4) the method of transgene delivery to the plant.

1.1 Choice of Fluorescent Protein

Most fluorescent proteins in current use are variants of either GFP (green fluorescent protein) obtained from the jellyfish *Aequoria victoria* or DsRed (red fluorescent protein) obtained from the coral *Discosoma* sp. [1]. Color variants of GFP and DsRed that cover almost the entire visible spectrum have been generated by mutation [2]. Besides differing in color, these variants differ in other properties such as brightness, stability, pH sensitivity, or ability to form multimers. The choice of color variant depends on the intended purpose. For example, experiments involving apoplastic localization should, given the low pH of the apoplast and the sensitivity of GFP to low pH, use DsRed or its derivatives, which are relatively insensitive to low pH, or an insensitive derivative of GFP. More importantly, color variants enable double or triple labeling, and more specifically, the co-localization (or otherwise) of a protein of unknown location with one of known location, to confirm a subcellular location inferred from the pattern of fluorescence observed within the cell. While some patterns of fluorescence, such as the net-like reticulate pattern of the endoplasmic reticulum, may be relatively definitive others, such as the punctate pattern of the Golgi apparatus and peroxisomes, may be ambiguous and therefore require confirmation by co-localization. Many other fluorescent protein variants have also been engineered for specific purposes such as sensing changes in pH (pHluorin), calcium concentration (cameleon) or redox status (RoGFP), testing for protein–protein interaction (bimolecular fluorescence complementation using split YFP), or visualizing protein movement (EosFP photoconvertible GFP), but these applications are beyond the scope of this chapter [3–5].

1.2 Location of the Fluorescent Protein Within the Fusion Protein

Typically target proteins are tagged with fluorescent proteins at the N- and/or C-termini, but can also be tagged internally. The choice of location depends on what is known or predicted about the target protein. If the structure of the protein is known, or boundaries between structural/functional domains can be predicted, then it may be possible to tag a protein internally without affecting its localization. If nothing is known or predicted, then the usual approach is to produce and test both N- and C-terminally tagged proteins. However, if the target protein has a predicted N- or C-terminal localization motif, such as an N-terminal myristoylation and palmitoylation motif or a C-terminal palmitoylation motif, then only a tag at the opposite terminus should be used. Above all, it should be remembered that tagging with fluorescent proteins can interfere with normal protein function or targeting.

Wherever possible, subcellular localization experiments using fluorescent protein tags should be accompanied by experiments addressing the function of the tagged proteins. If the tagged protein functions properly then it may be reasonable to assume if it is targeted to its proper subcellular location. Subcellular localization experiments can also be conducted with truncated target proteins to examine the role of possible localization motifs in more detail. Wherever possible, these experiments should be validated by experiments addressing the location of the full-length target protein.

1.3 The Level of Fusion-Protein Expression

The level of fusion-protein expression can be controlled to a large extent by the choice of promoter. Expression is usually driven either by a strong constitutive promoter or the gene's own (native) promoter. The main concern associated with strong constitutive promoters is saturation of localization mechanisms leading to leakage of tagged protein into other subcellular compartments. On the other hand, the native promoter may be inducible and therefore not expressed in the tissue used for visualization, may be weak, giving insufficient protein expression for visualization, or may be poorly defined. Wherever possible, localization experiments conducted using strong constitutive promoters for clear visualization of fusion proteins should be validated by experiments using native promoters. Alternatively, the issue of fusion-protein expression level can be addressed via the choice of transgene delivery mechanism as described below.

1.4 The Method of Transgene Delivery to the Plant

Three main methods are used for delivery of transgenes encoding plant proteins tagged with fluorescent proteins into plant cells: production of transgenic plants, using disarmed *Agrobacterium tumefaciens*, or transgenic roots, using *A. rhizogenes*; infiltration of plant leaves using *A. tumefaciens* (agroinfiltration, [6]); and particle bombardment, usually of leaves but potentially any plant tissue. Each method has its own advantages and disadvantages. Generation of binary vector constructs, their transfer to *A. tumefaciens* and plant regeneration, are time-consuming and labor-intensive steps performed over a period of several months to generate transgenic plants (apart from *Arabidopsis*). Agroinfiltration represents a significant improvement in turnaround time but still requires the generation of binary vector constructs and their transfer to *A. tumefaciens*. Both methods are limited by species recalcitrant to transformation by *A. tumefaciens* and the generation of transgenic plants may be further limited by difficulties in plant regeneration.

Particle bombardment only requires vector construction in *Escherichia coli* and is not limited by the amenability of a particular plant species to transformation with *A. tumefaciens*, or by plant regeneration. Particle bombardment can be applied to most plant species and has a short turnaround time with as little as a week elapsing from commencement of vector construction or transgene modification to observation of subcellular location. In transgenic plants, differences in protein expression, such as tissue-specific distribution or level of protein expression are controlled mainly by the choice of promoter but may vary according to position effects and transgene copy number. With a constitutive promoter, transgenic plants will show a steady-state level of protein accumulation, whereas agroinfiltrated and bombarded leaves will show protein accumulation starting from zero and rising over the time course of the experiment. Temporal and spatial differences in expression allow subcellular localization to be assessed over a range of expression levels from very low to very high. If necessary, the level of protein expression can be further varied by altering the concentration of *A. tumefaciens* used in agroinfiltration experiments, or the concentration of plasmids used in bombardment experiments. With a constitutive promoter, transgenic plants will show confluent fusion-protein expression whereas particle bombardment will generate isolated cells showing fusion-protein expression. Although these transformed cells are far fewer in number, their isolation from one another can allow the cell periphery and the distribution of the fusion protein in relation to the cell periphery to be more easily visualized (useful for localization of plasma membrane or secreted proteins). Agroinfiltration shows expression both in isolated cells and in patches of confluent cells with the ratio of the two varying according to the efficiency of the transformation, the concentration of bacteria used for infiltration, and the plant species involved. For example, *Nicotiana benthamiana* can show almost confluent expression throughout the leaf indicating very efficient transformation via agroinfiltration. On the other hand, the stress caused by the continued presence of *Agrobacterium* in agroinfiltration experiments (often leading to chlorotic panels in *Agrobacterium* empty vector controls) or the cellular damage caused by penetration of a gold particle in bombardment experiments, could potentially affect the localization of some proteins.

This chapter describes the use of particle bombardment-mediated transient transformation of leaf epidermal cells to determine the subcellular localization of plant proteins. This rapid procedure can also be applied to the detection and analysis of recognition sites in plant proteins targeted for cleavage by pathogen effector proteases.

2   Materials

2.1   Plant Material

1. Grow your plant of interest. Choose young, healthy, and relatively "flat" leaves. Do not use stressed or old leaves, or leaves with too many hairs.

2.2 Stock Solutions and Equipment Required

1. Biolistic Particle Delivery System (e.g., Model PDS-1000/He, Bio-Rad 165-2257).

2. Gold particles (1-μm diameter, Bio-Rad 165-2263).

3. Macrocarrier (Bio-Rad 165-2335).

4. Rupture Disc (1,350 psi, Bio-Rad 165-2330).

5. Stopping Screen (Bio-Rad 165-2336).

6. 2.5 M $CaCl_2$. Store at 4 °C.

7. 0.1 M spermidine (Sigma S0266). Store at –80 °C.

2.3 Preparation of Gold Particles

1. Weigh 50 mg gold particles in a 1.5 ml Eppendorf tube and suspend in 1 ml sterile MilliQ water by vortexing for 30 s.

2. Centrifuge at $10,000 \times g$ for 1 min and remove supernatant.

3. Resuspend gold particles in 100 % ethanol by vortexing for 30 s.

4. Centrifuge at $10,000 \times g$ for 1 min and remove supernatant.

5. Repeat **items 3** and **4** two more times.

6. Resuspend gold particles in 1 ml sterile MilliQ water.

7. Store gold particles in 1 ml water in a screw-cap tube at 4 °C.

2.4 Construction of Plasmid DNA for Coating Gold Particles

Construction of vectors for particle bombardment is a critical point for successful detection of transiently expressed GFP-tagged proteins (Fig. 1). Any preferred method for the construction of vectors can be used, but there are several important points to bear in mind when considering vector design. Key considerations for the construction of plasmid DNA for particle bombardment are:

1. Do not use vectors over 10 kb such as those often used for *Agrobacterium*-mediated transformation, as a large vector size drastically reduces the number of transformed epidermal cells. Ideally, use universal cloning vectors around 3 kb, such as pUC19 or pBluescript.

2. Try short, well-defined, strong, constitutive promoters, like the CaMV 35S promoter, first. Native promoters can be used, but there may be a more limited chance of success for the reasons described above.

3. If overexpression of a target gene is stressful or induces death of leaf epidermal cells in the target plant, transformed cells showing GFP fluorescence will rarely be observed and if observed are likely to show abnormal patterns of fluorescence associated with stress or induction of cell death (Fig. 1, RPP8-100N:GFP which contains part of the TIR domain). In this event, try shorter (partial) versions of the protein of interest to avoid any cell disruptive effects due to the activity of the expressed protein.

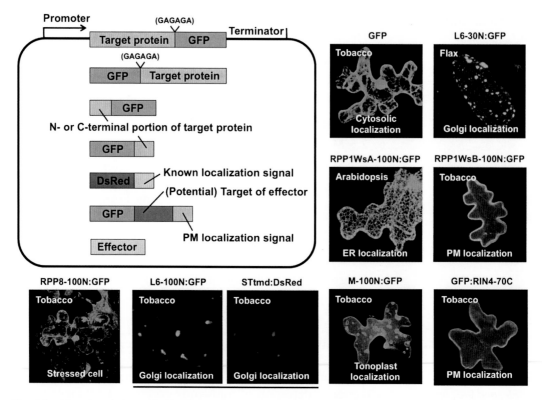

Fig. 1 Construction of plasmids used for particle bombardment (*top left*) and examples of localization patterns observed for GFP or DsRed fusion-protein constructs following particle bombardment of tobacco, Arabidopsis, or flax leaf epidermal cells (*right and bottom panels*). The *top left panel* shows a diagrammatic summary of *E. coli* plasmid (*black line*) constructs containing various transgenes (*colored boxes*) inserted between plant promoter and terminator sequences. The *right and bottom panels* show epidermal cells expressing GFP fused with the N-terminal 30 or 100 amino acids of plant disease resistance proteins or C-terminal 70 amino acids of RIN4 (an Arabidopsis protein target by multiple *Pseudomonas syringae* effectors and guarded by the RPS2 and RPM1 resistance proteins). These GFP fusion proteins show different subcellular localizations including plasma membrane (PM), the Golgi apparatus, tonoplast, and endoplasmic reticulum (ER) in leaf epidermal cells of various plants. STtmd:DsRed was co-expressed with L6-100N:GFP to confirm localization to the Golgi apparatus. STtmd, transmembrane domain and signal anchor sequences of rat 2,6-sialyl transferase. GFP alone localizes to the cytosol and nucleus

4. Try both N- and C-terminal GFP-tagging of target protein if there is no predicted localization signal. A short linker can be added between GFP and the target protein to allow for natural folding of the target protein and to reduce any steric hindrance. We generally have a six amino acid linker (Gly-Ala-Gly-Ala-Gly-Ala) between GFP and target proteins (Fig. 1).

5. Linear DNA fragments (promoter-GFP-tagged protein-terminator) amplified by PCR can be used for particle bombardment, but the efficiency is significantly lower than that obtained using plasmid DNA.

2.5 Vectors for Detecting Localization Signals in Target Proteins

To identify the localization signal in a target protein, try a series of shortened N- or C-terminal portions of the target protein tagged with GFP (or other fluorescent protein). Once the minimal sequence requirement is determined, site-directed mutations can be applied to analyze the amino acid residues required for localization. To confirm the subcellular localization of a target protein, co-expression with DsRed (or other fluorescent protein) tagged with a known localization signal can be used to look for co-localization (e.g., Fig. 1, L6-100N:GFP with STtmd-DsRed). Some examples of localization signal sequences suitable for co-localization experiments are: plasma membrane, N-terminal myristoylation/palmitoylation site (e.g. PBS1-30N, [7]); ER, C-terminal HDEL [8]; Golgi apparatus, N-terminal tagging with STtmd (transmembrane domain and signal anchor sequences of rat 2,6-sialyl transferase) ([9], Fig. 1); tonoplast, M-30N ([7], Fig. 1); peroxisomes, C-terminal SKL [10]; mitochondria, C-terminal tagging with β-ATPase or CPN-60 [11]; nuclei, N-terminal tagging with SV40-NLS (nuclear localization sequence) [12].

2.6 Vectors for Detecting the Proteolytic Activity of Effectors

The combination of a PM (plasma membrane) localization signal with a plant effector-target protein (or effector-target peptide motif) can be used as a sensor for the detection of effector proteins with proteolytic activity (Fig. 2) [13]. Potential effector-target proteins (or shortened variants) can be placed between GFP and a PM localization signal (Fig. 2). In the absence of the effector, the sensor protein is observed on the PM (Fig. 1, GFP:RIN4-70C), but co-expression with the effector protease releases the PM-localized GFP into the cytosol (Fig. 2, GFP:RIN4-70C with AvrRpt2). Site-directed mutations in the target protein can then be used to identify the amino acid residues essential for proteolytic cleavage by the effector protease.

3 Methods

3.1 Coating Gold Particles with Plasmid DNA

1. Suspend 25 μl (1.25 mg) of gold particles in water (prepared in Subheading 2.2) in 1.5 ml Eppendorf tube by vortexing for 30 s.

2. Add approx. 800 ng (in 5 μl) of plasmid DNA and mix by vortexing for 30 s. For co-bombardment experiments, add 400 ng of each plasmid.

3. Add 2.5 μl of 2.5 M CaCl$_2$ and mix by vortexing for 30 s.

4. Add 10 μl of 0.1 M spermidine and mix by vortexing for 3 min.

5. Centrifuge at $5,000 \times g$ for 10 s and remove supernatant.

6. Add 180 μl of ethanol and mix by vortexing for 30 s.

7. Centrifuge at $5,000 \times g$ for 10 s and remove supernatant.

8. Add 30–50 μl of ethanol and mix by vortexing. Make sure that there is no clumping of gold particles after the vortexing.

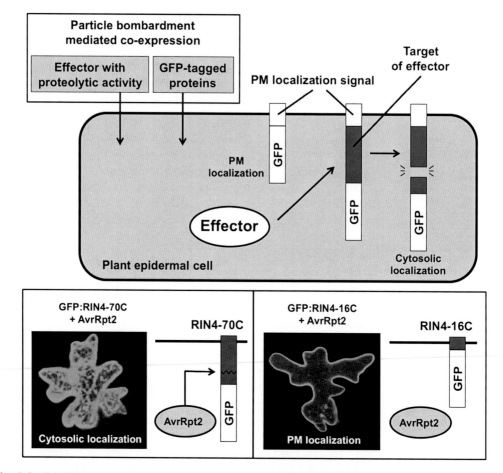

Fig. 2 Particle bombardment-mediated identification of target sites for the proteolytic activity of pathogen effectors. GFP:RIN4-70C (*bottom left panel*) or GFP:RIN4-16C (*bottom right panel*) was co-expressed with AvrRpt2 in epidermal cells of tobacco leaves. Micrographs showing a cytosolic pattern of GFP fluorescence indicate the cleavage of PM-localized GFP:RIN4-70C (*see* Fig. 1) by AvrRpt2. As GFP:RIN4-16C lacks the cleavage site, co-expression with AvrRpt2 has no effect on PM localization of GFP:RIN4-16C

3.2 Particle Bombardment

Either an ordinary particle gun system (PDS-1000/He) or the Helios gene gun system (Bio-Rad) can be used for particle bombardment. Please consult the instructions provided by the manufacturer for detailed procedures and safety precautions. Here we describe the basic procedure for particle bombardment with the Bio-Rad biolistic particle delivery system model PDS-1000/He (Fig. 3).

1. Place leaf (or several small leaves) of the target plant (abaxial side up) in a plastic Petri dish on top of filter paper. For small leaves (e.g., rosette leaves of *Arabidopsis*), use multiple leaves for one shot (Fig. 4) fixed with cellophane tape to avoid them being blown off when they are bombarded. Do not use leaves with moisture droplets on their abaxial side.

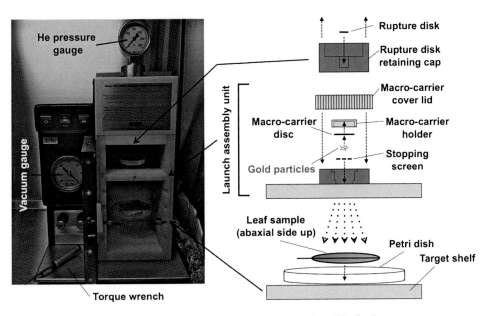

Fig. 3 Components of the biolistic PDS-1000/He particle delivery system (Bio-Rad)

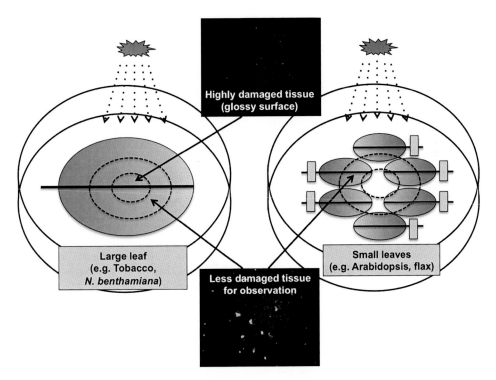

Fig. 4 Arrangement of large or small leaves in Petri dish on target shelf (Fig. 3). Low-magnification fluorescence micrographs of tobacco leaf tissue 9 h after bombardment with GFP, showing poor transformation of epidermal cells at the central impact area (*top panel*) and efficient transformation of epidermal cells in the surrounding area (*bottom panel*). Epidermal cells expressing GFP (*green*) are seen against a background of chlorophyll autofluorescence (*red*)

2. Turn on gun chamber (left red switch), vacuum pump, and He gas supply.

3. Wash macrocarrier discs (1 disc per shot) in 100 % ethanol and dry them by leaning up against the side of a Petri dish.

4. Insert dried macrocarrier disc in the macrocarrier holder. Place 5–10 µl of DNA-coated gold particles in ethanol (prepared in Subheading 3.1) in the center of macrocarrier disc. Mix the particle suspension well before removing each aliquot. Allow gold particles to dry completely.

5. Wash rupture discs (1 disc per shot) in 100 % 2-propanol and dry them by leaning up against the side of a Petri dish.

6. Insert rupture disc into rupture disc retaining cap and set it onto the end of the gas acceleration tube at the top of chamber using a torque wrench.

7. Place the stopping screen in Launch assembly unit and then place the inverted macrocarrier holder (gold particles down side) on top of the unit. Screw the macrocarrier cover lid over the inverted macrocarrier holder.

8. Slot launch assembly unit into upper shelf position 0 in chamber.

9. Place leaf sample in plastic Petri dish (without lid) on target shelf and slot into shelf position 3 or alternative shelf position depending on the leaf sample used.

10. Close the front door of the gun chamber and push the middle red switch upward to create a vacuum in the gun chamber.

11. When vacuum reaches –28 in. of Hg, hold the right red switch upward until the rupture of the disc, then release the finger from the switch and push the middle red switch to release vacuum from gun chamber.

12. When vacuum is back to 0, open front door of the gun chamber and collect the leaf sample.

3.3 Confocal Microscopy Following Particle Bombardment

1. Keep bombarded leaves at 23 °C in a humid (100 % humidity) chamber for 6–9 h (*see* **Note 1**).

2. Check the efficiency of transformation by fluorescence microscopy on a low-resolution dissecting microscope (Fig. 4). Usually, the very center of bombarded leaves is highly damaged and contains fewer GFP-labeled epidermal cells than the surrounding areas. Label the location of efficiently transformed areas with a marker pen (*see* **Note 2**).

3. Observe the localization of GFP-tagged protein by confocal laser-scanning microscopy (*see* **Note 3**). Use appropriate setting for the detection of fluorescence proteins. GFP, excitation peak 489 nm, emission peak 508 nm; DsRed, excitation peak 558 nm, emission peak 583 nm (other variants will have different excitation and emission peaks that should be sourced from the literature, [2]).

4   Notes

1. The best temperature for incubation of bombarded leaves may vary from 23 °C depending on the leaf sample used.

2. Do not incubate bombarded leaves for longer than 24 h as subcellular localization detected at 6–9 h after bombardment may be lost after 24 h.

3. At 6–9 h after bombardment, epidermal cells with various intensities of GFP fluorescence will be found. As overexpression of GFP-tagged proteins can cause localization artifacts, observations should include cells with both high and low levels of GFP expression.

References

1. Lippincott-Schwartz J, Snapp E, Kenworthy A (2001) Studying protein dynamics in living cells. Nat Rev Mol Cell Biol 2:444–456

2. Shaner NC, Patterson GH, Davidson MW (2007) Advances in fluorescent protein technology. J Cell Sci 120:4247–4260

3. Choi WG, Swanson SJ, Gilroy S (2012) High-resolution imaging of Ca2+, redox status, ROS and pH using GFP biosensors. Plant J 70:118–128

4. Kerppola TK (2006) Design and implementation of bimolecular fluorescence complementation (BiFC) assays for the visualization of protein interactions in living cells. Nat Protoc 1:1278–1286

5. Wu S, Koizumi K, Macrae-Crerar A, Gallagher KL (2011) Assessing the utility of photoswitchable fluorescent proteins for tracking intercellular protein movement in the *Arabidopsis* root. PLoS One 6:e27536

6. Kapila J, De Rycke R, Van Montagu M, Angenon G (1997) An *Agrobacterium-mediated* transient gene expression system for intact leaves. Plant Sci 122:101–108

7. Takemoto D, Rafiqi M, Hurley U, Lawrence GJ, Bernoux M, Hardham AR, Ellis JG, Dodds PN, Jones DA (2012) N-terminal motifs in some plant disease resistance proteins function in membrane attachment and contribute to disease resistance. Mol Plant Microbe Interact 25:379–392

8. Haseloff J, Siemering KR, Prasher DC, Hodge S (1997) Removal of a cryptic intron and subcellular localization of green fluorescent protein are required to mark transgenic *Arabidopsis* plants brightly. Proc Natl Acad Sci U S A 94:2122–2127

9. Boevink P, Oparka K, Santa Cruz S, Martin B, Betteridge A, Hawes C (1998) Stacks on tracks: the plant Golgi apparatus traffics on an actin/ER network. Plant J 15:441–447

10. Kragler F, Lametschwandtner G, Christmann J, Hartig A, Harada JJ (1998) Identification and analysis of the plant peroxisomal targeting signal 1 receptor NtPEX5. Proc Natl Acad Sci U S A 95:13336–133341

11. Logan DC, Leaver CJ (2000) Mitochondria-targeted GFP highlights the heterogeneity of mitochondrial shape, size and movement within living plant cells. J Exp Bot 51:865–871

12. Chiu W-L, Niwa Y, Zeng W, Hirano T, Kobayashi H, Sheen J (1996) Engineered GFP as a vital reporter in plants. Curr Biol 6:325–330

13. Takemoto D, Jones DA (2005) Membrane release and destabilization of *Arabidopsis* RIN4 following cleavage by *Pseudomonas syringae* AvrRpt2. Mol Plant Microbe Interact 18:1258–1268

Chapter 8

Purification of Fungal Haustoria from Infected Plant Tissue by Flow Cytometry

Diana P. Garnica and John P. Rathjen

Abstract

A hallmark of biotrophy in many fungal plant pathogens is the formation of a specialized pathogenic structure called the haustorium from infectious hyphae. This is the major parasitic structure, where nutrients are taken up from the host and pathogenicity factors are exported to the host tissue. Obligate biotrophic fungi can typically be cultured in vivo only to a limited extent and do not produce haustoria under these conditions. This has hampered the application of classic molecular biology techniques to haustoria-forming pathogens. The lectin Concanavalin A (Con A), which binds specifically to sugars present on the exterior of rust haustoria, was first used in a column-based affinity purification procedure in 1992 (Hahn and Mendgen, Protoplasma 170:95–103, 1992). Here we describe a new technique where we combine initial gradient purification of haustoria with flow-sorting based on labeling of haustoria with fluorescent Con A. Our method allows haustorial isolation with purity above 98 % and yields ten times more isolated haustoria in a single experiment than the previous procedure.

Key words Rust fungi, *Puccinia*, Haustoria, Density gradient, Flow cytometry

1 Introduction

Obligate fungal parasites source nutrients from living cells of their plant hosts through a sophisticated cellular structure termed the haustorium [1]. After penetration of host tissue by the fungus, the invasive hyphae contact host cells and start to differentiate. The fungus perforates the host cell wall and expands within, invaginating the host plasma membrane but not breaching it, to form a single-lobed or multilobed structure. The haustorium remains separated from the host cell by the extrahaustorial membrane (EHM), which is derived from and contiguous with the host plasma membrane, but appears to be differentiated from it [2]. On the fungal side of the EHM is a gel-like structure called the extra-haustorial matrix, composed largely of polysaccharides, which is bordered by the fungal plasma membrane on its interior surface. The haustorium remains connected to the fungal body through a

Paul Birch et al. (eds.), *Plant-Pathogen Interactions: Methods and Protocols*, Methods in Molecular Biology, vol. 1127, DOI 10.1007/978-1-62703-986-4_8, © Springer Science+Business Media New York 2014

neck structure which contacts the EHM, thus sealing the extra-haustorial matrix from the plant cell. Molecular exchange between the fungus and plant cells must occur across the extrahaustorial matrix; however, the structure and function of this interface is poorly understood [2].

Haustoria are not only feeding structures; they induce structural changes in the host cell including cytoskeletal rearrangements, nuclear migration, and chromatin condensation [3], and there is evidence that they influence host cell metabolism [4, 5]. Furthermore, they deliver essential virulence molecules called "effectors" into the extrahaustorial matrix, several of which are subsequently translocated into host cells [6, 7]. As such, the haustorium is a site of concerted host-pathogen interaction, and describing its functions is essential to understanding biotrophy. Despite its importance, its inaccessibility and the inability to culture it in vitro has constrained experimentation. Hahn and Mendgen [8] found that lectins showed differential affinity for the haustoria of *Uromyces* spp. and *Puccinia* spp., with Concanavalin A (Con A) showing the highest affinity. Additional observations led to the conclusion that Con A recognizes α-linked mannoside residues on the haustorial wall and the extrahaustorial matrix, since the EHM appears to be lost during isolation of haustoria from leaf tissue. Affinity of Con A for surface sugars is the basis of a scheme to purify haustoria by column-based chromatography from crude tissue extracts [8]. However, this purification method has two major disadvantages: the high level of contamination by chloroplasts which far outnumber haustoria, and the low yield of haustoria. Moreover, although the method has been used for other rust fungi [9], it failed to purify the haustoria of the powdery mildew species *Sphaeroteca fuliginea*, *Erysiphe pisi*, and *E. graminis* f.sp. *hordei*, suggesting that haustoria of these species differ in their external composition. Alternative separation techniques such as density gradient centrifugation have been used to isolate haustoria of powdery mildew fungi [2, 10, 11]. However, none of these techniques achieve high levels of purity, which is necessary for subsequent applications such as transcriptome sequencing, proteomics analysis, and metabolomic studies.

Recently, Takahara et al. [12] described a method for fluorescent vital staining of the intracellular hyphae of the pathogen *Colletotrichum higginsianum* from homogenates of infected Arabidopsis leaves, which could then be purified by fluorescence-activated cell sorting (FACS). Here we describe the purification of wheat stripe rust haustoria by Percoll density gradients combined with FACS based on affinity staining of the haustoria with fluorescent Con A. The method generates high yields of essentially pure haustoria that are suitable for downstream analyses that demand high purity.

2   Materials

Prepare all solutions using sterile MilliQ water and analytical grade reagents.

2.1   Haustoria Isolation from Plant Tissue

1. Infected plant material (*see* **Note 1**).
2. Ethanol 70 %. Prepare 250 ml and store at 4 °C.
3. Two liters of chilled MilliQ water.
4. Sodium hypochlorite 2 % (v/v), prepare fresh.
5. Sterile paper towel.
6. Waring blender.
7. Homogenization buffer: 0.2 M sucrose, 20 mM MOPS pH 7.2, and 0.2 % (v/v) β mercaptoethanol, added freshly. Prepare 250 ml of this buffer and store at 4 °C.
8. Isolation buffer (1× IB): 0.2 M sucrose, 20 mM MOPS pH 7.2. Prepare 500 ml of this buffer and store at 4 °C.
9. Isolation buffer (10× IB): 2 M sucrose, 0.2 M MOPS pH 7.2. Prepare 100 ml of this buffer and store at 4 °C.
10. Two clean and sterile 1 L Erlenmeyer flasks.
11. Ten Oak Ridge round-bottom centrifuge tubes, 45 ml capacity, with polypropylene screw closure.
12. Two nylon meshes (~20 cm × 30 cm), one of pore size 100 μm and the other of 20 μm (*see* **Note 2**).
13. Four polycarbonate round-bottom tubes, 45 ml capacity, without lids.
14. Percoll (GE healthcare life sciences, 17-0891-01).
15. Refrigerated benchtop centrifuge with swingout rotor.
16. Concanavalin A, Alexa Fluor® 488 Conjugate (Molecular Probes® C11252, *see* **Note 3**).
17. Rotary mixer.
18. Glass vial with plastic lid, 6 ml capacity.

2.2   Purification of Haustoria by Flow Cytometry

1. Dickinson BD FACSARIA II cell sorter (BD Biosciences), or similar.
2. 5 ml glass assay tubes.
3. Phosphate buffered saline (1× PBS): 8 g NaCl, 0.2 g KCl, 1.44 g Na_2HPO_4, 0.24 g KH_2PO_4, dissolved in 1 L of MilliQ water, final pH 7.4, and sterilized by autoclaving.

3   Methods

Carry out all procedures at 4 °C unless otherwise specified.

3.1   Haustoria Isolation from Infected Tissue Using Percoll Gradients

Steps including the timing of tissue harvesting (*see* **Note 1**), the amount of infected tissue to be processed, the pore size of the meshes, and Con A staining time, have to be determined empirically according to the pathosystem under investigation.

1. Harvest 20–25 g of heavily infected tissue (leaf pieces of ~6 cm length), 8–9 days after infection (or 1 day before sporulation) (*see* **Note 1**).

2. To remove external contaminating organisms, wash the tissue with tap water several times, then incubate for 3 min in 2 % sodium hypochlorite. Wash with tap water three times or until sodium hypochlorite is completely removed, then incubate the tissue for 1 min in chilled 70 % ethanol. Wash tissue with chilled MilliQ water several times to remove the alcohol. Dry the tissue as much as possible with sterile paper towel.

3. Using a Waring blender, homogenize the infected plant material in 150 ml of homogenization buffer at maximum speed for 25 s.

4. Filter the homogenate through a 100 μm nylon mesh by gravity flow. Recover the solid particles retained on the mesh and return them to the blender, add the remaining 100 ml of homogenization buffer and blend for 15 s at 18,000 rpm. Pass through the 100 μm mesh to remove cell debris and combine the two filtrates.

5. Pass the filtrate (~250 ml) through the 20 μm mesh by gravity flow and distribute the new filtrate in six chilled Oak Ridge round-bottom centrifuge tubes, 45 ml capacity (*see* **Note 2**).

6. Centrifuge the tubes at $1,080 \times g$ for 15 min at 4 °C, using a centrifuge with swingout rotor. In the meantime, prepare Percoll-30 solution: 24 ml Percoll, 30 ml MilliQ water, and 6 ml 10× IB.

7. Resuspend each pellet very gently in 2 ml of ice-cold 1× IB using a 1 ml pipette; combine all of the resuspended pellets and bring the total volume to 20 ml with 1× IB. Mix the resuspended pellets (20 ml) with the Percoll-30 solution to a total volume of 80 ml. Split the mixture into four 45 ml polycarbonate round-bottom tubes.

8. Centrifuge at $25,000 \times g$ for 30 min at 4 °C without braking.

9. Carefully remove the tubes from the centrifuge, and draw off the first 10 ml from each tube very slowly using a 10 ml pipette and an automatic pipette controller under low suction speed (*see* **Note 4**). Dilute the pooled 40 ml haustorial fraction 1:10

into 360 ml of 1× IB. Mix well and distribute the mixture into ten chilled 45 ml Oak Ridge round-bottom centrifuge tubes.

10. Centrifuge the tubes at $1080 \times g$ for 15 min at 4 °C, using a centrifuge with swingout rotor (**steps 11–13** are optional, *see* **Note 5**). If following **steps 11–13** resuspend the pellets in 10 ml of 1× IB, otherwise go to **step 14**.

11. Prepare Percoll-25 solution: 10 ml Percoll, 17 ml MilliQ water, and 3 ml 10× IB. Mix this solution with the resuspended pellets for a final volume of 40 ml.

12. Centrifuge at $25,000 \times g$ for 30 min at 4 °C without braking in two 45 ml polycarbonate round-bottom tubes. Draw off the first 10 ml from each tube very slowly using a 10 ml pipette and an automatic pipette controller under low suction speed. Dilute the pooled haustorial fractions 1:10 in 180 ml of 1× IB. Mix well and distribute this mixture into chilled 45 ml Oak Ridge round-bottom centrifuge tubes.

13. Centrifuge the tubes at $1080 \times g$ for 15 min at 4 °C, using a centrifuge with swingout rotor.

14. Resuspend the pellets in 1× IB to a final volume of 4 ml, and transfer the suspension to a glass vial with plastic lid, 6 ml capacity.

15. Add 200 µl of 1 mg/ml Con A-Alexa 488 (Invitrogen), cover the container with foil, and mix gently on a rotary mixer for 20 min at room temperature (or for 45 min at 4 °C).

16. Pellet the haustoria at $4,000 \times g$ in a benchtop centrifuge for 5 min at 4 °C. Remove the supernatant and wash the pellet twice with 1× IB. Resuspend in a final volume of 4 ml of 1× IB; keep sample on ice in the dark and proceed directly to FACS sorting.

3.2 Flow Cytometry of Pre-isolated Haustoria

This section describes the use of flow cytometry to sort haustoria based their fluorescence after labeling with Con A-Alexa 488. The Alexa 488 fluor is excited using a 13 mW 488 nm solid-state laser (Sapphire, Coherent Inc. Santa Clara, CA) and detected using a 502 nM longpass as well as a 530/30 nM bandpass filter. Chlorophyll autofluorescence is excited with the same 488 nm laser and detected using a 655 longpass and 695/40 bandpass filter.

1. Dilute the haustorial sample 1:10 with chilled 1× PBS to adjust the concentration of the chloroplast–fungal cell mixture to approximately $2.5–5.0 \times 10^6$ particles/ml.

2. Set up and optimize the cell sorter (*see* **Note 6**). For fungal haustoria (5–20 µm in diameter): 100 µM nozzle, 20 psi sheath pressure.

3. Use 1× PBS as sheath fluid, run cells using the lowest sample pressure with a resulting sample rate of 5,000–10,000 events per second. Collect sorted cells in 5 ml glass tubes containing 1× PBS, keeping them at 4 °C, and agitating them periodically at 300 rpm to prevent settling.

4. Set fluorescence and scatter parameters. Identify and exclude chlorophyll-containing particles based on chlorophyll autofluorescence. Identify and select the population of Con A-positive particles by Alexa 488 fluorescence (*see* Fig. 1a).

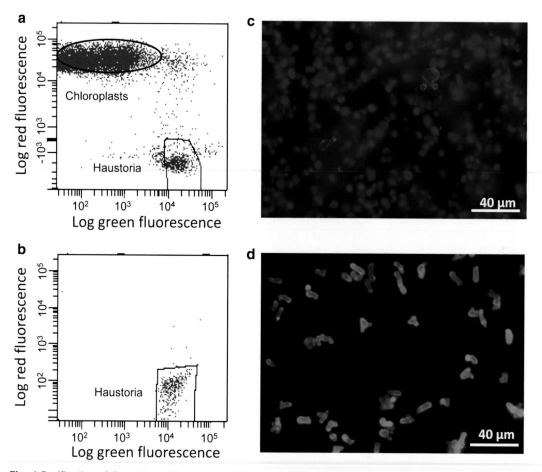

Fig. 1 Purification of *Puccinia striiformis* f.sp. *tritici* haustoria by FACS, and microscopic analysis of pre- and post-sorted samples. Haustoria were partially purified by Percoll gradients as described, then stained with Con A-Alexa 488 prior to sorting. (**a**) Flow-cytometric analysis of pre-isolated haustoria showing the discrimination of two populations based on red chloroplast autofluorescence and green Alexa 488 fluorescence from labeled haustoria. (**b**) Flow-cytometric analysis of a sorted sample to test the efficiency of the sorting process, average purity of 98 %. (**c**) Microscopic image of a sample prior to sorting, with a haustoria:chloroplast ratio of about 1:80. (**d**) Microscopic image of a sample post-sorting, with a haustoria:chloroplast ratio of about 49:1. In (**a**) and (**b**) *plots* displaying *flow cytometry* data, haustoria are represented by the *green dots* and chloroplasts by the *blue dots*

Use forward and side scatter information to identify and exclude doublets by comparing the forward scatter height and width and the side scatter height and width. To optimize the forward scatter signal, use a neutral density 2 filter in front of the forward scatter detector.

5. Confirm sort purity by running a small amount of the sorted cells to determine the percentage of positive events for the sort parameters (Fig. 1b) (*see* **Notes 7** and **8**).

6. Pool the samples from FACS purification and centrifuge them at $1,080 \times g$ for 15 min at 4 °C. Discard the supernatant and snap-freeze the final pellet in liquid nitrogen, and store it at −80 °C until required.

4 Notes

1. Harvest tissue as late as possible before sporulation to increase the number of haustoria; however, it should be done at least 1 day before pustule eruption. In wheat plants infected with *P. striiformis* and maintained at 18 °C post-infection, this time is usually 8–9 days.

2. The pore size of the meshes is critical for reducing contaminating particles before FACS. Different fungal species display a diverse range of haustoria size, thus small-scale isolations to test different pore sizes and microscopic analyses are necessary to optimize the filtration step.

3. The stock solution was prepared as per manufacturer's instructions but without addition of sodium azide.

4. Haustoria from different fungal species exhibit slightly different density characteristics. It is highly recommended to establish the density level at which the majority of haustoria resolve. Small-scale density tests can be performed by following the protocol above until **step 8**, Subheading 3.1. Once the tubes are removed from the centrifuge, aliquots of 1 ml are taken carefully with a wide-bore pipette tip, and the number of haustoria can be counted in each aliquot using a hemocytometer.

5. **Steps 11** and **12** are included specifically to reduce the complexity of the sample prior to flow cytometry. However, depending on the final use of the sorted haustoria, these two steps can be omitted.

6. The procedure of setting up a flow cytometer varies depending on the machine and needs to be performed by appropriately trained personnel.

7. Typically, 9×10^5–1.5×10^6 purified haustorial cells can be purified from 15 to 20 g of wheat tissue heavily infected with *P. striiformis*.

8. Vital tests should be applied at this stage to verify the viability of haustoria post-sorting. A high percentage of viable cells could be very important for some subsequent applications as isolation of intact RNA or proteins. In our studies we used CellTracker™ Orange CMRA (Invitrogen, Catalog number C34551) successfully following the manufacturer's instructions, and obtained ≥98 % viable haustoria post-sorting.

References

1. Szabo LJ, Bushnell WR (2001) Hidden robbers: the role of fungal haustoria in parasitism of plants. Proc Natl Acad Sci USA 98:7654–7655

2. Micali CO, Neumann U, Grunewald D, Panstruga R, O'Connell R (2011) Biogenesis of a specialized plant–fungal interface during host cell internalization of Golovinomyces orontii haustoria. Cell Microbiol 13:210–226

3. Heath MC (1997) Signalling between pathogenic rust fungi and resistant or susceptible host plants. Ann Bot Lond 80:713–720

4. Voegele RT, Mendgen K (2003) Rust haustoria: nutrient uptake and beyond. New Phytol 159:93–100

5. Stergiopoulos I, de Wit PJGM (2009) Fungal effector proteins. Annu Rev Phytopathol 47:233–263

6. Kemen E, Kemen AC, Rafiqi M, Hempel U, Mendgen K, Hahn M, Voegele RT (2005) Identification of a protein from rust fungi transferred from haustoria into infected plant cells. Mol Plant Microbe Interact 18: 1130–1139

7. Rafiqi M, Gan P, Ravensdale M, Lawrence G, Ellis J, Jones D, Hardham A, Dodds PN (2010) Internalization of flax rust avirulence proteins into flax and tobacco cells can occur in the absence of the pathogen. Plant Cell 22(6): 2017–2032

8. Hahn M, Mendgen K (1992) Isolation by cona binding of haustoria from different rust fungi and comparison of their surface qualities. Protoplasma 170:95–103

9. Catanzariti AM, Dodds PN, Lawrence GJ, Ayliffe MA, Ellis JG (2006) Haustorially expressed secreted proteins from flax rust are highly enriched for avirulence elicitors. Plant Cell 18:243–256

10. Dekhuijzen HM (1966) The isolation of haustoria from cucumber leaves infected with powdery mildew. Netherlands J Plant Pathol 72:1–11

11. Manners JM, Gay JL (1977) The morphology of haustorial complexes isolated from apple, barley, beet and vine infected with powdery mildews. Physiol Plant Pathol 11:261–262

12. Takahara H, Dolf A, Endl E, O'Connell R (2009) Flow cytometric purification of Colletotrichum higginsianum biotrophic hyphae from Arabidopsis leaves for stage-specific transcriptome analysis. Plant J 59: 672–683

Part III

From Assays of Pathogen Virulence to Effector Function

Chapter 9

Functional Characterization of Nematode Effectors in Plants

Axel A. Elling and John T. Jones

Abstract

Secreted effectors represent the molecular interface between the nematode and its host plant. Studies that aimed at deciphering molecular plant–nematode interactions are hampered by technical hurdles that prevent the generation of transgenic nematodes. However, RNA interference (RNAi) has proven to be a valuable tool to specifically knock-down nematode effector genes, both ex planta and in planta. Plant-mediated RNAi of nematode genes not only facilitates functional characterization of effectors but also lends itself to a novel control strategy against plant-parasitic nematodes. Here, we describe currently used methods to silence genes in plant-parasitic cyst and root-knot nematodes.

Key words RNAi, Effector, *Meloidogyne*, *Globodera*, *Heterodera*, Plant–nematode interaction

1 Introduction

Plant-parasitic nematodes display a variety of interactions with their hosts. They can be sedentary or browsing and can be endoparasites or ectoparasites. The interactions of some nematodes, including most of the migratory ectoparasitic species, are limited to simple grazing on root cells. However, some nematode species, including cyst nematodes and root-knot nematodes, the most economically important species, are biotrophic and induce profound changes in the roots of their hosts. Second-stage juveniles (J2) of these species hatch from eggs in the soil and locate and penetrate host roots. They then migrate through the root to cells near the vascular cylinder where they induce the formation of a feeding site. Root-knot nematodes induce repeated rounds of mitosis in the absence of cytokinesis leading to the formation of large multinucleate giant cells. Cyst nematodes induce syncytia that are formed by the breakdown of cell walls and fusion of adjacent protoplasts. In both cases the nematodes need to keep the resulting structure alive for the remainder of their life cycle, a period of up to 6 weeks

Paul Birch et al. (eds.), *Plant-Pathogen Interactions: Methods and Protocols*, Methods in Molecular Biology, vol. 1127, DOI 10.1007/978-1-62703-986-4_9, © Springer Science+Business Media New York 2014

in the case of cyst nematodes. The biology of a wide range of plant-parasitic nematodes has recently been reviewed in detail [1].

The interactions of biotrophic nematodes with their hosts are mediated by effectors (defined here as any secreted nematode factor that manipulates the host plant to the benefit of the nematode). Nematode effectors are produced in two sets of esophageal gland cells—subventral and dorsal—and secreted into the host through the stylet, a hollow protrusible spear-like mouthpart. Nematode effectors have several biological functions in the interaction between host and pathogen including softening of the cell wall during migration, induction of the nematode feeding site, and suppression of host defenses.

As a result of genome sequencing [2–4], expressed sequence tag (EST) (e.g. [5, 6]), and microarray projects [7], candidate effectors have been identified from many different plant-parasitic nematodes. In some cases a testable biological function can be derived from the sequence itself. For example, a variety of cell wall degrading enzymes are secreted by plant-parasitic nematodes [8] and these can be identified on the basis of sequence similarity and their biochemical properties can be tested in in vitro assays. However, in many cases effectors are pioneers that have no similarity to functionally defined proteins from other species. Ascribing function to these sequences is particularly challenging. Here we review the most commonly applied in planta methods that are currently used for analyzing the function of these proteins. These include:

(a) RNA interference (RNAi) by soaking of J2s.

(b) In planta RNAi.

(c) Phenotyping of nematodes grown on genetically modified plants over-expressing effectors or effector targets, or plants in which effector targets are knocked out.

Methods (a) and (b) will provide information about the phenotype caused by silencing a nematode gene of interest. Method (c) describes the most commonly used protocols for growing nematodes on transgenic plants and scoring effects of the transgene insertion on the nematode. This protocol is applicable to plants produced as described in method (b) but can also be used to assess the effects of growing nematodes on plants that overexpress effectors.

2 Materials

2.1 RNAi by Soaking

1. Cloned cDNA of the gene that is to be silenced.

2. Six primers for each gene to be silenced (see Fig. 1), each at 10 µM concentration (see Note 1).

3. PCR reagents (10× Taq buffer, 25 mM $MgCl_2$, 2 mM dNTP mix, F, R and T7F, and T7R primers (Fig. 1), template cDNA, Taq DNA polymerase, thermal cycler).

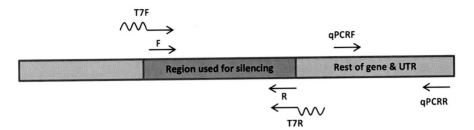

Fig. 1 Schematic diagram showing PCR primers required for RNAi by soaking. F and R-primers for amplification of the region of the gene to be used for the silencing experiment

4. Equipment and reagents for agarose gel electrophoresis.

5. DNA purification kit (e.g., Qiagen PCR Purification Kit).

6. MegaScript RNAi kit (Ambion).

7. 0.5 M Octopamine 0.5 M.

8. 300 mM spermidine.

9. 5 % gelatin.

10. 0.25× M9 buffer (1×M9 = 43.6 mM Na_2HPO_4, 22 mM KH_2PO_4, 18.7 mM NH_4Cl, 8.6 mM NaCl).

11. Rotator.

12. Host plants for infections.

13. Materials for qPCR or semi-quantitative PCR.

2.2 In Planta RNAi

1. Cloned cDNA of the gene that is to be silenced.

2. pHANNIBAL vector (http://www.pi.csiro.au) (*see* **Note 2**).

3. pART27 binary vector (http://www.pi.csiro.au).

4. TA cloning vector kit (e.g., TOPO TA from Invitrogen or pGEM-TEasy from Promega).

5. Primers for sense fragment: 5′-*Xho*I-gene-specific sequence (forward primer) and 5′-*Eco*RI-gene-specific sequence (reverse primer).

6. Primers for antisense fragment: 5′-*Xba*I-gene-specific sequence (forward primer) and 5′-*Hind*III-gene-specific sequence (reverse primer).

7. T7 and SP6 sequencing primers.

8. PCR purification, gel purification, and miniprep kits.

9. PCR reagents and thermal cycler.

10. Restriction enzymes.

11. T4 ligase and buffer.

12. LB medium (liquid and solidified with agar).

13. Primers to confirm hairpin constructs: (P-5: 5'-GGGATGAC GCACAATCC-3'; P-3: 5'-GAGCTACACATGCTCAGG-3'; I-5:5'-ATAATCATACTAATTAACATCAC-3' I-3:5'-TGATA GATCATGTCATTGTG-3') [12].

14. PCR tubes and 1.5 mL Eppendorf tubes.

15. Ice.

16. Pipettes.

17. Petri dishes, 12-well tissue culture dishes.

2.3 Phenotyping of Nematodes

1. *Arabidopsis* seeds.

2. Bleach, Tween-20, 100 % ethanol, sterile water.

3. Rotator.

4. 1.5 mL Eppendorf tubes.

5. 12-well tissue cultures dishes.

6. Parafilm.

7. Modified Knop's media (solidified).

8. #60, #200, #500 test sieves (Taylor).

9. 70 % Sucrose.

10. Baermann pan.

11. 0.01 % Mercuric chloride.

12. Low-melting point agarose.

13. Acid fuchsin for staining nematodes in roots (250 mL glacial acetic acid; 750 mL distilled water; 3.5 g acid fuchsin). Destaining solution-acidified glycerol (30 mL glycerol containing a few drops of concentrated HCl).

14. Phloxine B for staining root-knot nematode egg masses (0.10 g/L).

3 Methods

3.1 RNAi by Soaking

1. Amplify the region to be used for silencing of your gene in two separate PCR reactions. In one use the T7F primer in combination with the R primer, in the other use the F primer in combination with the T7R primer (*see* Fig. 1). This will generate two PCR products, one with the T7 promoter sequence at the 5' end of the coding strand and one with the T7 promoter sequence at the 5' end of the antisense strand. Several micrograms of PCR product will be required for the downstream steps so it is advisable to do several (up to 8) repeats of each PCR reaction.

2. Check that a single PCR product has been produced by running a small aliquot of each reaction on a 1.5 % agarose gel (*see* **Note 3**).

3. Purify the PCR products using your preferred kit. Combine each of the T7F/R reactions and each of the F/T7R reactions to maximize the concentration of each of the two PCR products.

4. Produce the two strands of RNA in separate reactions using the MegaScript kit following the manufacturer's instructions. Briefly, the purified PCR products are incubated with ATP, CTP, GTP, and UTP along with the T7 enzyme mix supplied. The reaction is left overnight at 37 °C, rather than for the 2–4 h recommended in the kit instructions, in order to maximize yield.

5. Anneal the two strands by mixing the two reactions, heating to 75 °C for 5 min and then allowing to cool at room temperature. A small amount of the reaction products can be checked on an agarose gel at this stage if desired—dsRNA will migrate slightly slower than a DNA marker of the same size (the template DNA PCR product is ideal for this purpose).

6. Remove DNA template and purify the dsRNA as indicated in the Megascript kit. The dsRNA should then be quantified using a spectrophotometer.

7. Set up the soaking experiment. Enough nematodes (several thousand J2) need to be used to allow replicated infection studies and to allow extraction of mRNA for cDNA synthesis and semi-quantitative PCR. Centrifuge the nematodes for 2 min at top speed in a benchtop centrifuge and resuspend in a solution containing 0.25× M9 buffer, 50 mM octopamine, 3 mM spermidine, 0.05 % gelatin, and 2 μg/μL dsRNA. The volume used for this step will depend on the number of nematodes being used and the yield of dsRNA; for PCN 200 μL is a good starting point. Wrap the tube in foil and leave on a rotator for 24 h at room temperature. It is essential to use a control in which nematodes are soaked in dsRNA from a non-endogenous gene (e.g., green fluorescent protein), given the high concentrations of dsRNA that are used in these experiments.

8. Remove a small amount of the soaking liquid after the soaking is complete and run on an agarose gel as above to confirm that degradation of the dsRNA has not occurred during soaking. Wash the nematodes three times in 0.25× M9 buffer.

9. Dilute the soaking liquid to a total volume of 1 mL 0.25× M9. Count the numbers of nematodes present in three 5 μL aliquots from this 1 mL and calculate the concentration of nematodes. Remove enough nematodes for infection studies (a minimum of five 100 nematode aliquots are required). Centrifuge and freeze the remaining nematodes; these will be used for mRNA extraction.

10. For infection studies use a minimum of five plants and infect each with 100 nematodes. Score numbers of nematodes developing into mature females. Precise details will vary depending on the pathosystem being tested but try to minimize the size

of the root system in order to make it easier to locate the nematodes. For detailed studies it may also be desirable to stage the nematodes at set time points after infection (weekly for 4–6 weeks) in order to determine the life stage at which any effect of the dsRNA treatment is occurring. Nematodes can be stained as described in Subheading 3.3 below.

11. Extract mRNA from the remaining nematodes and assess mRNA levels of the target gene in comparison to a control gene in the samples soaked in the target dsRNA and the dsRNA for the non-endogenous control gene. This can be done by qPCR or using a semi-quantitative method as described in ref. 9.

3.2 In Planta RNAi

pHANNIBAL vector (*see* **Note 2**).

1. Select target region of adequate length in nematode gene of interest (*see* **Note 4**).

2. Set up separate PCRs to amplify sense and antisense arms of nematode gene of interest (*see* Fig. 2).

3. Purify PCR products using PCR purification kit.

4. Clone purified PCR products into TA cloning vector (e.g., pGEM-TEasy) and transform into *E. coli*. Plate transformed *E. coli* on LB plates with ampicillin (100 mg/L) and blue-white selection with IPTG and X-gal.

5. Pick white colonies and grow in liquid LB media with ampicillin (100 mg/L) overnight. Recover plasmid by miniprep and sequence insert using T7 and SP6 primers.

6. Digest confirmed clones with *Xho*I and *Eco*RI (sense arm) or *Xba*I and *Hind*III (antisense arm) to release insert. Purify released inserts using gel purification kit.

7. Digest pHANNIBAL sequentially with *Xho*I and *Eco*RI to clone digested sense arm insert and *Xba*I and *Hind*III to clone digested antisense arm insert (*see* Fig. 2).

8. Digest pHANNIBAL (now containing both sense and antisense arm inserts) with *Not*I to release ihpRNA cassette and clone cassette into *Not*I site in pART27 binary vector.

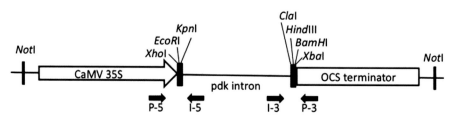

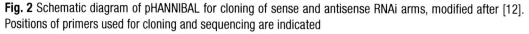

Fig. 2 Schematic diagram of pHANNIBAL for cloning of sense and antisense RNAi arms, modified after [12]. Positions of primers used for cloning and sequencing are indicated

9. Transform pART27 into *E. coli* and spread cells on LB plates with spectinomycin (100 mg/L). Pick colonies and grow in liquid LB with spectinomycin (100 mg/L) overnight. Recover pART27 plasmid by miniprep.

10. Digest pART27 with *Bgl*II (linearizes plasmid by cutting once in pHANNIBAL pdk intron) to confirm clones.

11. Set up PCRs to amplify sense arm (using primers P-5 and I-5) and antisense arm (using primers P-3 and I-3). The respective products will be 250 bp longer than the sense and antisense arms.

12. Purify PCR products and sequence to confirm inserts.

13. Transform pART27 binary vector with ihpRNA cassette into appropriate *Agrobacterium tumefaciens* strain (e.g., GV3101).

14. Transform *Arabidopsis* using floral dip method [10] and select transformants on kanamycin (100 mg/L).

15. Select homozygous T3 plants from at least ten independently transformed lines for phenotyping assays.

3.3 Phenotyping of Nematodes

1. Surface sterilize T3 seeds of transformed *Arabidopsis* lines and respective wild-type controls. Place seeds into a 1.5 mL Eppendorf tube and add 10 % bleach and 0.01 % Tween-20. Place on rotator for 3 min.

2. Spin down seeds, remove supernatant, and wash with 70 % ethanol. Put back on rotator for 1 min.

3. Spin down seeds, remove supernatant, and wash seeds three times with sterile de-ionized water.

4. Transfer single seeds into separate wells of a 12-well dish containing 1.5 mL modified Knop's media (solidified).

5. Seal plates with Parafilm and place in growth chamber (16 h light/8 h dark). Grow plants 7–14 days before inoculating with surface-sterilized nematodes.

6. For root-knot nematodes, collect nematode eggs using a #60/#200/#500 sieve tower. To clean eggs and remove contaminating plant tissue and soil particles, resuspend egg slurry in 20 mL water and transfer to a 50 mL tube. Add 20 mL of 70 % sucrose and shake. Overlay with 10 mL water and centrifuge in a swing bucket centrifuge at $380 \times g$ for 5 min. Eggs and juvenile nematodes will concentrate at the interface that will form. Collect eggs by pipetting and immediately wash with water to remove sucrose on a #500 test sieve. Rinse eggs into a 50 mL tube once they have been washed.

7. Hatch purified eggs in a modified Baermann pan (*see* **Note 5**). If working with cyst nematodes, start the procedure here with sterile or cleaned cysts.

8. Collect hatched J2 by centrifuging in a swing bucket centrifuge at $380 \times g$ for 3 min.

9. Remove supernatant and wash once with 0.001 % Tween-20 and twice with sterile water.

10. Surface sterilize J2 by adding 0.01 % mercuric chloride for 4 min. Collect J2 by centrifuging in a swing bucket centrifuge at $380 \times g$ for 3 min. Remove supernatant and wash three times with sterile water.

11. Resuspend nematodes in 1.5 % low-melting point agarose.

12. Pipette 200–500 J2 into each well and seal plate with Parafilm. Return to growth chamber and grow until desired timepoints of experiment have been reached.

13. Stain nematodes in roots with acid fuchsin and count/measure nematodes and/or nematode infection sites (syncytia, galls). Place infected roots in 50 mL sterile distilled water to which 1 mL acid fuchsin stain has been added, bring up to boiling point on a hotplate or in a microwave oven. Rinse in several changes of sterile distilled water. To destain, place the roots in 20–30 mL acidified glycerol and bring up to boiling point on a hotplate or in a microwave oven. Allow to cool before observing under a microscope.

14. If staining of root-knot nematode egg masses is desired to facilitate gall counting, incubate each root system in phloxine B (0.10 g/L) for 10–15 min. Rinse three times with sterile distilled water. Egg masses will stain bright pink, whereas plant tissue will be clear to slight pink.

15. Additional phenotypic analyses such as root length measurements may be added depending on the question being addressed.

4   Notes

1. Careful consideration should be given to the region of the gene chosen for silencing. You may wish to use a region that is highly specific to your gene of interest. Alternatively, if functional redundancy is likely to be an issue, a region of the gene that is conserved and that will target silencing of all related sequences can be chosen. It is essential that the qPCR primers are targeted at a region outside the region chosen for silencing, as substantial quantities of dsRNA will remain in the sample, making assessment of transcript levels using this region impossible. siRNAs have also been used for silencing nematode genes. For a review of this technology *see* ref. 11.

2. Several vectors have been used for in planta RNAi of nematode genes (*see* Table 1). However, pHANNIBAL has been used successfully more often than other vector systems to date and is readily available, which is why we focus on pHANNIBAL

Table 1
Selected in planta RNAi studies

Gene	Nematode species	Plant species	Delivery system	Promoter	Intron/spacer	Target region (nt)	References
Splicing factor	*M. incognita*	Tobacco	pHANNIBAL	35S	Pdk	349	[13]
Integrase	*M. incognita*	Tobacco	pHANNIBAL	35S	Pdk	624	[13]
3B05	*H. schachtii*	Arabidopsis	pHANNIBAL	35S	Pdk	233	[14]
4G06	*H. schachtii*	Arabidopsis	pHANNIBAL	35S	Pdk	153	[14]
8H07	*H. schachtii*	Arabidopsis	pHANNIBAL	35S	Pdk	82, 338	[14]
10A06	*H. schachtii*	Arabidopsis	pHANNIBAL	35S	Pdk	103	[14]
16D10	*M. arenaria, M. incognita, M. hapla, M. javanica*	Arabidopsis	pHANNIBAL	35S	Pdk	42, 271	[15]
Major sperm protein	*H. glycines*	Soybean	pATMSPi	ACT2	Major sperm protein sequence	231	[16]
L-Lactate hydrogenase	*M. incognita*	Soybean (hairy roots)	pRAP17	FMV-sgt	Unspecified, chloramphenicol resistance gene	400	[17]
Tyrosine phosphatase	*M. incognita*	Soybean (hairy roots)	pRAP17	FMV-sgt	Unspecified, chloramphenicol resistance gene	400	[17]
Mitochondrial stress-70 protein precursor	*M. incognita*	Soybean (hairy roots)	pRAP17	FMV-sgt	Unspecified, chloramphenicol resistance gene	400	[17]
ATP synthase beta-chain mitochondrial precursor	*M. incognita*	Soybean (hairy roots)	pRAP17	FMV-sgt	Unspecified, chloramphenicol resistance gene	400	[17]

(continued)

Table 1
(continued)

Gene	Nematode species	Plant species	Delivery system	Promoter	Intron/spacer	Target region (nt)	References
Small ribosomal protein 3a	H. glycines	Soybean (hairy roots)	pRAP17	FMV-sgt	Unspecified, chloramphenicol resistance gene	200–500 (not specified)	[18]
Small ribosomal protein 4	H. glycines	Soybean (hairy roots)	pRAP17	FMV-sgt	Unspecified, chloramphenicol resistance gene	200–500 (not specified)	[18]
Synaptobrevin	H. glycines	Soybean (hairy roots)	pRAP17	FMV-sgt	Unspecified, chloramphenicol resistance gene	200–500 (not specified)	[18]
Splicosomal SR protein	H. glycines	Soybean (hairy roots)	pRAP17	FMV-sgt	Unspecified, chloramphenicol resistance gene	200–500 (not specified)	[18]
Fib-1	H. glycines	Soybean (hairy roots)	pANDA35HK	35S	GUS	170	[19]
Y25C1A.5	H. glycines	Soybean (hairy roots)	pANDA35HK	35S	GUS	292	[19]
Cpn-1	H. glycines	Soybean (hairy roots)	pANDA35HK	35S	GUS	408	[20]
Y25	H. glycines	Soybean (hairy roots)	pANDA35HK	35S	GUS	294	[20]
Prp-17	H. glycines	Soybean (hairy roots)	pANDA35HK	35S	GUS	289	[20]
MjTis11	M. javanica	Tobacco	pHANNIBAL	35S	Pdk, uidA	772	[21]
MjTis11	M. javanica	Tobacco	pHANNIBAL	TobRB7	Pdk, uidA	772	[21]
Miduox	M. incognita	Tomato (hairy roots), Arabidopsis	pFGC5941	35S	CHS	554	[22]
Mispc3	M. incognita	Tomato (hairy roots), Arabidopsis	pFGC5941	35S	CHS	477	[22]

in this protocol. Detailed information about cloning with pHANNIBAL can be found in ref. 12. Other vector systems will require modifications to the materials and methods described here.

3. It is essential that a single band is produced in the PCR reactions. If this is not the case, adjust annealing temperature and/or $MgCl_2$ concentration and repeat the reaction.

4. Using pHANNIBAL, 42–624 bp have proven successful in RNAi experiments targeting nematode genes in planta; other vectors may require different lengths. Ensure that there are no off-target effects on the host or pathogen by BLAST searches if sufficient genome is available. In general, sequences with 20 nt or more that are identical between the RNAi construct and a non-target gene will result in off-target silencing. If trying to target a gene family in the nematode, select a target region that is conserved in all members of the gene family. If only a specific gene of a gene family is to be targeted, make sure to select a unique region.

5. Cyst nematodes may require addition of chemicals to induce hatching. Soybean cyst nematode can be stimulated to hatch with 3.14 mM $ZnSO_4$. Other cyst nematodes can be hatched in solutions of root exudates. These are made by placing the roots of a plant into a beaker containing 250 mL distilled water, leaving overnight and then filtering the liquid. Root diffusates can be stored at 4 °C for several months. Root-knot nematodes generally do not require stimulants in order to hatch.

Acknowledgments

The James Hutton Institute receives funding from the Scottish Government. Work on RNAi in the Elling laboratory has been funded by the United States Department of Agriculture, Washington State Department of Agriculture, Washington State Potato Commission, Idaho Potato Commission, and Washington Grain Commission. PPNS No. 0612, Department of Plant Pathology, College of Agricultural, Human, and Natural Resource Sciences, Agricultural Research Center, Project No. WNP00744, Washington State University, Pullman, WA 99164-6430, USA.

References

1. Jones JT, Gheysen G, Fenoll C (2011) Genomics and molecular genetics of plant-nematode interactions. Springer, New York

2. Abad P, Gouzy J, Aury JM et al (2008) Genome sequence of the metazoan plant-parasitic nematode *Meloidogyne incognita*. Nat Biotechnol 26:909–915

3. Opperman CH, Bird DMK, Williamson VM et al (2008) Sequence and genetic map of *Meloidogyne hapla*: a compact nematode genome for plant parasitism. Proc Natl Acad Sci U S A 105:14802–14807

4. Kikuchi T, Cotton JA, Dalzell JJ et al (2011) Genomic insights into the origin of parasitism

in the emerging plant pathogen *Bursaphelenchus xylophilus*. PLoS Pathog 7:e1002219

5. Gao B, Allen R, Maier T, Davis EL, Baum TJ, Hussey RS (2003) The parasitome of the phytonematode *Heterodera glycines*. Mol Plant Microbe Interact 16:720–726

6. Jones JT, Kumar A, Pylypenko LA, Thirugnanasambandam A, Castelli L, Chapman S, Cock PJ, Grenier E, Lilley CJ, Phillips MS, Blok VC (2009) Identification and functional characterisation of effectors in expressed sequence tags from various life cycle stages of the potato cyst nematode *Globodera pallida*. Mol Plant Pathol 10:815–828

7. Elling AA, Mitreva M, Gai X, Martin J, Recknor J, Davis EL, Hussey RS, Nettleton D, McCarter JP, Baum TJ (2009) Sequence mining and transcript profiling to explore cyst nematode parasitism. BMC Genomics 10:58

8. Haegeman A, Jones JT, Danchin E (2011) Horizontal gene transfer in nematodes: a catalyst for plant parasitism? Mol Plant Microbe Interact 24:879–887

9. Chen Q, Rehman S, Smant G, Jones JT (2005) Functional analysis of pathogenicity proteins of the potato cyst nematode *Globodera rostochiensis* using RNAi. Mol Plant Microbe Interact 18:621–625

10. Bechtold N, Ellis J, Pelletier G (1993) In-planta Agrobacterium-mediated genetransfer by infiltration of adult *Arabidopsis-thaliana* plants. Compt Rend Acad Sci Iii Sci Vie Life Sci 316:1194–1199

11. Dalzell JJ, McMaaster S, Fleming CC, Maule AG (2010) Short interfering RNA-mediated gene silencing in *Globodera pallida* and *Meloidogyne incognita* infective stage juveniles. Int J Parasitol 41:91–100

12. Wesley S, Liu Q, Wielopolska A, Ellacott G, Smith N, Singh S, Helliwell C (2003) Custom knock-outs with hairpin RNA-mediated gene silencing. Methods Mol Biol 236:273–286

13. Yadav BC, Veluthambi K, Subramaniam K (2006) Host-generated doubles stranded RNA induces RNAi in plant-parasitic nematodes and protects the host from infection. Mol Biochem Parasitol 148:219–222

14. Sindhu AS, Maier TR, Mitchum MG, Hussey RS, Davis EL, Baum TJ (2009) Effective and specific *in planta* RNAi in cyst nematodes: expression interference of four parasitism genes reduces parasitic success. J Exp Bot 60:315–324

15. Huang G, Allen R, Davis EL, Baum TJ, Hussey RS (2006) Engineering broad root-knot resistance in transgenic plants by RNAi silencing of a conserved and essential root-knot nematode parasitism gene. Proc Natl Acad Sci U S A 103:14302–14306

16. Steeves RM, Todd TC, Essig JS, Trick HN (2006) Transgenic soybeans expressing siRNAs specific to a major sperm protein gene suppress *Heterodera glycines* reproduction. Funct Plant Biol 33:991–999

17. Ibrahim HMM, Alkharouf NW, Meyer SLF, Aly MAM, El-Din AEKYG, Hussein EHA, Matthews BF (2011) Post-transcriptional gene silencing of root-knot nematode in transformed soybean roots. Exp Parasitol 127:90–99

18. Klink VP, Kim KH, Martins V, MacDonald MH, Beard HS, Alkharouf NW, Lee SK, Park SC, Matthews BF (2009) A correlation between host-mediated expression of parasite genes as tandem inverted repeats and abrogation of development of female *Heterodera glycines* cyst formation during infection of *Glycine max*. Planta 230:53–71

19. Li J, Todd TC, Trick HN (2010) Rapid in planta evaluation of root expressed transgenes in chimeric soybean plants. Plant Cell Rep 29:113–123

20. Li J, Todd TC, Oakley TR, Lee J, Trick HN (2010) Host-derived suppression of nematode reproductive and fitness genes decreases fecundity of *Heterodera glycines* Ichinohe. Planta 232:775–785

21. Fairbairn DJ, Cavallaro AS, Bernard M, Mahalinga-Iyer J, Graham MW, Botella JR (2007) Host-delivered RNAi: an effective strategy to silence genes in plant-parasitic nematodes. Planta 226:1525–1533

22. Charlton WL, Harel HYM, Bakhetia M, Hibbard JK, Atkinson HJ, McPherson MJ (2010) Additive effects of plant expressed double-stranded RNAs on root-knot nematode development. Int J Parasitol 40:855–864

Chapter 10

Silencing of Aphid Genes by Feeding on Stable Transgenic *Arabidopsis thaliana*

Alexander D. Coleman, Marco Pitino, and Saskia A. Hogenhout

Abstract

Aphids are economically important pests that predominantly feed from the plant phloem. Genome, transcriptome, and proteome data are being generated for these insects, and predicted secreted proteins in aphid saliva have been identified. These secreted proteins are candidate effectors that may modulate plant processes and aid aphid colonization of plants. The next step is to develop post-genomics strategies to study the functions of identified aphid genes. One such strategy is to express aphid effector genes in planta to assess whether aphid effectors alter plant development and aphid survival and fecundity. A second strategy is to knock down the expression of aphid target genes by RNA interference (RNAi). In this chapter, we describe how to knock down aphid gene expression using plant-mediated RNAi. This strategy is useful for assessing the contribution of aphid effectors to aphid colonization of plants.

Key words Aphid effectors, Virulence, Plant–aphid interactions, Silencing, Aphid gene knock down, Post-genomics tools, RNA interference (RNAi)

1 Introduction

Aphids are sap-feeding insects that predominantly feed from the plant phloem. These insects are economically important pests of crops. Aphids and related species in the order Hemiptera transmit the majority of described plant viruses and can also cause extensive feeding damage. In addition, several hemipteran insects have developed resistance to the most widely used pesticides currently on the market.

A number of research groups worldwide have generated genomics resources for aphids, including genome, transcriptome, and proteome information. This has led to the identification of genes potentially involved in regulating aphid development and aphid-plant interactions [1–4]. For example, predicted secreted proteins in aphid saliva are likely delivered in the plant during aphid feeding [5]. Hence, aphid secreted saliva proteins are candidate effector proteins.

Paul Birch et al. (eds.), *Plant-Pathogen Interactions: Methods and Protocols*, Methods in Molecular Biology, vol. 1127, DOI 10.1007/978-1-62703-986-4_10, © Springer Science+Business Media New York 2014

The next step is to develop tools that will allow the functions of identified aphid genes to be studied. Such techniques are essential for confirming predicted functions of aphid effectors and other insect genes, including those involved in the transmission of plant viruses, and may lead to the identification of novel strategies to control aphids.

We have developed two strategies to investigate aphid gene functions ([5–7]; Chapter 11). The first strategy is to express the aphid effector genes in planta to assess whether aphid effectors alter plant development or aphid survival and fecundity. Two methods to study aphid effector function in planta were developed. The first method uses agroinfiltration to transiently express aphid effector genes in plant leaves. This method is useful for all plant species amenable to agroinfiltration, including *Nicotiana benthamiana*, and may be used in leaf-disc assays to assay aphid gene functions at higher throughput. Details of this method are provided in Chapter 11. In the second method, stable transgenic *Arabidopsis thaliana* plants constitutively expressing the aphid effector genes are generated. Aphid genes may be stably expressed in the majority of plant cells using the 35S promoter or predominantly in plant phloem using, for example, the AtSuc2 promoter.

A second strategy developed for studying aphid gene functions is to knock down aphid gene expression by plant-mediated RNA interference (RNAi). Double-stranded RNAs (dsRNAs) corresponding to aphid target genes are expressed in plants and acquired by aphids during feeding, resulting in partial silencing of aphid target gene expression. The dsRNAs may be expressed transiently using agroinfiltration of plant leaves or expressed stably under control of 35S or AtSuc2 promoters in transgenic plants. RNAi can be achieved in aphids via microinjection of dsRNAs into the aphid body or feeding of dsRNAs from artificial diets [8–12]. However, these methods require removal of aphids from their natural habitat, the plant, and require a high level of insect handling. Moreover, microinjection results in high mortality rates and smaller aphid species are less suitable for injection. Thus, the plant-mediated RNAi strategy has several advantages.

In this chapter, we describe how to generate stable transgenic plants for plant-mediated RNAi of aphids, including how to (1) clone constructs into the pJawohl8:RNAi plasmid, (2) generate stable transgenic plants, (3) perform aphid survival and fecundity assays, (4) carry out qRT-PCRs to study gene expression knock down in aphids, and (5) perform northern blots. The pJawohl8:RNAi constructs may also be used in transient expression experiments (Chapter 11). We developed the plant-mediated RNAi technique for the green peach aphid (GPA), *Myzus persicae*, because this species can be readily reared on the model plants *A. thaliana* and *N. benthamiana*. In addition, GPA is one of the most economically important insect pests as it colonizes over 400 plant species in diverse plant families and has developed resistance to the majority of pesticides.

2   Materials

2.1   Total RNA Extraction from Plants/Aphids and Quality Control

1. RNaseZap (Life Technologies, Paisley, UK).

2. TRIzol Reagent (Life Technologies).

3. Chloroform.

4. Isopropanol.

5. 75 % EtOH.

6. RNase-free water (Qiagen, Hilden, Germany).

7. Nanodrop Spectrophotometer ND2000 (Thermo Scientific, Loughborough, UK).

8. RNA gel: 1 % agarose gel in 1× BPTE buffer [10× buffer consists of 100 mM piperazine-N,N'-bis(2-ethanesulfonic acid) (PIPES), 300 mM Bis(2-hydroxyethyl)-amino-tris(hydroxymethyl)-methane (Bis-Tris), 10 mM EDTA, pH 6.5].

9. Glyoxal loading dye (Life Technologies).

2.2   Cloning of Constructs into pJawohl8-RNAi

1. Competent bacterial strains: DH5α (Life Technologies); GV3101 containing pMP90RK plasmid [13].

2. Plasmid vectors: pDONR™207 (Life Technologies); pJawohl8-RNAi (I.E. Somssich, Max Planck Institute for Plant Breeding Research, Germany).

3. Selective media (DH5α + pDONR™207): Luria Broth (LB) containing 7 μg/mL Gentamicin.

4. Selective media (GV3101 containing pMP90RK + pJawohl8-RNAi): Luria Broth (LB) containing 25 mg/L Kanamycin, 25 mg/L Gentamicin, 50 mg/L Rifampicin, and 25 mg/L Carbenicillin.

5. Sequencing: Big Dye v3.1 (Life Technologies).

6. Cloning system: Gateway® Recombination Cloning Technology (Life Technologies).

2.3   Growing of Plants, Transformation, and Selection of T2 Homozygous Lines

1. *Arabidopsis thaliana*: Col-0 background.

2. BASTA: phosphinothricin spray (120 mg/L solution).

3. BASTA selective seedling growth media: 0.8 % Murashige and Skoog (MS) medium supplemented with 20 mg/mL BASTA for selection.

4. Compost: Scotts Levington F2.

5. Controlled-environment conditions: 18 °C, 10 h light, 60 % humidity.

2.4   Northern Blotting to Detect siRNAs of Transgene

1. 15 % Polyacrylamide/7 M urea/20 mM MOPS pH 7.0, 5 mL gel: 2.1 g urea/1.5 mL sterile water; 1.85 mL of 40 % acrylamide/bisacrylamide solution 19:1; 0.25 mL of 10 % 3-(N-morpholino)propanesulfonic acid (MOPS); 50 μL of

10 % ammonium persulfate (APS, freshly made); 2.5 μL of tetramethylethylenediamine (TEMED).

2. Gel running buffer (10× 200 mM MOPS): 0.2 M MOPS; 20 mM sodium acetate (CH₃COONa); 10 mM ethylenedi-aminetetraacetic acid (disodium EDTA) sterilized by filtration through a 0.45 μM filter, and stored at room temperature protected from light.

3. Stopmix: 5 mM EDTA; 0.1 % bromophenol blue; 0.1 % xyleno cyanol; 95 % formamide.

4. MicroRNA marker: MicroRNA marker (New England Biolabs (NEB), Hitchin, UK).

5. Semi-dry blotter: Trans-blot™ (Bio-Rad, Hempstead, UK).

6. Blotting membrane: Amersham Hybond-N membrane (GE Healthcare, Little Chalfont, UK).

7. Cross-linking solution (for 12 mL—enough for two gels): 122.5 μL 1-methylimidazole (12.5 M) (Sigma-Aldrich, Gillingham, UK); 10 mL dH₂O; 0.373 g 1-Ethyl-3-(3-dimethylaminopropyl)carbodiimide (EDC, 0.2 M) (Sigma-Aldrich). Set volume to 12 mL and adjust pH to 8.0 by the addition of 1 M HCl.

8. Hybridization solution: Ambion ULTRAhyb® Ultrasensitive Hybridization Buffer (Life Technologies).

9. Gel extraction kit: QIAquick Gel Extraction Kit (Qiagen).

10. DNA labeling: Ambion Klenow fragment (Life Technologies) with [α-32P] Deoxycytidine triphosphate (dCTP).

11. Washing solution: 0.2× Saline Sodium Citrate buffer (SSC) (diluted from 20× stock solution consists of 3 M sodium chloride and 300 mM trisodium citrate, pH 7.0); 0.1 % Sodium dodecyl sulfate (SDS).

12. Stripping solution: 10 mM tris(hydroxymethyl)amino-methane (TRIS)/HCl pH 8.0; 5 mM EDTA; 0.1 % SDS.

13. Blot imaging: Phosphor storage plates (GE Healthcare); Typhoon™ 9200 scanner hardware (GE Healthcare); ImageQuant™ software (GE Healthcare).

2.5 Insect Bioassays

1. Green peach aphid (GPA) *Myzus persicae* (RRes genotype O) maintained on Chinese cabbage (*Brassica rapa*) [5–7].

2. Sealed experimental cages: Clear, acrylic cages to house 4–8 plants (Fig. 1a).

3. Controlled-environment conditions: 18 °C, 8 h light, 48 % humidity.

4. Insect fecundity assays: Sealed experimental cages containing the entire plant in single pots (10 cm diameter) (Fig. 1b).

Fig. 1 Cages used for aphid rearing (**a**) and insect fecundity assays (**b**)

2.6   Statistical Analysis of Fecundity and qRT-PCR Data

1. Statistical packages: GenStat (VSNi, Hemel Hempstead, UK).
2. qRT-PCR software: CFX manager (Bio-Rad).
3. Genorm: qBASEplus (Biogazelle, Zwijnaarde, Belgium).

2.7   qRT-PCR Analyses

1. DNase: RQ1 RNase-Free DNase (Promega, Southampton, UK).
2. 10 mM dNTP mix (made from individual dNTPs): dATP, 10 mM (Life Technologies, 100 mM); dCTP, 10 mM (Life Technologies, 100 mM); dGTP, 10 mM (Life Technologies, 100 mM); dTTP, 10 mM (Life Technologies, 100 mM).
3. Reverse Transcriptase: Moloney Murine Leukemia Virus (M-MLV) Reverse Transcriptase Kit (Life Technologies) with Oligo (dT) (500 µg/mL) (Life Technologies).
4. qRT-PCR ready mix: SYBR Green Jumpstart TAQ R/Mix 500rxn (Sigma-Aldrich).
5. qRT-PCR plate: Thermo-Fast Plate 96-Well PCR Non-Skirted White (Thermo Scientific).
6. Adhesive film: Adhesive PCR Film (Thermo Scientific).
7. qRT-PCR instrument: CFX connect™ machine (Bio-Rad).

3   Methods

3.1   Total RNA Extraction from Plants/Aphids and Quality Control (See Note 1)

1. Wipe working area and implements coming into contact with samples with RNaseZap.
2. Grind plant/insect samples in liquid nitrogen to a fine powder.

3. Allow samples to come to room temperature for 2 min and add 1 mL of TRIzol Reagent per 1 mg of tissue.

4. Incubate the homogenate for 5 min (*see* **Note 2**).

5. Add 0.2 mL of chloroform per 1 mL of TRIzol, mix thoroughly for 15 s. Incubate at room temperature for 10 min.

6. Centrifuge at $17,900 \times g$ for 15 min at 4 °C (*see* **Note 3**).

7. Transfer the aqueous phase to a fresh tube. Add 0.5 mL isopropanol per 1 mL of TRIzol. If upper layer does not seem clear, again add 0.2 mL of chloroform per 1 mL of TRIzol and repeat **step 6**.

8. Mix and incubate at room temperature for 10 min.

9. Centrifuge at $17,900 \times g$ for 10 min at 4 °C (*see* **Note 4**).

10. Remove supernatant and add 1 mL of cold 75 % ethanol. Resuspend pellet and mix vigorously.

11. Centrifuge at $17,900 \times g$ for 5 min at 4 °C.

12. Pour off ethanol wash. Spin down to collect last remaining ethanol and pipette off. Briefly air dry sample for 5 min.

13. Dissolve sample in 30–50 μL RNase-free water.

14. Test concentration and quality of RNA using a Nanodrop Spectrophotometer. An A_{260}/A_{280} ratio of 2.0 ± 0.1 indicates a clean preparation of RNA.

15. Visualize total RNA on RNA gel. Prepare 1 μg total RNA per sample using 1:1 volume of glyoxal loading dye, mix and heat samples to 50 °C for 30 min prior to loading (*see* **Note 5**).

16. Store RNA samples at –20 °C for short term or –80 °C for long term.

3.2 Cloning of Constructs into pJawohl8-RNAi

1. Extract total RNA from GPA adults using the total RNA extraction protocol (described above) and make cDNA (as described in Subheading 3.8, **steps 1–4**).

2. Amplify target gene coding sequences from GPA cDNA by PCR with specific primers containing additional attb1 and attb2 linkers for cloning with Gateway® system (*see* **Note 6**).

3. Introduce PCR product into pDONR™207 plasmid to create an entry clone using Gateway® BP reaction according to the manufacturer's instructions.

4. Transform plasmid DNA into electrocompetent DH5α by electroporation and culture on selective media plates overnight at 37 °C; 1 μL of BP reaction is sufficient for electroporation.

5. Sequence subsequent clones to verify correct size and sequence of inserts using overlapping forward and reverse Sanger sequence reactions.

6. Introduce verified inserts into the pJawohl8-RNAi binary destination vector using Gateway® LR reaction according to the manufacturer's instructions (*see* **Note 7**).

7. Introduce constructs into *Agrobacterium tumefaciens* strain GV3101 containing pMP90RK plasmid by electroporation. Grow at 28 °C for 2 days in selective media until colony formation (1 μL of LR reaction is usually sufficient for electroporation).

8. Pick four or five colonies per construct and identify positives by PCR using one gene specific and one vector specific primer.

9. Grow liquid cultures for transformation of *Arabidopsis thaliana*.

3.3 Growing of Plants, Transformation, and Selection of T2 Homozygous Lines

1. Transform pJawohl8:RNAi constructs into *Arabidopsis thaliana* ecotype Col-0 using the floral dip method [14] and harvest seed from the dipped plants.

2. Sow seeds and spray seedlings with BASTA to select for transformants.

3. Germinate T2 seeds on BASTA selective seedling growth media. Plants showing a ratio of 3:1 dead/alive (evidence of single insertion) segregation should be taken forward to the T3.

4. Germinate seed on MS + BASTA and select lines with 100 % survival ratio (homozygous).

5. Confirm the presence of each construct insert by PCR and sequencing. Select three independent lines for each construct for use in insect bioassays.

3.4 Northern Blotting to Detect siRNAs Derived from the Transgene

1. Harvest approximately 2 g of whole, 2 week-old *Arabidopsis thaliana* T3 transgenic seedlings.

2. Extract total RNA from leaves/seedlings using method described in Subheading 3.1.

3. Prepare samples as follows:

   (a) Aliquot 15–30 μg of total RNA per sample; the RNA should be in a volume of approximately 7–10 μL.

   (b) Add the same volume (7–10 μL) of stopmix buffer to each sample.

   (c) Denature the samples 1 min at 90 °C or 5 min at 65 °C and store on ice until the loading of the gel.

4. Run samples on 15 % polyacrylamide gel using 1× MOPS running buffer:

   (a) Pre-run gel at 100 V for 30 min.

   (b) Load RNA samples & MicroRNA marker onto gel (*see* **Note 8**).

(c) Run the gel for approximately 2 h at 80–100 V, until the bromophenol blue reaches the bottom.

(d) Stain the gel for 5 min in 1 μg/mL EtBr to test gel quality.

5. Blot gel to a membrane using semi-dry transfer cell at 130 mA per small gel for 35 min; cut one corner off the membrane for orientation.

6. Cross-link RNA to the membrane by incubating the membrane for 2 h at 60 °C using the cross-linking solution. Briefly wash membrane twice with distilled water for 10 min.

7. Pre-hybridize membrane with hybridization buffer for 2 h at 37 °C.

8. Prepare DNA probe corresponding to insert:

(a) Perform PCR on aphid cDNA using flanking primers for insert region.

(b) Run completed reaction on 1 % agarose gel.

(c) Excise band from gel.

(d) Perform gel purification as per kit instructions.

9. Label DNA probes using Klenow fragment with [α-32P] dCTP according to manufacturer's instructions to generate highly specific probes.

10. Add probe to hybridization solution surrounding the membrane and hybridize for 16–24 h at 37 °C (do not pipette the probe directly onto the membrane).

11. Wash membrane twice for 30 min at 37 °C with washing solution.

12. Wrap samples in clear plastic (i.e., cling-film or Saran wrap) and expose to phosphor storage plate with cassette provided.

13. Detect signals after 3–6 day exposure by scanning using phosphor scanner and analyze using the associated software.

14. Strip blots for re-probing by soaking in stripping buffer at 100 °C for 5 min.

15. To control for equal loading of RNA amounts, hybridize blots with a probe corresponding to U6 (snRNA 59-GCTAATC TTCTCTGTATCGTTCC-39) [15].

3.5 Insect Bioassays

1. Sow T3 seed in compost.

2. After 10–14 days, transfer seedlings to single pots (10 cm diameter) and place in environmental growth room.

3. Confine five GPA adults to each single 4-week-old *Arabidopsis* lines in sealed experimental cages containing the entire plant (Fig. 1b); expose insects to three independent lines per construct and control plants expressing dsRNA derived from GFP (dsGFP).

4. After 2 days, remove all adults, leaving nymphs on plants.

5. Count the number of remaining adults and offspring present on the 10th, 14th and 16th day; remove counted nymphs.

6. Repeat experiment three times to create data from three independent biological replicates with four plants per line per replicate.

7. For assaying gene knock down by qRT-PCR (*see* below): At the 16th day, collect three batches of five adult aphids from each dsRNA-expressing line after exposure to plants for 16 days and flash freeze the aphid samples in liquid nitrogen.

3.6 Statistical Analysis of Fecundity Data

1. Calculate mean number of nymphs produced on each line ± standard error.

2. Import data into statistical software e.g., Genstat.

3. Check data for approximate normal distribution by visualizing residuals.

4. Employ a classical linear regression analysis using a generalized linear model (GLM). A Poisson distribution should be used and select "number of nymphs" as the response variate.

5. Use aphid nymph production on 4 plants per treatment as independent data points in statistical analyses using biological replicate as a variable (*see* **Note 9**).

3.7 Design of qRT-PCR Primers and Reference Gene Set

1. Design primers (*see* **Note 10**) for target gene sequences and for a minimum of eight reference (house-keeping) gene sequences.

2. Test efficiency of each primer set by performing dilution series of GPA cDNA (1:1; 1:10; 1:100; 1:1,000). Represent each cDNA dilution in 2–4 technical replicates per sample for each primer set. Plot Threshold Cycle ($C(t)$) values (y axis) against Log dilution (x axis) and calculate gradient of line (m). Assess primer percentage efficiency in a spreadsheet (e.g., Microsoft Excel) according to the following formula: $=100 \times POWER(10, 1/m)/2$. A primer efficiency of 100 ± 10 % is ideal.

3. Use Genorm software to identify which of the eight or more reference genes are most stable between aphid treatments and select three or more reference genes to include in gRT-PCRs to investigate knock down of aphid target genes.

3.8 qRT-PCR Analyses to Investigate Knock Down of Aphid Target Genes

1. Extract total RNA from GPA exposed to test plants (Subheading 3.5, **step 7**).

2. Use 1–5 μg total RNA for each sample and set up DNase digestion reaction according to the manufacturer's instructions.

3. Purify RNA samples with RNeasy Mini Kit as per the kit instructions.

4. Make first-strand cDNA using M-MLV reverse transcriptase plus Oligo (dT) according to the manufacturer's instructions.

5. Dilute cDNA samples to 100 μL with distilled water and mix well.

6. Lay out 96-well plate for qRT-PCR reaction as follows:

   (a) Each sample should be represented by gene of interest plus 1–4 reference genes depending on results of GENORM analysis; a minimum of two reference genes is recommended.

   (b) Each sample should be represented by two or three technical replicates.

   (c) Each reaction should contain 3 μL of cDNA, 0.5 μL of each specific primers (10 pmol/mL), and 10 μL of 2×SYBR Green (Bio-Rad) in a final volume of 20 μL.

   (d) Seal plate using adhesive film.

   (e) Run plate in qRT-PCR instrument using the following PCR program for all PCR reactions: 90 °C for 3 min, followed by 40 cycles of 95 °C for 30 s, 60 °C for 30 s, 72 °C for 30 s followed by 10 min at 72 °C to end.

3.9 Statistical Analysis of qRT-PCR Data

1. Calculate Threshold Cycle (C(t)) values using qRT-PCR software.

2. Analyze relative gene expression data using $2^{-\Delta\Delta^C T}$ method [16]; this will provide normalized C(t) values for difference in cDNA amount using reference gene C(t) values.

3. Import normalized transcript values into statistics software and analyze results for significant difference with Student's t-test in which fold changes can be calculated by comparing the normalized transcript levels of target genes for GPA fed on transgenic plants expressing dsRNA corresponding to the target gene compared to aphids fed on dsGFP transgenic (control) plants.

4   Notes

1. Carry out all procedures at room temperature unless otherwise stated.

2. This allows the complete dissociation of nucleoprotein complexes.

3. Following centrifugation, the mixture will separate into three phases: a lower red phenol–chloroform phase, an interphase, and colorless upper aqueous phases. RNA remains exclusively in upper phase whereas proteins/DNA will remain in the lower two phases.

4. RNA precipitate will form a gelatinous white pellet on the side of the tube.

5. Discrete, thick 28S, and 18S ribosomal RNA (rRNA) gel bands at an approximate mass ratio of 2:1 are indications of high integrity.

6. Design primers to amplify >70 bp of gene of interest, inclusion of gene UTRs is preferred.

7. This will generate plasmids containing target gene fragments as inverted repeats.

8. The RNA marker consists of three synthetic single-stranded RNA oligonucleotides of 17, 21, and 25 residues on gels and hybridizes on blots with corresponding microRNA probe to determine size of siRNA between 21 and 23 nucleotides.

9. There should be no significant difference between the (three or more) biological replicates.

10. Primer design tips: primer melting temperature should be between 58 and 60 °C, amplicon length between 50 and 200 nucleotides, ensure high transcript specificity.

References

1. Harmel N, Létocart E, Cherqui A, Giordanengo P, Mazzucchelli G, Guillonneau F, De Pauw E, Haubruge E, Francis F (2008) Identification of aphid salivary proteins: a proteomic investigation of *Myzus persicae*. Insect Mol Biol 17:165–174

2. Carolan JC, Fitzroy CIJ, Ashton PD, Douglas AE, Wilkinson TL (2009) The secreted salivary proteome of the pea aphid *Acyrthosiphon pisum* characterised by mass spectrometry. Proteomics 9:2457–2467

3. IAGC (2010) Genome sequence of the pea aphid *Acyrthosiphon pisum*. PLoS Biol 8:e1000313

4. Carolan JC, Caragea D, Reardon KT, Mutti NS, Dittmer N, Pappan K, Cui F, Castaneto M, Poulain J, Dossat C et al (2011) Predicted effector molecules in the salivary secretome of the pea aphid (*Acyrthosiphon pisum*): a dual transcriptomic/proteomic approach. J Proteome Res 10:1505–1518

5. Bos JIB, Prince D, Pitino M, Maffei ME, Win J, Hogenhout SA (2010) A functional genomics approach identifies candidate effectors from the aphid species Myzus persicae (green peach aphid). PLoS Genet 6:e1001216

6. Pitino M, Coleman AD, Maffei ME, Ridout CJ, Hogenhout SA (2011) Silencing of aphid genes by dsRNA feeding from plants. PLoS One 6:e25709

7. Pitino M, Hogenhout SA (2013) Aphid protein effectors promote aphid colonization in a plant species-specific manner. Mol Plant Microbe Interact 26:130–139

8. Mutti NS, Park Y, Reese JC, Reeck GR (2006) RNAi knockdown of a salivary transcript leading to lethality in the pea aphid, *Acyrthosiphon pisum*. J Insect Sci 6:1–7

9. Jaubert-Possamai S, Le Trionnaire G, Bonhomme J, Christophides GK, Rispe C et al (2007) Gene knockdown by RNAi in the pea aphid *Acyrthosiphon pisum*. BMC Biotechnol 7:8

10. Mutti NS, Louis J, Pappan LK, Pappan K, Begum K et al (2008) A protein from the salivary glands of the pea aphid, *Acyrthosiphon pisum*, is essential in feeding on a host plant. Proc Natl Acad Sci USA 105:9965–9969

11. Shakesby AJ, Wallace IS, Isaacs HV, Pritchard J, Roberts DM et al (2009) A water-specific aquaporin involved in aphid osmoregulation. Insect Biochem Mol Biol 39:1–10

12. Whyard S, Singh AD, Wong S (2009) Ingested double-stranded RNAs can act as species-specific insecticides. Insect Biochem Mol Biol 39:824–832

13. Hellens R, Mullineaux P, Klee H (2000) Technical focus: a guide to Agrobacterium binary Ti vectors. Trends Plant Sci 5: 446–451

14. Bechtold N, Ellis J, Pelletier G (1993) In-planta agrobacterium-mediated gene-transfer by infiltration of adult *Arabidopsis-thaliana* plants. Compt Rendus Acad Sci III Sci Vie Life Sci 316:1194–1199

15. Lopez-Gomollon S, Dalmay T (2011) Detecting sRNAs by northern blotting. MicroRNAs in development: methods and protocols. pp 25–38

16. Livak KJ, Schmittgen TD (2001) Analysis of relative gene expression data using real-time quantitative PCR and the 2(T) (-delta delta C) method. Methods 25: 402–408

Chapter 11

Leaf-Disc Assay Based on Transient Over-Expression in *Nicotiana benthamiana* to Allow Functional Screening of Candidate Effectors from Aphids

Patricia A. Rodriguez, Saskia A. Hogenhout, and Jorunn I.B. Bos

Abstract

Aphids, like plant pathogens, are known to form close associations with their host. While probing and feeding, these insects deliver effectors inside the host, which are thought to be involved in suppression of host defenses and/or the release of nutrients. With increasing availability of aphid genome and transcriptome sequencing data, effectors can now be identified using bioinformatics- and proteomics-based approaches. The next step is then to apply functional assays relevant to plant–aphid interactions to identify effector activities. This chapter describes an effective and medium-throughput screen for the identification of effectors that affect aphid fecundity upon in planta over-expression. This assay will allow the identification of aphid effectors with a role in aphid virulence and can be adapted to other plant species amenable to agroinfiltration as well as to other assays based on transient expression, such as RNAi.

Key words Aphid effectors, Leaf-disc assay, Fecundity, Virulence, Plant–aphid interactions

1 Introduction

Aphids are economically important pests that, like other hemipterans, feed from the plant phloem using specialized mouthparts, called stylets. Despite having a mobile lifestyle, aphids need to maintain a close association with their host for extended periods of time to be able to survive and reproduce successfully. Although aphid stylets follow a mainly extracellular pathway en route to the phloem, they also briefly probe most cells along this pathway providing for direct contact with the cytoplasm of the plant cells [1]. While probing and feeding, aphids secrete saliva into the host-stylet interface. Aphid saliva has been implicated in suppression of host defense responses such as the clogging of plant sieve elements [2]. This suggests that the saliva contains molecules that function as effectors to manipulate host defenses and/or promote the release of nutrients. The recent availability of aphid genome and transcriptome

Paul Birch et al. (eds.), *Plant-Pathogen Interactions: Methods and Protocols*, Methods in Molecular Biology, vol. 1127, DOI 10.1007/978-1-62703-986-4_11, © Springer Science+Business Media New York 2014

sequence data has facilitated the identification of predicted secreted salivary proteins using bioinformatics pipelines. These data have also facilitated proteomics approaches for the identification of proteins present in aphid saliva [3–6]. These salivary proteins are hypothesized to play a major role in plant–aphid interactions and are considered candidate aphid effectors. Indeed, we previously performed a functional genomics approach, which identified a number of aphid candidate effectors that shared functional features with plant pathogen effectors in that they suppressed host defenses and/or affected aphid virulence [5]. As part of the functional genomics approach, we developed a novel assay that allowed screening of a large number of candidate aphid effectors for their ability to affect aphid virulence on *Nicotiana benthamiana*. Using this assay, we found that *Myzus persicae* effectors Mp10 and Mp42 reduced aphid fecundity, whereas MpC002 increased aphid fecundity upon over-expression *in planta*. Although we have developed this assay for use in *N. benthamiana*, it can be adapted to other plant species that are amenable to agroinfiltration. An alternative approach to over-express effectors is to generate transgenic lines as described in chapter 10 (Aphid RNAi Chapter), although this will reduce the throughput. The leaf-disc assay here can also be easily adapted to, for example, test the effects of aphid gene silencing on virulence by transiently over-expressing dsRNA [7, 8].

2   Materials

2.1   Over-Expression of Effectors by Agroinfiltration

1. *Agrobacterium tumefaciens* strain GV3101 carrying a plant expression vector encoding the protein of interest.
2. Luria-Bertani (LB) medium and agar plates supplemented with appropriate antibiotics for selection of *Agrobacterium* carrying the plant expression vector.
3. Incubators and shakers set at 28 °C.
4. High-speed tabletop centrifuge.
5. Agroinfiltration buffer: 10 mM $MgCl_2$, supplemented with acetosyringone (125 μM final concentration) directly before use.
6. Spectrophotometer.
7. Needleless 1 mL syringes.
8. 4- to 6-week old *N. benthamiana* plants.
9. Glasshouse chamber set at 20–23 °C and 16 h light.

2.2   Preparation of 24-Well Plates Containing Leaf Discs

1. 24-well plate.
2. Caps of 5 mL Polypropylene round-bottom test tubes (BD Falcon, Cat. No. 352063).
3. Scissors.

4. Fine mesh.

5. Glue gun with glue sticks.

6. Leaf-disc cutter/cork borer, number 7.

7. Forceps.

8. Water agar, 1 %.

9. 25 mL disposable pipettes.

2.3 Aphid Transfer and Counting of Progeny

1. *Myzus persicae* aphid colony maintained on *Nicotiana tabacum* plants.

2. Fine brushes.

3. Scoring sheets.

4. Plastic petri dishes.

5. Dissecting microscope.

6. Growth room set at 18 °C and 16 h light.

2.4 Statistical Analyses

• Computer with statistical data analyses software to perform ANOVA.

3 Methods

We generate expression constructs by using Gateway compatible 35S-based expression vectors such as the pB7WG2 vector. Genes of interest are cloned without the signal peptide encoding sequence to allow expression of effectors inside the host cell cytoplasm. Expression levels of proteins of interest can be examined by western blotting by including epitope tags in the expression construct as long as the addition of tags does not interfere with effector activity. It is recommended that over-expression assays are performed in *N. benthamiana* plants prior to initiating the leaf-disc assays to determine if the proteins of interest induce a visible phenotype that will interfere with the leaf-disc assays (e.g., leaf crinkling or cell death).

3.1 Over-Expression of Effectors by Agroinfiltration

1. Grow *Agrobacterium* strains carrying the vector constructs in 5 mL liquid LB medium overnight in a shaker at 28 °C with appropriate antibiotic selection. To select for *Agrobacterium* strain GV3101 rifampicin 50 μg/mL can be included, plus the appropriate antibiotic to select for the vector construct.

2. Centrifuge the overnight culture for 8 min at $2,656 \times g$ in a tabletop centrifuge to pellet the *Agrobacterium* cells. Resuspend the cells in an appropriate amount of agroinfiltration buffer and dilute to a final OD_{600} of 0.3.

3. Infiltrate *Agrobacterium* solutions into the underside of *N. benthamiana* leaves using a needleless syringe and prevent

causing wound damage. Infiltrate an area that will allow leaf tissue collection with cork borer No. 7 while excluding the mark left by the syringe. Use a permanent marker pen to outline the infiltrated leaf area (*see* **Notes 1** and **2**).

3.2 Preparation of 24-Well Plates Containing Leaf Discs

1. Cut the top off 5 mL polypropylene round-bottom test tube caps (these are caps made from a soft plastic that is flexible and can be cut with scissors), to create a small tube. Use glue to attach a fine mesh over one of the openings. Remove excess glue using scissors. Prepare 24 caps to cover all wells in the plate.

2. Pipette 1 mL of melted 1 % water agar into the individual wells using a 25 mL disposable pipette. Let the agar cool and set. Cover the plate to prevent drying out of the agar (*see* **Note 3**).

3. One day after agroinfiltration of *N. benthamina* plants, cut leaf discs using a No. 7 cork borer. Carefully place the leaf discs upside down onto the agar (with the underside surface upward) using forceps (*see* **Note 4**).

3.3 Aphid Transfer and Counting of Aphid Progeny

1. Transfer four first-instar nymphs from a healthy aphid colony to the individual leaf disc using a fine brush (*see* **Note 5**).

2. Ensure that transferred nymphs are in place and have survived the transfer process.

3. Seal individual wells with the mesh-covered caps to prevent escape of aphids, while maintaining air exchange. This is done by pushing the caps with the mesh-side down into the individual wells (Fig. 1a, b).

4. Place plate in an insect culture room at 18 °C and under 16 h of light (or in a slightly colder environment to reduce the reproduction rate if desirable and depending on the aphid species/genotype) (*see* **Note 6**).

5. Six days after placing the nymphs in the 24-well plate, transfer the (now adult) aphids to a new 24-well plate with fresh leaf discs prepared as described in Subheading 3.2. Record the number of surviving aphids per well on a scoring sheet (*see* **Note 7**).

6. Six days after the previous transfer (on day 12), transfer the aphids for a second time to a new 24-well plate with freshly infiltrated leaf discs prepared as described in Subheading 3.2. Depending on the aphid species and temperature, the aphids may have produced nymphs by this stage. Record the number of adult aphids and nymphs per well on a scoring sheet. Only the adult aphids are transferred to the new plate.

7. Record the number of adult aphids and nymphs on day 14 and day 17 after setting up the initial 24-well plate on a scoring sheet. Remove any newly produced nymphs after each count.

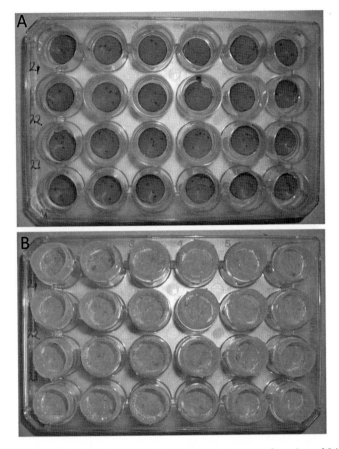

Fig. 1 The 24-well plate setup for aphid virulence assays. (**a**) Overview of 24-well plate setup with *Nicotiana benthamiana* leaf discs and first-instar nymphs. (**b**) Overview of 24-well plate sealed with individual caps to prevent aphid escape and allow air exchange

3.4 Statistical Analyses

1. Calculate the aphid survival rates by adding up the number of surviving aphids (the initial four nymphs placed on each leaf disc) of the six wells per effector/control treatment.

2. Calculate the average production rates per effector/control treatment by dividing the number of nymphs on day 12 by the number of adults on day 6 (calculated per well), dividing the number of nymphs on day 14 by the number of adults on day 12, and dividing the number of nymphs on day 17 by the number of adults on day 14.

   Thus, the average reproduction rate is calculated as follows:

$$\text{Average production rate} = \text{Average}\left(\frac{\text{Number of nymphs on day 12}}{\text{Number of adults on day 6}}\right)$$
$$+ \text{Average}\left(\frac{\text{Number of nymphs on day 14}}{\text{Number of adults on day 12}}\right)$$
$$+ \text{Average}\left(\frac{\text{Number of nymphs on day 17}}{\text{Number of adults on day 14}}\right).$$

3. Calculate the total nymph production per adult aphid per effector/control treatment by combining the average production rates throughout the experiment.

4. Perform one-way ANOVA using statistical analyses software, such as GenStat, to determine whether differences in survival and reproduction rates are significant among treatments. Effector/control construct is selected as treatment and the replicates as block.

4   Notes

1. *N. benthamiana* leaves are easy to infiltrate due to leaf size and the tissue texture. Always ensure that plants are grown under optimal conditions. Choose leaves for infiltration that are young (second to third leaf from the top of the plant) to ensure high levels of protein expression.

2. In order to reduce the effects of leaf-to-leaf variation, it is recommended to express a set of three effectors plus a vector control side-by-side on the same leaf. Infiltrate six individual leaves from different plants with each effector/control set to be tested. Assays are typically performed in at least three biological replicates.

3. Be sure to have plates ready with agar when collecting leaf materials so that leaves/leaf discs do not dry out.

4. Each effector construct will be represented by six individual leaf discs per replicate and will thus fill one row in the plate. This means that one 24-well plate can accommodate three effector constructs plus one control construct.

5. Nymphs need to be selected from a healthy aphid colony and handled with the greatest care to ensure they survive transfer to the 24-well plates. Some practice may be needed in selecting and handling aphids prior to setting up large-scale experiments to ensure good survival rates after transfer.

6. The temperature and humidity may need adjustment depending on the aphid species and/or genotype. Ideally, the temperature for performing reproduction assays should be such that aphid reproduction rates are relatively low, as it is more likely to detect small effects on aphid reproduction when aphids are not performing optimally.

7. *Agrobacterium*-mediated expression is transient and lasts for up to 5–7 days. Levels of protein expression can be assessed by western blotting, if desirable, using epitope tagged proteins.

References

1. Tjallingii WF (2006) Salivary secretions by aphids interacting with proteins of phloem wound responses. J Exp Bot 57:739–745

2. Will T, Tjallingii WF, Thönnessen A, Van Bel AJE (2007) Molecular sabotage of plant defence by aphid saliva. Proc Natl Acad Sci U S A 104:10536–10541

3. Harmel N, Létocart E, Cherqui A, Giordanengo P, Mazzucchelli G, Guillonneau F, De Pauw E, Haubruge E, Francis F (2008) Identification of aphid salivary proteins: a proteomic investigation of Myzus persicae. Insect Mol Biol 17:165–174

4. Carolan JC, Fitzroy CIJ, Ashton PD, Douglas AE, Wilkinson TL (2009) The secreted salivary proteome of the pea aphid Acyrthosiphon pisum characterised by mass spectrometry. Proteomics 9:2457–2467

5. Bos JIB, Prince D, Pitino M, Maffei ME, Win J, Hogenhout SA (2010) A functional genomics approach identifies candidate effectors from the aphid species Myzus persicae (green peach aphid). PLoS Genet 6: e1001216

6. Carolan JC, Caragea D, Reardon KT, Mutti NS, Dittmer N, Pappan K, Cui F, Castaneto M, Poulain J, Dossat C et al (2011) Predicted effector molecules in the salivary secretome of the pea aphid (Acyrthosiphon pisum): a dual transcriptomic/proteomic approach. J Proteome Res 10:1505–1518

7. Pitino M, Coleman AD, Maffei ME, Ridout CJ, Hogenhout SA (2011) Silencing of aphid genes by dsRNA feeding from plants. PLoS One 6:e25709

8. Pitino M, Hogenhout SA (2013) Aphid protein effectors promote aphid colonization in a plant species-specific manner. Mol Plant Microbe Interact 26:130–139

Chapter 12

A Growth Quantification Assay for *Hyaloperonospora arabidopsidis* Isolates in *Arabidopsis thaliana*

Daniel F.A. Tomé, Jens Steinbrenner, and Jim L. Beynon

Abstract

There is a considerable interest in determining the role of individual oomycete effectors in promoting disease. Widely used strategies are based on manipulating effector-expression levels in the pathogen and by over-expressing particular effectors in the host by genetic transformation. In the case of the oomycete, *Hyaloperonospora arabidopsidis* (*Hpa*) genetic manipulation is not yet possible, so over-expression of predicted effectors in stably transformed Arabidopsis lines is used to investigate their capability for promoting virulence. Here, we describe a technique for quantifying pathogen growth based on the counting of asexual reproductive structures called sporangiophores in the compatible interaction between the *Hpa* isolate Noks1 and the Col-0 Arabidopsis accession.

Key words Oomycete, *Hyaloperonospora arabidopsidis*, Effector, Susceptibility, Sporangiophores, Pathogen growth

1 Introduction

Plants have evolved a very efficient defense system to protect them from plant pathogens that inhabit every ecosystem. This surveillance system detects slowly evolving pathogen-associated molecular patterns (PAMPs), pathogen produced molecules conserved among many organisms, and activates defense responses (PAMP-triggered immunity, PTI) in a timely manner [1]. Successful pathogens overcome PTI by deploying secreted proteins, so-called effectors, which target the different cellular components involved in mediating PTI. This is usually sufficient to suppress the PTI response to a level that enables colonization of the host and to allow the establishment of a compatible interaction. In turn, plants have evolved a second layer of defense termed effector-triggered immunity (ETI), which is mediated by resistance (R) proteins [2] that recognize the presence of effectors resulting in a localized cell death hypersensitive response (HR). This can happen upon direct interaction between the R-protein and the effector [3, 4] or by the

Paul Birch et al. (eds.), *Plant-Pathogen Interactions: Methods and Protocols*, Methods in Molecular Biology, vol. 1127, DOI 10.1007/978-1-62703-986-4_12, © Springer Science+Business Media New York 2014

detection of changes to a virulence target caused by the effector [5]. An interaction is termed incompatible when detection of the pathogen by the host via PTI or ETI results in suppression of pathogen growth and reproduction. In addition to effector proteins, pathogens can also secrete non-proteinaceous toxins/chemicals to manipulate host hormonal signaling pathways and further suppress ETI [6].

Oomycetes are fungal-like organisms that are actually more closely related to brown algae and diatoms than to fungi. This group comprises a number of aggressive plant pathogens that successfully suppress plant defenses [7, 8] and can cause severe crop losses [9]. Unlike bacteria, oomycetes lack a type-three secretion system (TTSS) to directly deliver effectors inside the host cell to suppress PTI and ETI. Instead, oomycete effectors are secreted in specialized feeding structures, called haustoria, and translocated into the host cell by mechanisms which are still under debate [10, 11]. It is believed that oomycetes suppress plant defense by employing effectors in this way.

Hpa (formerly *Peronospora parasitica*) is an obligate biotrophic oomycete of the model plant Arabidopsis that causes what is commonly referred to as downy mildew [12]. The genome sequence of *Hpa* was published in 2010 revealing 134 predicted RxLR effectors [13], a relatively low number compared to other oomycete species [14, 15]. The high genetic diversity among *Hpa* isolates has enabled the discovery of many "<u>r</u>esistance to <u>P</u>eronospora <u>p</u>arasitica" (RPP) loci that encode specific R-proteins [16]. In contrast, only four *Hpa* avirulence effector genes have been identified and cloned [17–20], showing that we are still far from understanding the relative contributions of oomycete effectors toward suppressing host defenses or triggering ETI responses.

A recent study employed an effector delivery system based on the TTSS of *Pseudomonas syringae* to investigate the role of individual effectors in suppressing plant immunity [21]. The authors show that some *Hpa* effectors suppressed classic PTI responses, such as the reactive oxygen burst and callose deposition, and promoted bacterial growth. Although a useful analytic technique this does not assess the consequence of effector production *in planta* on the growth and development of the pathogen from which the effectors originate. Pathogen growth can be measured directly or indirectly depending on the pathogen. For bacteria, it is common to measure bacterial growth by counting the numbers of colony forming units in leaf material upon serial dilution plating. For necrotrophic pathogens, measuring the lesion size on a host leaf is widely used because it correlates with pathogen growth. The biotrophic nature of *Hpa* makes these approaches impractical. Therefore, *Hpa* growth is sometimes measured indirectly by counting the number of trypan blue stained oospores inside infected leaves or measuring the total amount of asexual spores on the surface of infected leaves [22]. However, these assays can sometimes give inaccurate results, such as the underestimation of asexual spore numbers due to spore loss during plant manipulation. In this chapter, we describe a method for measuring *Hpa* growth based on sporangiophore counting of the virulent *Hpa* isolate Noks1 on Col-0. It covers the revival and maintenance of the pathogen, steps that can be slightly complicated due to its obligate lifestyle and inability to grow on synthetic media similarly to other laboratory model organisms. This method can be used to accurately quantify small changes in *Hpa* growth, making it ideal for investigating putative roles in defense suppression of individual *Hpa* RxLR effectors constitutively expressed following stable integration in the *Arabidopsis* genome.

2   Materials

1. *Hpa* asexual spores from Noks1 isolate. Heavily sporulating seedlings from a compatible interaction are stored in small plastic tubes and kept at –80 °C.

2. Plant materials:

   (a) Col-0.

   (b) *Ws-eds1* [23].

   (c) Stable transgenic lines in Col-0 background constitutively expressing *Hpa* RxLR effectors generated as described previously [21].

3. Ice box.

4. Sterile MilliQ water.

5. 70 % (v/v) ethanol.

6. Miracloth mesh (EMD Millipore, Billerica, Massachusetts, USA).

7. Microscope and hemocytometer.

8. Dissecting microscope with adequate light source.

9. Fine forceps and scissors.

10. Gilson pipettes (P10 and P1000) and tips.

11. Consumables:

    (a) BD Falcon 50 ml conical tubes.

    (b) 40 cell tray insert (P40).

    (c) Seed trays with holes.

    - Standard size: 35 × 20cm (William Sinclair Horticulture Ltd., Lincoln, UK).

    - Intermediate size: 21 × 17cm (William Sinclair Horticulture Ltd., Lincoln, UK).

Fig. 1 Different plant materials and propagators used when working with *Hpa*. (**a**) Small propagator used for reviving isolates from frozen stocks on a single pot of a P40 tray containing *Wsneds1* seedlings. This propagator does not require sealing with adhesive tape for saturated humidity conditions. (**b**) Intermediate size propagator used for regular maintenance of *Hpa* on six pots of a compatible accession. (**c**) Regular size propagator used on experimental P40 trays

    (d) Colored plastic labels.

    (e) Electrical adhesive tape.

    (f) Disposable gloves.

    (g) Aluminum foil.

12. Clear small propagators (*see* Fig. 1a, Bell Boy Mini Propagators, Crowders, UK).

13. Clear propagator lids (iGrowphonics, Sandy, Bedfordshire, UK, but other transparent lids will do) for intermediate and standard seed trays (*see* Fig. 1b,c).

14. Equipment:

    (a) Class 2 (CL2) cabinet or flow-hood.

    (b) Two plant growth cabinets set for 10 h photoperiod, 100 μE/cm, 60 % humidity. Have one cabinet set at constant 18 °C and the other at constant 20 °C.

    (c) Spraying equipment:

       • Sparmax Professional Mini Piston Airbrush Compressor.

       • Badger 350-4 Airbrush Kit (Everything Airbrush, Poole, Dorset, UK). Familiarize yourself with all the components of the kit and how to use it.

3   Methods

3.1   Reviving Hpa Isolates from Frozen Stocks and Regular Subculturing

1. Fill a P40 tray with soil and let it gradually absorb water from below, via holes in the bottom of the P40 tray insert.

2. Sow *Ws-eds1* seed in these P40 trays at high density (>50 seeds per P40 tray insert), wrap the trays in aluminum foil and vernalize for 48 h at 4 °C. Cover tray with a propagator lid (Fig. 1c) and grow seeds at 20 °C and 60 % humidity in a 10 h photoperiod cabinet for 7 days. This is the material used for reviving all *Hpa* isolates (*see* **Note 1**).

3. Take a frozen *Hpa* stock, add 2 ml of ice-chilled sterile MilliQ water and resuspend the asexual spores by vigorous shaking or using a vortexer (*see* **Note 1**).

4. In the CL2 cabinet, carefully pipette 5 μl droplets of the spore suspension on the cotyledons of 7-day-old *Ws-eds1* seedlings. Make sure every seedling has at least one drop on it (*see* **Notes 2** and **3**).

5. Place the pot inside a small transparent propagator (Fig. 1a), previously sterilized with 70 % ethanol and rinsed with sterile MilliQ water, and place in a separate growth chamber at 10 h photoperiod, 60 % humidity, constant 18 °C for 7 days.

6. After 7 days, harvest infected seedlings. Using scissors and forceps sterilized with 70 % ethanol, cut the seedlings bellow the cotyledons and transfer into a BD Falcon 50 ml tube, taking care not to carry over any soil.

7. Add 5 ml of ice-chilled sterile MilliQ water and resuspend the asexual spores by vigorous shaking or using a vortexer.

8. Filter the suspension by gravity through a Miracloth mesh into a fresh BD Falcon 50 ml tube.

9. Adjust compressor pressure to 1 bar. Connect compressor to airbrush gun attached to the bottle using the hose and wash the bottle from the inside, by passing through 70 % ethanol followed by sterile MilliQ water.

10. Load the filtered spore suspension into the spray bottle and evenly spray six new P40 modules of 7-day-old *Ws-eds1* seedlings.

11. Place these in an intermediate size propagator (21 × 17 cm, Fig. 1b), previously sterilized with 70 % ethanol and rinsed with sterile MilliQ water, and seal the edges with electrical adhesive tape. Seal vents with tape if the propagator has any.

12. Place the propagator in growth cabinet at 10 h photoperiod, 60 % humidity and constant 18 °C for 7 days, after which there will be enough spores to spray several experimental trays.

13. Repeat **steps 6–12** every 7 days to maintain the pathogen isolate(s) available for experiments (*see* **Note 4**).

3.2 Experimental Trays Layout and Spraying

1. Fill P40 trays with soil and let them gradually absorb water from below, via holes in the bottom of the P40 tray insert. Do not leave trays for longer than 15 min, as this leads to excessive water uptake making it ideal for contaminating pathogens to grow. These trays will become the experimental trays containing controls and stable expressing transgenic lines (*see* **Note 5**).

2. On a P40 tray, carefully sow 15–20 seeds per pot. Do not sow more as this could result in bigger plants shadowing smaller ones resulting in more variable results.

3. Sow *Ws-eds1* seed on the outside of each tray (Fig. 2a). These are used as controls for even spraying and should always show high sporulation. Condensation inside the propagator lids tends to form droplets that drop on the outside modules, so these should not be used for sowing plants to be scored.

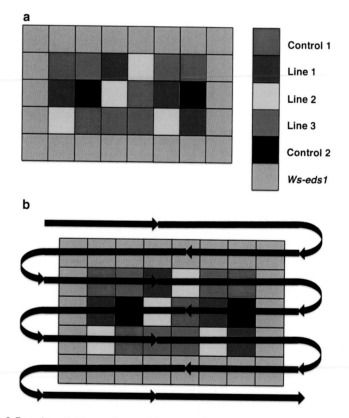

Fig. 2 Experimental tray setup and tray spraying. (**a**) A P40 tray is used to test three lines and two controls. Control 1 (Col-0 GUS, baseline control) and the three lines are randomly sown and color-coded. The remaining pots (Control 2) can be used for a highly sporulation control (HaRxL14 O.E. lines). (**b**) Spraying of an experimental tray. Each row should be sprayed once, switching to the next row when spraying outside the tray as indicated by the *black arrows*. Repeat three more times after rotating the tray 90° before each spraying

4. Of the 18 pots available in the center of a P40 tray, randomly distribute equal number of pots of controls and lines being tested (*see* **Note 6**, Fig. 2a). Color-code the pots to eliminate counting bias and inter-pot variability that can arise from their position in the tray. These include indiscernible differences in light, humidity, soil moisture, seed density, and water droplets on the surface of the leaves.

5. Wrap the trays in aluminum foil and vernalize for at least 48 h at 4 °C in the dark.

6. Grow plants for 14 days in a 10 h photoperiod cabinet, 100 μE/cm, at 20 °C and 60 % humidity covering the tray with a clear propagator lid and sealing the edges with adhesive tape (Fig. 1).

7. Harvest Noks1 spores from the 6-pot seedlings 7 days after spraying as described in Subheading 3.1. After filtering the spore suspension through Miracloth, keep the spores on ice. The time between filtering the spores and spraying should be kept to a minimum because spore viability decreases overtime.

8. Quantify the spores using a hemocytometer. Adjust spore concentration to 30,000/ml. We found this concentration to produce reproducible results.

9. Place experimental trays in the CL2 cabinet and spray the whole tray with 12 ml in one continuous movement following a predefined pattern at a distance of 10 cm (Fig. 2b). Rotate tray 90° and spray again. Repeat until tray has been sprayed from all four sides (*see* **Note 7**).

10. Seal propagator lid to tray with adhesive tape as well as the vents on top of the propagator (*see* **Note 8**).

11. Place sealed trays in the dedicated growth cabinet (10 h photoperiod, 100 μE/cm, 60 % humidity and constant 18 °C). Do not water the tray after spraying.

12. After 3 days open propagator vents by removing the adhesive tape (*see* **Note 8**), but do not remove the tape sealing the tray to the propagator lid.

13. Wait another day and count sporangiophores on 15 individual seedlings in each pot using tweezers and a dissecting microscope at 200–300× magnification (*see* **Notes 9–12**). Count a total of 60 seedlings from each line (*see* **Notes 13–14**). High sporulation on the *Ws-eds1* seedlings sown on the outside pots of a P40 tray shows that the whole tray was evenly sprayed (Fig. 2a).

14. If using an incompatible *Hpa* isolate, wait at least 2 days after opening the vents before counting (*see* **Note 15**).

15. When finished, dispose the tray and count seedlings in an autoclave bag.

16. Compare the average sporangiophore count of each O.E. line to the control of the same tray using the Student's *t*-test. The actual averages will vary from experiment to experiment but the HaRxLR14 O.E. control should always show a higher average than Col-0 GUS O.E. (baseline control). *See* Fig. 3a for some representative examples with HaRxLR14 O.E. T$_4$ lines (*see* **Notes 16–17**).

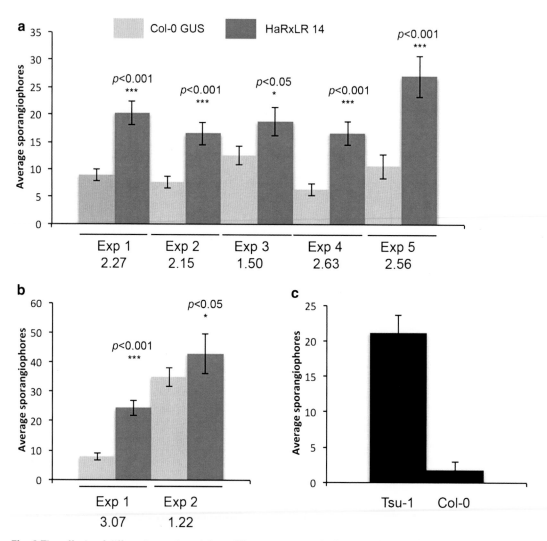

Fig. 3 The effects of different experimental conditions on sporangiophore numbers in Col-0 GUS and HaRxL14 O.E. lines. (**a**) Average Noks1 sporangiophore numbers for Col-0-GUS and HaRxL14 O.E. lines sprayed with 30,000 spores/ml of Noks1 from five independent experiments. (**b**) Average Noks1 sporangiophore numbers after 4 and 5 days, respectively. Both trays were sprayed with the same inoculum and incubated next to each other. (**c**) Emoy2 sprayed at 50,000 spores/ml on compatible Tsu-1 and incompatible Col-0 accessions. Counting was performed 5 and 6 days after spraying on Tsu-1 and Col-0, respectively. Error bars show mean values +/− STE. The *p*-value was determined using Student's *t*-test. The mean Noks1 sporangiophore fold increase in the HaRxL14 O.E. is shown below each experiment number

4 Notes

1. When reviving a frozen stock, it is necessary to use *Ws-eds1* seedlings because these plants are hyper-susceptible to *Hpa* hence greatly increasing the chances of successfully reviving frozen stocks. Seven days after reviving the frozen stock, a few seedlings will show heavy sporulation. However, there are not enough spores for experimental use and a further round of subculturing is necessary to bulk up the spore numbers. The maximum viability of frozen stocks is not known, but we have successfully revived 20-year-old stocks.

2. *Hpa* is a naturally occurring pathogen of a non-crop plant and hence does not require a special license to work with in the UK. CL2 cabinets should be used as they minimize contaminations of seedling pots with other plant pathogens, in particular when working with the hyper-susceptible *Ws-eds1* seedlings and of other *Arabidopsis* plants.

3. If a CL2 cabinet is not available a regular flow-hood will suffice, taking great care to sterilize all surfaces before use and between different isolates of *Hpa*. All infected material (gloves, soil, seedlings, and P40 tray insert) should be disposed of in autoclave bags and used propagators and trays should be thoroughly washed with 70 % ethanol and water.

4. For continuous maintenance of the pathogen, spray six P40 tray inserts (Fig. 1b) containing 7-day-old seedlings grown in high density of a compatible *Arabidopsis* accession with fresh spores every 7 days. It is recommended that at least two separate propagator trays be used in parallel to reduce the risk of a contamination eliminating an ongoing *Hpa* culture. If working with more than one *Hpa* isolate at a time, revive each isolate from frozen stocks every 3 months to minimize the risk of cross contamination between isolates. This can result in a mixed population with more than one isolate. For confirmation of the isolate identity, sequencing of the highly polymorphic effector *ATR13* (primer pair ATR13-2F 5′-CAA TGC GCC TTG TTC ACG CGG TAC and ATR13-2R 5′-CAA CTA CTG TCT GTC AAG GGC AGCC) from infected seedlings used to generate the frozen stocks is recommended [17].

5. Each experimental tray must always have its own control lines included to eliminate tray-to-tray variability. For the Noks1 isolate, we use a Col-0 transgenic line over-expressing (O.E.) the GUS protein, which was generated using the same plasmid as the Col-0 effector O.E. lines, as the baseline control. Sporangiophore numbers on these plants will be treated as the baseline for *Hpa* infection on each particular tray, as GUS does not enhance or decrease *Hpa* sporulation. Some Col-0 effector O.E. lines have previously been shown to allow higher

sporulation levels than the Col-0 line over-expressing GUS [21], and so are used as a positive control for an increased sporulation levels. We regularly use the O.E. line for HaRxL14 due to its reproducible phenotype [21], but other lines can also be used. If a positive control for reduced sporulation levels is desired, the HaRxLL60 O.E. line can be used [21]. In addition, it is best to randomize the locations of the controls and lines being tested in each tray (Fig. 2a).

6. Color-code the pots to eliminate counting bias and inter-pot variability that can arise from their position in the tray. These include indiscernible differences in light, humidity, soil moisture, seed density, and water droplets on the surface of the leaves. It is best if the user doing the potting is not the same one doing the sporangiophore counting and so does not know the color-code.

7. For increased infection rates it is best to spray experimental trays in the final 2 h of the cabinet's photoperiod, as the degree of pathogen growth is dependent on the circadian rhythm of the host plant [24].

8. It is important to properly seal the trays after spraying. This helps maintain saturated humidity conditions inside the tray promoting spore germination and infection of the plant. It also prevents airborne spores of other pathogens from contaminating the tray, as well as *Hpa* spores escaping into the growth cabinet. Opening the propagator vents causes a gradual decrease in the tray humidity and droplets on the leaf surface promoting sporangiophore growth.

9. Sporangiophore counting can be a tedious process, especially if numbers are very high per seedling. An adequate light source and a dark background will make them stand out and easier to count. Try not to be disturbed while counting as not to lose count or use a hand held button counter. Avoid counting where an air current is present, as this will cause the seedlings to move, making counting difficult. One tray takes 2–4 h to count, depending on sporangiophore numbers and experience of the user.

10. Count sporangiophores on the whole seedling. This includes cotyledons, true leaves, petioles, and hypocotyl. Sporangiophores are mostly concentrated on the abaxial side of leaves and cotyledons due to the higher stomata density. They are also predominant in the cotyledons and the first two true leaves, although the subsequent true leaves might have a few due to their tiny size at the time of spraying.

11. When scoring sporangiophores, care is required when scoring the adaxial side of true leaves as their trichomes can sometimes be confused for sporangiophores, especially when spores have been shed. Do not score seedlings where leaves were

accidentally removed/damages when being removed from the pot using the fine forceps.

12. Sporangiophores can have different shapes and sizes. Bushy structures can sometimes appear on the abaxial side. These are sporangiophores with short stalks. Thin stalks can be immature sporangiophores or sporangiophores that have shed their spores, hence should also be counted. Sporangiophore morphology can also vary slightly between *Hpa* isolates. We noticed that Emoy2 has bigger and better defined structures than Noks1.

13. Sporangiophore numbers can vary greatly between seedlings, even in the same pot. Therefore, it is necessary to count sporulation on 15 seedlings per pot to enable statistical analysis. If following the example of Fig. 2a, a total of 60 seedlings will be counted per line, which usually produces good statistics. Abnormally, small seedlings should be discounted as they are at a different growth stage resulting in lower sporulation counts.

14. Unless using an incompatible *Hpa* isolate, it is not recommended to count ≥5 days after spraying due to the high numbers of sporangiophores per seedling that can exceed 100. The accuracy of the counting decreases when counting such numbers due to the high density of individual sporangiophores. Also, it may be difficult to assess differences in sporulation levels between a more susceptible line and the baseline control at later time points, when this control line reaches the maximum number of sporangiophores (Fig. 3b).

15. Some accession-isolate combinations are deemed incompatible. This happens when there is recognition of a specific effector by a cognate *R*-gene, triggering HR, and restricting the pathogen's growth and sporulation [25]. There are different degrees of recognition phenotypes, depending on the robustness of the response [26]. Most show full resistance (no sporulation), but reduced sporulation is possible when the resistance response is slow. When this is the case it is possible to perform the same pathogenicity test as the one described in the paper using an incompatible isolate-accession combination, such as Emoy2 on Col-0. For such a test a few changes to the protocol are called for to maximize the number of sporangiophores to be counted. First, seedlings should be 10 days old at the time of spraying. Secondly, 50,000 spores/ml concentration should be used. Thirdly, sprayed seedlings should be incubated for 5–6 days before being counted to give enough time for all sporangiophores to mature. We found that with these changes, average Emoy2 sporangiophores numbers in Col-0 were less than 2 per seedling, significantly less than when using the compatible accession Tsu-1 (Fig. 3c). This approach might be useful for determining if an effector has a specific role in suppressing the *RPP* gene-dependent HR. Finally, the protocol described in this chapter can also be used to measure growth differences

in an *Hpa* isolate on an incompatible isolate-accession combination, providing that the adequate *rpp* knockout (KO) background for that accession was used to generate the effector O.E. lines as described previously [21].

16. Generation of O.E. stable transgenic lines requires at least third-generation transformant (T₃) seed to ensure stable expression of the transgene [21]. We tested if it was possible to speed up the screening of *Hpa* effector O.E. lines by screening second-generation transformants (T₂) seed instead of waiting for T₃. Experimental trays with T₂ seed were soaked in BASTA (glufosinate ammonium) 50 μg/ml instead of water. This allowed germination of all seeds but only seedlings expressing the *bar* gene, together with the *Hpa* effector gene, would survive. When plants were treated this way, we repeatedly noticed that the susceptible phenotype of the positive controls HaRxL14 O.E. was no longer evident. Our observations suggest that BASTA interferes with *Hpa* growth and so should not be used for screening T₂ seed using this assay. Finally, other herbicides have been found to have negative effects on the growth of several plant pathogens both in vivo and in vitro [27] and likewise should also be avoided.

17. The protocol described in this chapter can also be used with any other Arabidopsis lines as long as the appropriate controls are used. This is especially true for SALK KO lines [28], as there is a large body of evidence showing specific interactions between *Hpa* effectors and *Arabidopsis* proteins [29]. By using SALK KO lines for effector interacting proteins with this method one can systematically search for new genes that play a critical role in immune responses, even if they have been initially described as having unrelated functions, such as well characterized primary metabolism proteins [30, 31].

Acknowledgments

D.F.A.T. and J.S. are funded by the UK Biotechnology and Biological Sciences Research Council grant BB/G015066/1 to J.L.B.

References

1. Jones JDG, Dangl JL (2006) The plant immune system. Nature 444:323–329

2. Dodds PN, Rathjen JP (2010) Plant immunity: towards an integrated view of plant–pathogen interactions. Nat Rev Genet 11: 539–548

3. Frederick RD, Thilmony RL, Sessa G, Martin GB (1998) Recognition specificity for the bacterial avirulence protein AvrPto is determined by Thr-204 in the activation loop of the tomato Pto kinase. Mol Cell 2:241–245

4. Jia Y, McAdams SA, Bryan GT, Hershey HP, Valent B (2000) Direct interaction of resistance gene and avirulence gene products confers rice blast resistance. EMBO J 19: 4004–4014

5. Mackey D, Holt BF III, Wiig A, Dangl JL (2002) RIN4 interacts with *Pseudomonas syringae* type III effector molecules and is required for RPM1-mediated resistance in *Arabidopsis*. Cell 108:743–754

6. Geng X, Cheng J, Gangadharan A, Mackey D (2012) The coronatine toxin of pseudomonas syringae is a multifunctional suppressor of Arabidopsis defense. Plant Cell Online 24: 4763–4774

7. Birch PRJ, Boevink PC, Gilroy EM, Hein I, Pritchard L, Whisson SC (2008) Oomycete RXLR effectors: delivery, functional redundancy and durable disease resistance. Curr Opin Plant Biol 11:373–379

8. Hein I, Gilroy EM, Armstrong MR, Birch PRJ (2009) The zig-zag-zig in oomycete-plant interactions. Mol Plant Pathol 10:547–562

9. Haverkort AJ, Boonekamp PM, Hutten R, Jacobsen E, Lotz LAP, Kessel GJT, Visser RGF, Vossen EAG (2008) Societal costs of late blight in potato and prospects of durable resistance through cisgenic modification. Potato Res 51:47–57

10. Kale SD (2011) Oomycete and fungal effector entry, a microbial Trojan horse. New Phytol 193:874–881

11. Wawra S, Belmonte R, Lo bach L, Saraiva M, Willems A, van West P (2012) Secretion, delivery and function of oomycete effector proteins. Curr Opin Microbiol 15:633–636

12. Coates ME, Beynon JL (2010) Hyaloperonospora arabidopsidis as a pathogen model. Annu Rev Phytopathol 48:329–345

13. Baxter L, Tripathy S, Ishaque N, Boot N, Cabral A, Kemen E, Thines M, Ah-Fong A, Anderson R, Badejoko W, Bittner-Eddy P, Boore JL, Chibucos MC, Coates M, Dehal P, Delehaunty K, Dong S, Downton P, Dumas B, Fabro G, Fronick C, Fuerstenberg SI, Fulton L, Gaulin E, Govers F, Hughes L, Humphray S, Jiang RHY, Judelson H, Kamoun S, Kyung K, Meijer H, Minx P, Morris P, Nelson J, Phuntumart V, Qutob D, Rehmany A, Rougon-Cardoso A, Ryden P, Torto-Alalibo T, Studholme D, Wang Y, Win J, Wood J, Clifton SW, Rogers J, Van Den Ackerveken G, Jones JDG, Mcdowell JM, Beynon J, Tyler BM (2010) Signatures of adaptation to obligate biotrophy in the Hyaloperonospora arabidopsidis genome. Science 330:1549–1551

14. Haas BJ, Kamoun S, Zody MC, Jiang RHY, Handsaker RE, Cano LM, Grabherr M, Kodira CD, Raffaele S, Torto-Alalibo T, Bozkurt TO, Ah-Fong AMV, Alvarado L, Anderson VL, Armstrong MR, Avrova A, Baxter L, Beynon J, Boevink PC, Bollmann SR, Bos JIB, Bulone V, Cai G, Cakir C, Carrington JC, Chawner M, Conti L, Costanzo S, Ewan R, Fahlgren N, Fischbach MA, Fugelstad J, Gilroy EM, Gnerre S, Green PJ, Grenville-Briggs LJ, Griffith J, Grünwald NJ, Horn K, Horner NR, Hu C-H, Huitema E, Jeong D-H, Jones AME, Jones JDG, Jones RW, Karlsson EK, Kunjeti SG, Lamour K, Liu Z, Ma L, MacLean D, Chibucos MC, McDonald H, McWalters J, Meijer HJG, Morgan W, Morris PF, Munro CA, O'Neill K, Ospina-Giraldo M, Pinzón A, Pritchard L, Ramsahoye B, Ren Q, Restrepo S, Roy S, Sadanandom A, Savidor A, Schornack S, Schwartz DC, Schumann UD, Schwessinger B, Seyer L, Sharpe T, Silvar C, Song J, Studholme DJ, Sykes S, Thines M, van de Vondervoort PJI, Phuntumart V, Wawra S, Weide R, Win J, Young C, Zhou S, Fry W, Meyers BC, van West P, Ristaino J, Govers F, Birch PRJ, Whisson SC, Judelson HS, Nusbaum C (2009) Genome sequence and analysis of the Irish potato famine pathogen Phytophthora infestans. Nature 461: 393–398

15. Jiang RHY, Tripathy S, Govers F, Tyler BM (2008) RXLR effector reservoir in two Phytophthora species is dominated by a single rapidly evolving superfamily with more than 700 members. Proc Natl Acad Sci U S A 105:4874–4879

16. Slusarenko AJ, Schlaich NL (2003) Downy mildew of Arabidopsis thaliana caused by Hyaloperonospora parasitica (formerly Peronospora parasitica). Mol Plant Pathol 4:159–170

17. Allen RL, Bittner-Eddy PD, Grenville-Briggs LJ, Meitz JC, Rehmany AP, Rose LE, Beynon JL (2004) Host-parasite coevolutionary conflict between Arabidopsis and downy mildew. Science 306:1957–1960

18. Rehmany AP, Gordon A, Rose LE, Allen RL, Armstrong MR, Whisson SC, Kamoun S, Tyler BM, Birch PRJ, Beynon JL (2005) Differential recognition of highly divergent downy mildew avirulence gene alleles by RPP1 resistance genes from two Arabidopsis lines. Plant Cell Online 17:1839–1850

19. Bailey K, Cevik V, Holton N, Byrne-Richardson J, Sohn KH, Coates M, Woods-Tor A, Aksoy HM, Hughes L, Baxter L, Jones JDG, Beynon J, Holub EB, Tor M (2011) Molecular cloning of ATR5Emoy2 from *Hyaloperonospora arabidopsidis*, an avirulence determinant that triggers RPP5-mediated defense in *Arabidopsis*. MPMI 24:827–838

20. Goritschnig S, Krasileva KV, Dahlbeck D, Staskawicz BJ (2012) Computational prediction and molecular characterization of an oomycete effector and the cognate arabidopsis resistance gene. PLoS Genet 8:1–12

21. Fabro G, Steinbrenner J, Coates M, Ishaque N, Baxter L, Studholme DJ, Körner E, Allen RL, Piquerez SJM, Rougon-Cardoso A, Greenshields D, Lei R, Badel JL, Caillaud M-C, Sohn K-H, Van Den Ackerveken G, Parker JE, Beynon J, Jones JDG (2011) Multiple candidate effectors from the oomycete pathogen Hyaloperonospora arabidopsidis suppress host plant immunity. PLoS Pathog 7:1–18

22. Caillaud M-C, Piquerez SJM, Fabro G, Steinbrenner J, Ishaque N, Beynon J, Jones JDG (2012) Subcellular localization of the Hpa RxLR effector repertoire identifies a tonoplast-associated protein HaRxL17 that confers enhanced plant susceptibility. Plant J 69:252–265

23. Parker JE, Holub EB, Frost LN, Falk A, Gunn ND, Daniels MJ (1996) Characterization of *eds1*, a mutation in arabidopsis suppressing resistance to *Peronospora parasitica* specified by several different *RPP* genes. Plant Cell Online 8:2033–2046

24. Wang W, Barnaby JY, Tada Y, Li H, Tör M, Caldelari D, Lee D-U, Fu X-D, Dong X (2011) Timing of plant immune responses by a central circadian regulator. Nature 470:110–114

25. van der Biezen EA, Freddie CT, Kahn K, Parker JE, Jones JDG (2002) Arabidopsis RPP4 is a member of the RPP5 multigene family of TIR-NB-LRR genes and confers downy mildew resistance through multiple signaling components. Plant J 29:1–13

26. Holub EB (2008) Natural history of Arabidopsis thaliana and oomycete symbioses. Eur J Plant Pathol 122:91–109

27. Kortekamp A (2011) Unexpected Side Effects of Herbicides: Modulation of Plant-Pathogen Interactions, Herbicides and Environment, Dr Andreas Kortekamp (Ed.), InTech, Available from: http://www.intechopen.com/books/herbicides-and-environment/unexpected-side-effects-of-herbicides-modulation-of-plant-pathogen-interactions

28. Alonso JM, Alonso JEM, Stepanova AN, Leisse TJ, Kim CJ, Chen H, Shinn P, Stevenson DK, Zimmerman J, Barajas P, Cheuk R, Gadrinab C, Heller C, Jeske A, Koesema E, Meyers CC, Parker H, Prednis L, Ansari Y, Choy N, Deen H, Geralt M, Hazari N, Hom E, Karnes M, Mulholland C, Ndubaku R, Schmidt I, Guzman P, Aguilar-Henonin L, Schmid M, Weigel D, Carter DE, Marchand T, Risseeuw E, Brogden D, Zeko A, Crosby WL, Berry CC, Ecker JR (2003) Genome-wide insertional mutagenesis of Arabidopsis thaliana. Science 301:653–657

29. Mukhtar MS, Carvunis A-R, Dreze M, Epple P, Steinbrenner J, Moore J, Tasan M, Galli M, Hao T, Nishimura MT, Pevzner SJ, Donovan SE, Ghamsari L, Santhanam B, Romero V, Poulin MM, Gebreab F, Gutierrez BJ, Tam S, Monachello D, Boxem M, Harbort CJ, Mcdonald N, Gai L, Chen H, He Y, Vandenhaute J, Roth FP, Hill DE, Ecker JR, Vidal M, Beynon J, Braun P, Dangl JL (2011) Independently evolved virulence effectors converge onto hubs in a plant immune system network. Science 333:596–601

30. Moonil K, Lim J-H, Ahn CS, Park K, Kim GT, Kim WT, Pai H-S (2006) Mitochondria-associated hexokinases play a role in the control of programmed cell death in Nicotiana benthamiana. Plant Cell Online 18: 2341–2355

31. Chivasa S, Tomé DFA, Hamilton JM, Slabas AR (2011) Proteomic analysis of extracellular ATP-regulated proteins identifies ATP synthase β-subunit as a novel plant cell death regulator. Mol Cell Proteomics 10:1–13

Chapter 13

Simple Quantification of *In Planta* Fungal Biomass

Michael Ayliffe, Sambasivam K. Periyannan, Angela Feechan, Ian Dry, Ulrike Schumann, Evans Lagudah, and Anthony Pryor

Abstract

An accurate assessment of the disease resistance status of plants to fungal pathogens is an essential requirement for the development of resistant crop plants. Many disease resistance phenotypes are partial rather than obvious immunity and are frequently scored using subjective qualitative estimates of pathogen development or plant disease symptoms. Here we report a method for the accurate comparison of total fungal biomass in plant tissues. This method, called the WAC assay, is based upon the specific binding of the plant lectin *w*heat germ *a*gglutinin to fungal *c*hitin. The assay is simple, high-throughput, and sensitive enough to discriminate between single *Puccinia graminis* f.sp *tritici* infection sites on a wheat leaf segment. It greatly lends itself to replication as large volumes of tissue can be pooled from independent experiments and assayed to provide truly representative quantification, or, alternatively, fungal growth on a single, small leaf segment can be quantified. In addition, as the assay is based upon a microscopic technique, pathogen infection sites can also be examined at high magnification prior to quantification if desired and average infection site areas are determined. Previously, we have demonstrated the application of the WAC assay for quantifying the growth of several different pathogen species in both glasshouse grown material and large-scale field plots. Details of this method are provided within.

Key words Plant, Fungus, Quantification, Chitin, Disease, *Puccinia*

---

1 Introduction

Effective plant disease resistance is a prerequisite for stable food production [1–3]. An essential requirement for the development of disease resistant plant cultivars is an accurate assessment of the plant's resistance status. Many plant resistance phenotypes are partial rather than obvious immunity making scoring of resistant and susceptible plants often a challenging proposition. Plant disease assessment is frequently based upon qualitative estimates of pathogen growth [4, 5] or alternatively by the progression of plant disease symptoms [6]. Such scoring systems are inherently subjective and, to be meaningful, require considerable pathology experience [7].

Paul Birch et al. (eds.), *Plant-Pathogen Interactions: Methods and Protocols*, Methods in Molecular Biology, vol. 1127, DOI 10.1007/978-1-62703-986-4_13, © Springer Science+Business Media New York 2014

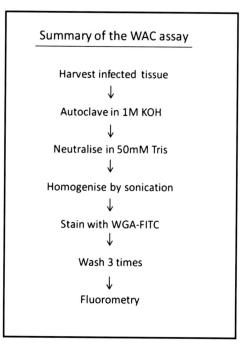

Fig. 1 Summary of the WAC assay

Here we report a simple method for the accurate quantification of pathogen development in plants. Previously we have demonstrated the application of this method for quantifying wheat infection with cereal rust pathogens (*Puccinia graminis*, *P. striiformis* and *P. triticina*) and infection of grape-vines with *Erysiphe necator* [8]. This assay uses the lectin wheat germ agglutinin (WGA) conjugated to a fluorophore such as fluorescein isothiocyanate (FITC) for quantification of fungal biomass of infected plant leaf tissue. WGA has been used extensively for plant pathology studies due to its high binding specificity for chitin, a major component of the cell wall of many fungal species [9–12]. Chitin consists of polymers of *N*-acetyl glucosamine, and WGA specifically binds to trimers of this repeat unit [13, 14]. In the assay, fungal biomass is quantified by the binding of WGA-FITC to fungal chitin present in processed infected plant tissues and subsequent fluorometry (summarized in Fig. 1). Consequently, we have named this *w*heat germ *a*gglutinin *c*hitin assay as WAC.

The WAC assay has a number of advantages for quantifying plant infection by fungal pathogens. It greatly lends itself to experimental replication as large volumes of tissue from multiple independent infection assays can be pooled and processed to provide a truly representative sample. The process can be stopped at many points throughout the protocol and processed tissue can be stored for long periods of time prior to assaying. There is no laborious grinding of samples and no nucleic acid extraction required. It is high-throughput and is not prone to sample cross contamination like PCR-based quantification. The assay is sensitive enough to discriminate between single rust pathogen infection sites on a wheat leaf and is likely to be applicable to many plant pathosystems, given the presence of chitin in the cell walls of most fungal species.

When establishing the WAC assay for a new pathosystem it is recommended that several preliminary steps are undertaken. From previous analysis it was apparent that minor changes in the protocol were necessary for wheat rust pathogens compared with grapevine powdery mildew (described below). In addition, background fluorescence was apparent in *Arabidopsis* tissues that appeared to be age-related making this assay less attractive in this species [8]. A microscopic analysis of tissue and tissue homogenates when establishing the protocol in a new pathosystem is therefore advisable. However, this is a very straightforward process as the WAC assay is based upon a microscopic staining process [15].

For demonstrative purposes, in the following protocol we have specified the conditions required for WAC assay analysis of wheat plants infected with cereal rust. Other pathosystems will obviously have different growth requirements for both host plants and pathogen species. However, the subsequent manipulations for the WAC assay will remain the same.

2   Materials

1. Wheat germ agglutinin conjugated to fluorescein isothiocyanate (WGA-FITC) (Sigma-Aldrich). The stock solution is made to a final concentration of 1 mg/ml in water. Aliquots (200 µl) are frozen at –20 °C until use.

2. 1 M KOH solution containing 0.1 % Silwet L-77 (Lehle Seeds, Texas) used for clearing plant tissue during autoclaving.

3. 50 mM Tris–HCl pH 7.0 used for neutralizing tissue after autoclaving and subsequent washing of samples.

4. A probe sonicator (e.g., Braun 1,000 l) or polytron blender.

5. 200 µl PCR tubes in strips of 12 tubes (e.g., Ultraflux 200 µl 12 strip tubes, Scientific Specialties Inc., USA).

6. Black 96-well plates suitable for fluorometry (e.g., Corning 96-well, flat bottom, black polystyrene plates, Corning Inc. USA).

7. A fluorometer with 485 nm adsorption and 535 nm emission wavelengths e.g., Wallac Victor 1420 multilabel counter (Perkin and Elmer Life Science).

8. 50 ml screw cap Falcon tubes.

9. A bench top centrifuge that takes 96-well plates.

10. Purified crab shell chitin particles (Sigma Aldrich).

11. A microscope with 20× resolution and capable of FITC detection (e.g., Zeiss Axioimager, Zeiss, Germany).

12. 15 cm pots of compost soil mix (1:1), 1 g/l Aquasol (Yates, Australia) and Osmocote (Scotts, USA).

3 Methods

3.1 Establishing the WAC Assay with a Chitin Standard Curve

When establishing the WAC assay for the first time, a useful control experiment is to produce a chitin standard curve by supplementing uninfected tissue of the host plant with a serial dilution of purified chitin particles.

1. Purified crab shell chitin particles (Sigma Aldrich) are ground in a mortar and pestle to produce particles as fine as possible (see **Note 1**). The insoluble chitin particles are then resuspended in water to produce a 1 mg/ml suspension (see **Note 2**).

2. Uninfected wheat seedling leaf tissue is cut into approximately 3 cm segments, weighed and added to a 50 ml falcon tube. A maximum of 8 g can be added to a single tube and still enable subsequent manipulations to be undertaken in the same tube.

3. A solution of 1 M KOH containing 0.1 % Silwet L-77 is poured over the leaf tissue. Ensure that all the tissue is covered with this buffer (see **Note 3**).

4. Leaf samples are autoclaved at 121 °C and 15 psi for 20 min with the tube caps slightly loosened to allow pressure release.

5. After autoclaving, the KOH solution is poured off. As much as KOH solution is removed as possible. Tubes are filled up with 50 mM Tris–HCl pH 7.0, inverted and the buffer then immediately poured off. Tubes are refilled with 50 mM Tris–HCl pH 7.0 and left at room temperature for at least 20 min (see **Note 4**).

6. The buffer is poured off and 5 ml of 50 mM Tris–HCl pH 7.0 added for each gram fresh weight of leaf tissue present, i.e., a final tissue concentration of 200 mg/ml (see **Note 5**).

7. Leaf tissue samples are homogenized using a probe sonicator or small hand-held blender. Each sample generally takes around 1 min to process. It is important to obtain a suspension as uniform as possible.

8. 200 μl of wheat leaf homogenate is added to each of twenty-four 200 μl PCR tubes using a cut off pipette tip (see **Note 6**). Strips of 12 PCR tubes attached together are convenient for handling. Wheat leaf homogenate is always agitated prior to pipetting to ensure that a uniform suspension is sampled.

9. To each of four tubes containing leaf homogenate, 0, 5, 10, 20, 30, or 40 μl of chitin particle suspension (described in Subheading 3.1.1) is added. The chitin particle suspension is also agitated prior to pipetting and a cut off pipette tip used.

10. 10 μl of WGA-FITC (1 mg/ml) is added to each PCR tube and the sample mixed thoroughly by pipetting. Samples are stained for at least 15 min (*see* **Note 7**).

11. PCR tube strips are centrifuged in a bench top centrifuge at $600 \times g$ for 3 min (*see* **Note 8**).

12. After centrifugation the supernatant containing unbound WGA-FITC stain is removed from each sample using a 200 μl pipetteman.

13. Samples are resuspended in 200 μl of 50 mM Tris–HCl pH 7.0. This can be rapidly achieved using a 12 channel multipipette and moderately forceful expulsion of buffer from the pipette.

14. Samples are washed three times in 200 μl of 50 mM Tris–HCl pH 7.0 and then resuspended in 100 μl of the same buffer.

15. Samples are transferred to black 96-well plates suitable for fluorometry using a 12 channel multipipette with the ends of the pipette tips cut off.

16. Fluorescence of each sample is measured in a fluorometer (e.g., Wallac Victor 1420 multilabel counter) using 485 nm adsorption and 535 nm emission wavelengths and a 1.0 s measurement time.

17. Fluorescence values from replicated samples are averaged and subjected to standard statistical analyses. A typical chitin standard curve is shown in Fig. 2.

From previous experiments [8] we have shown that fluorescence values that range from 0 to 200,000 fluorescence units are directly proportional to the chitin content (Fig. 2). Fluorescence values that exceed 200,000 units fall outside this linear range and are no longer directly proportional to chitin content. This nonlinear relationship is caused by WGA-FITC stain becoming a limiting factor with increasing amounts of chitin in the sample [8]. For this reason samples that exceed 200,000 units are further diluted and re-assayed. From our experience, a twofold dilution of very heavily infected wheat leaf tissue homogenate is generally sufficient to maintain these samples within the linear range of fluorescence under the conditions specified.

3.2 Infection of Wheat Seedlings for Quantification of Puccinia graminis f.sp tritici Growth

1. Wheat seeds are planted in 15 cm pots containing a 1:1 mixture of soil and compost. Five pots of each wheat genotype are planted (*see* **Note 9**). Emergent seedlings are fertilized fortnightly with Aquasol (1 g/l), and Osmocot fertilizer beads are added to each pot.

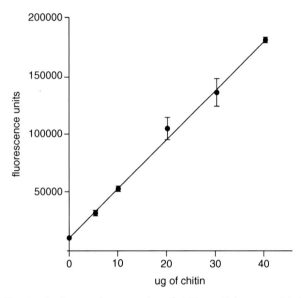

Fig. 2 Chitin standard curve. A suspension of chitin particles was added to uninfected wheat leaf tissue homogenate and processed via the WAC assay. Each data point represents the average fluorescence from four technical replicates

2. Seedlings are grown under a 16 h light/21 °C and 8 h dark/18 °C growth regime.

3. Seedlings are infected at the 4–5 leaf stage with wheat stem rust pathogen (*Puccinia graminis.* f.sp *tritici*), hereafter referred to as *Pgt*. Prior to infection, some seedlings are harvested as described in **points 6** and 7 as a zero time point for each genotype.

4. For *Pgt* infection seedling pots are placed in a plastic box containing 2 cm of water. A mixture of rust spores and talc powder (1:1) is sprayed onto the surface of the wheat seedlings, followed by a fine mist of water. The box is sealed with a plastic lid and incubated overnight at 18 °C.

5. Seedlings are removed the next day and returned to glasshouse growth conditions described in Subheading 3.2.2.

6. One seedling is harvested from each pot for each day of the time course. Seedlings from each genotype are pooled resulting in five seedlings per time point per genotype. All the aerial tissue of each seedling is harvested.

7. The total fresh weight of each pooled sample is recorded. Tissue is cut into 3 cm segments and placed in 50 ml Falcon tubes.

8. Tissue samples can then be either processed and viewed microscopically (Subheading 3.4) or have fungal infection quantified by the WAC assay (Subheading 3.6).

3.3 Growth and Infection of Adult Wheat Plants in the Field

1. Wheat seed is planted in field plots with approximately 50 seeds per meter row. Each wheat genotype is replicated in ten separate plots in the field.

2. Interspersed among each row of wheat cultivars of interest are rows of three fully susceptible wheat genotypes that serve as a source of natural *Pgt* inoculum throughout the growing season.

3. Mixed susceptible plots are infected at the mid-vegetative stage with *Pgt* urediniospores.

4. Flag leaves are harvested from wheat lines of interest from each plot after completion of grain filling. Ten flag leaves are harvested at random from each plot and the fresh weight recorded (*see* **Note 10**).

5. Harvested flag leaves are cut into 3 cm segments and placed into 50 ml Falcon tubes.

6. Tissue samples can either be processed and viewed microscopically (Subheading 3.4) or have fungal infection quantified by the WAC assay (Subheading 3.6).

3.4 Microscopic Analysis of Rust Pathogen-Infected Wheat Leaf Tissue

It is advisable when establishing the WAC assay, either for the first time or with a new pathosystem, to examine pathogen-infected tissue microscopically prior to the WAC assay to ensure that the WGA-FITC conjugate shows good binding specificity for the pathogen to be assayed and that the host tissue shows little background fluorescence (examples shown in Fig. 3a–e) (*see* **Note 11**). Microscopy of leaf homogenates is also valuable when establishing the technique to ensure that samples are relatively homogeneous, and background staining of tissue homogenate is not apparent (Fig. 3f, g). Fortunately, this is simple to do as the WAC assay is based upon a staining method for microscopy [15].

1. Leaf tissue from rust infected wheat plants is harvested and cut into approximately 3 cm pieces.

2. Tissue segments are placed into 10 ml screw capped tubes with enough 1 M KOH containing 0.1 % Silwet added to entirely cover the leaf segments.

3. Samples are autoclaved at 121 °C and 15 psi for 20 min with the caps loosened.

4. Buffer is gently poured off the leaf segments. Note that the tissue is fragile at this stage. The tubes are filled with 50 mM Tris–HCl pH 7.0. This solution is then immediately poured off and the tubes are again refilled with 50 mM Tris–HCl pH 7.0. Samples are left to neutralize for at least 20 min (*see* **Note 4**).

5. Buffer is poured off the leaf segments and enough 50 mM Tris–HCl pH 7.0 added to just cover the sample (*see* **Note 12**).

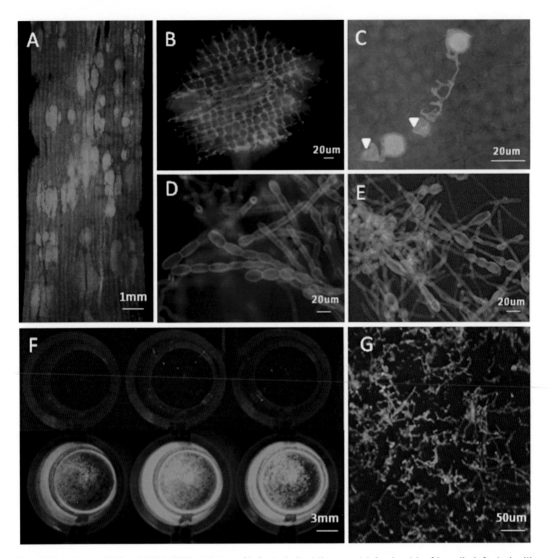

Fig. 3 Fungal specificity of WGA-FITC staining of infected plant tissues. (**a**) A wheat leaf heavily infected with *Pgt* and stained with WGA-FITC as described in Subheading 3.4. Extensive infection hyphae ramification is observed throughout the entire leaf segment while uredinia on the leaf surface can be seen as circular struc-tures. (**b**) A single *Pgt* uredinia on a wheat leaf showing urediniospore production and extensive ramification of infection hyphae throughout the apoplastic space between mesophyll cells. Note, little background staining of host tissue is apparent. (**c**) Two flax rust (*Melampsora lini*) infection sites on flax (*Linum usitatissimum*). Both spores have germinated on the leaf surface to produce a germ tube and an appressorium (marked with *arrow head*) over a stomate. (**d**) Condiophores and hyphae produced by *Blumeria graminis* f.sp. *tritici* (powdery mil-dew) on a wheat leaf. (**e**) Condiophores and hyphae produced by *Erysiphe necator* (powdery mildew) on a grape-vine leaf (*Vitis vinifera*). (**f**) Wheat leaf homogenates in a 96-well plate from uninfected (*top row*) and *Pgt*-infected (*bottom row*) tissue showing WGA-FITC staining. Triplicate WAC reactions are shown for each sample. (**g**) *Pgt*-infected wheat leaf homogenates viewed under blue light at 2× magnification. Filamentous fungal hyphae can be seen that are stained

20 μg of WGA-FITC (20 μl of a 1 mg/ml solution) is added per ml of 50 mM Tris–HCl buffer present.

6. Samples are allowed to stain for at least 15 min. Longer staining times result in better staining. Even very long staining times (e.g., overnight) do not cause a significant increase in background.

7. After staining, samples are removed from the tube for microscopy. This is best achieved by gently washing the samples into a glass petri dish using a squirt bottle of water. This will also dilute any unbound stain sufficiently for microscopy. Do not try to pull the samples with forceps as they are fragile.

8. Leaf segments are placed onto a microscope slide using a flat spatula. Water is gently sprayed directly onto the surface of the leaf segment using a squirt bottle. This causes the leaf samples to unfold and straighten out. Excess water is removed by blotting.

9. A cover slip is placed on the leaf samples which are then viewed under blue light excitation. Good resolution of all infection structures with little back ground staining of host tissue should be apparent (Fig. 3a–e).

3.5 Quantifying Average Infection Site Size Using the WAC Assay

A combination of microscopy and chitin quantification by the WAC assay enables the average size of rust infection sites to be quantified and compared between leaf segments derived from different wheat genotypes [8]. This is useful for comparing disease development between resistant and susceptible germplasm and enables both microscopic evaluation and quantification of infections sites from the same material. The total number of infection sites is counted microscopically on a leaf segment after WGA-FITC staining, and this same segment is then processed by the WAC assay. An average fluorescence per infection site can then be calculated. We have previously shown for *Pgt* that the average fluorescence is equivalent to the average infection site area (*see* **Notes 13** and **14**).

1. Rust infected wheat leaf segments are harvested (Subheadings 3.2 and 3.3) and processed and stained as described in Subheading 3.4.

2. The total number of infection sites on each leaf segment is counted under the microscope. It is essential that all sites are counted, i.e., both large and small sites (*see* **Note 15**).

3. After microscopy leaf segments from the same genotype are pooled into a tube, 1 ml of 50 mM Tris–HCl pH 7.0 is added per 2–3 leaf segments.

4. The tissue is sonicated using a probe sonicator.

5. 200 µl of tissue homogenate is added to a 200 µl PCR tube. At least three replicates are performed per sample.

6. 10 µl of WGA-FITC is added to each tube and samples mixed with vigorous pipetting. Samples are left to stain for at least 15 min (*see* **Note 16**).

7. **Steps 11–17** in Subheading 3.1 are then followed.

8. Average fluorescence per infection site is calculated by total sample fluorescence/number of infection sites present on the combined leaf segments. The total fluorescence of the sample is calculated by the average fluorescence of a 200 µl sample × total sample volume in ml/0.2.

3.6 Quantifying Pathogen Growth in Infected Tissue Samples by the WAC Assay

1. Samples harvested in Subheading 3.2 or 3.3 are covered with enough 1 M KOH containing 0.1 % Silwet to completely cover the sample.

2. Samples are autoclaved at 121 °C and 15 psi for 20 min with tube caps loosened.

3. The KOH solution is poured off and tubes filled with 50 mM Tris–HCl pH 7.0. This buffer is immediately poured off and replaced with 50 mM Tris–HCl pH 7.0. Samples are neutralized for at least 20 min (*see* **Note 11**).

4. The buffer is replaced with 5 ml of Tris–HCl pH 7.0 per gram fresh weight of tissue.

5. Samples are homogenized with a probe sonicator or polytron blender. Samples must be as homogeneous as possible.

6. Using a pipette tip with the end removed, 200 µl samples are aliquoted into 200 µl PCR tubes in strips of 12. Four to six replicates per sample are routinely undertaken. Samples are agitated between pipetting to ensure a homogeneous suspension is maintained.

7. 10 µl (1 mg/ml solution) of WGA-FITC is added to each tube. Samples are mixed by pipetting and stained at room temperature for at least 15 min (*see* **Note 7**).

8. Samples are centrifuged at $600 \times g$ for 3 min in a swing out bench top centrifuge.

9. As much supernatant as possible is removed from each sample with a pipetteman and samples then resuspend in 200 µl of 50 mM Tris–HCl pH 7.0.

10. Samples are washed three times with 200 µl of 50 mM Tris–HCl pH 7.0 and resuspended in 100 µl of 50 mM Tris–HCl pH 7.0 (*see* **Note 17**).

11. Fluorescence is quantified in a fluorometer using 485 nm adsorption and 535 nm emission wavelengths and 1.0 s measurement time. A time course of *Pgt* growth on two wheat genotypes is shown as an example in Fig. 4.

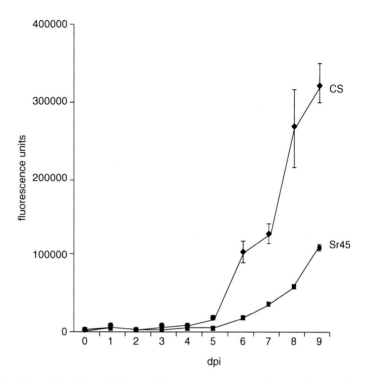

Fig. 4 Quantification of *Pgt* growth on two wheat genotypes. Chinese Spring wheat seedlings (CS) and Chinese Spring wheat seedlings with the *Sr45* gene introgressed as a chromosome 1D substitution from *Aegilops tauchii* were infected with *Pgt*. Ten seedlings were harvested for each time point and chitin accumulation compared with the WAC assay

4 Notes

1. As chitin is difficult to grind, the particles can be sieved through a very fine mesh to remove unground fragments. It is essential to make the chitin particles as small and uniform in size as possible. Large fragments will greatly interfere with subsequent quantification.

2. In subsequent pipetting steps of the chitin suspension, it is important to agitate the solution vigorously prior to each pipetting to ensure a homogeneous aliquot of the suspension is obtained.

3. Additional Silwet can be added if the tissue is particularly waxy and buoyant. It is important to make sure that the tissue is submerged in the buffer.

4. It is essential that the samples are neutralized with 50 mM Tris–HCl pH 7.0. The assay will not work under alkaline conditions. Samples can be tested with pH strips for confirmation of neutralization if needed.

5. Both uninfected and infected leaf samples can be stored for months at 4 °C after autoclaving. Do not store the samples at room temperature as fungal contamination can occur which will cause increased chitin levels. We have not tried freezing the samples, but it is likely that the sample would last indefinitely under these conditions.

6. The ends of pipette tips are removed for all pipetting of leaf homogenates and chitin samples throughout the procedure to ensure a uniform sample is aliquoted.

7. Samples can be stained for much longer periods (e.g., overnight) with no deleterious effects. However, it is essential that samples that are to be directly compared are stained at the same time and for the same length of time. Longer staining times can result in increased levels of fluorescence. Samples can be re-agitated by pipetting at various times throughout the staining step to assist staining.

8. A 96-well, 200 μl PCR plate makes a convenient holder for the PCR tube strips in the centrifuge.

9. Replication of infected material is essential for meaningful results to be obtained in quantitative assays. Uniform inoculation densities are not easy to obtain, so replication of material is needed to overcome potential variation arising from this limitation. Fortunately, the WAC assay is very amenable to pooling of leaf tissue.

10. We again stress the importance of replication for meaningful quantitative results in both seedling assays and field experiments. Good field results were obtained using ten replicated plots for each genotype. We have not determined the minimum number of field plots required for an experiment.

11. The following example highlights the value of microscopic analysis when establishing the WAC assay for a new pathosystem. Cereal rusts produce large amounts of growth within the plant apoplast whereas mildew pathogens produce large amounts of growth on the leaf surface. Microscopy of mildew infected leaves after processing indicated that external fungal structures frequently snapped off the leaf surface following autoclaving leaving only haustoria behind in the leaf tissue. It was likely that much of this external fungal growth would be lost without the inclusion of centrifugation steps prior to all buffer changes in the WAC assay described in Subheading 3.6. Consequently mildew infected leaf samples were centrifuged at $2,000 \times g$ for 10 min prior to all buffer changes described in **step 3** of Subheading 3.6. This additional step enabled good quantification of mildew pathogen growth by the WAC assay [8].

12. A minimum volume is used to reduce the amount of WGA-FITC required. If heavily infected leaf segments are present it may be necessary to add more stain. Higher stain concentrations do not seem to result in increased levels of background.

13. Average *Pgt* infection site area was calculated previously by photographing a representative number of individual infection sites on each leaf segment. The length (*L*) and width (*W*) of each infection site was measured from each photographic image using the polyline facility of the AnalySIS Life Science Professional program (Olympus, Australia). The area of each site was then calculated as $L/2 \times W/2 \times \pi$ (i.e., the area of an ellipse). This approach is considerably more laborious than determining average fluorescence per infection site [8].

14. The propensity of mildew infection structures to break away from the leaf surface as described in **Note 11** suggests that the assay described in Subheading 3.5 may not be applicable for mildew pathogens. In contrast the assay described in Subheading 3.5 was very successful in determining the average infection site size of cereal rust infection sites due to pathogen growth being predominately within the plant leaf [8].

15. For wheat rust, each spore germination event that led to the production of at least a substomatal vesicle was considered as a successful infection site. Spores that had germinated but did not reach this stage of development were not counted.

16. Even though samples are already WGA-FITC stained for microscopy, we restain to ensure that all chitin in the leaf homogenates is detected. It is possible that tissue homogenization increases stain accessibility when compared with intact leaf segments, although we have not tested this experimentally.

17. If establishing the WAC assay for the first time examine the infected tissue homogenates by bright field and blue light microscopy. Abundant homogenate should be present under bright field of which only a small proportion is strongly fluorescent under blue light (Fig. 3f, g).

Acknowledgments

The authors wish to thank the Australian Grains Research and Development Corporation for financial support, and Melanie Soliveres and Soma Chakraborty for technical support.

References

1. Strange RN (2005) Plant disease: a threat to global food security. Annu Rev Phytopathol 43:83–116

2. Ayliffe MA, Singh R, Lagudah ES (2008) Durable resistance to stem rust needed. Curr Opin Plant Biol 11:187–192

3. Fisher MC, Henk DA, Briggs CJ, Brownstein JS, Madoff LC, McCraw SL, Gurr SL (2012) Emerging fungal threats to animal, plant and ecosystem health. Nature 484:186–194

4. Gassner G, Starib W (1932) Die bestimmung der biologischen rassen de weizengelbrostes (*Puccinia glumarum* f.sp *tritici* (Schmidt.) Erikss.undHenn.) Arbeiten der Biologischen Reichsanstalt fur Land und Forstwirtschaft, Berlin, 20: 141–163

5. Stakman EC, Stewart DM, Loegering WQ (1962) Identification of physiologic races of *Puccinia graminis* var. *tritici*. Agricultural Research Service E617. United States Department of Agriculture, Washington, DC

6. Horsfall JG, Barratt RW (1945) An improved grading system for measuring plant disease. Phytopathology 35:655

7. Bock CH, Gottwald TR, Parker PE, Ferrandino F, Welham S, van den Bosch F, Parnell S (2010) Some consequences of using the Horsfall-Barratt scale for hypothesis testing. Phytopathology 100:1030–1041

8. Ayliffe M, Periyannan S, Feechan A, Dry I, Schumann U, Wang M-B, Pryor A, Lagudah E (2013) A simple method for comparing fungal biomass in infected plant tissues. Mol Plant Microbe Interact 26: 658–667

9. Mirelman D, Galun E, Sharon N, Lotan R (1975) Inhibition of fungal growth by wheat germ agglutinin. Nature 256:414–416

10. Molano J, Bowers B, Cabib E (1980) Distribution of chitin in the yeast cell wall. J Cell Biol 85:199–212

11. Chong J, Harder DE, Rohringer R (1985) Cytochemical studies on *Puccinia graminis* f. sp. *tritici* in a compatible wheat host. I. Walls of intercellular hyphal cells and haustorium mother cells. Can J Bot 63:1713–1724

12. O'Connell RJ, Ride JP (1990) Chemical detection and ultrastructural localisation of chitin in cells walls of *Colletotrichum lindemuthianum*. Physiol Mol Plant Pathol 37:39–53

13. Allen AK, Neuberger A, Sharon N (1973) The purification, composition and specificity of wheat germ agglutinin. J Biochem 131: 155–162

14. Roth J (1978) The biological functions of the lectins, 2. Wheat germ agglutinin. In: Roth J (ed) The lectins: molecular probes in cell biology and membrane research. VEB Gustav Fischer Verlag, Jena, pp 21–23

15. Ayliffe M, Devilla R, Mago R, White R, Talbot M, Pryor A, Leung H (2011) Non-host resistance of rice to rust pathogens. Mol Plant Microbe Interact 24:1143–1155

Chapter 14

Virus-Induced Gene Silencing and *Agrobacterium tumefaciens-Mediated* Transient Expression in *Nicotiana tabacum*

Zhao Zhang and Bart P.H.J. Thomma

Abstract

Virus-induced gene silencing (VIGS) is a rapid method for transient silencing of plant genes. In this chapter, we describe the methodology for *Tobacco rattle virus* (TRV)-based VIGS in *Nicotiana tabacum*. In combination with subsequent co-expression of the tomato immune receptor Ve1 and the corresponding *Verticillium* effector Ave1 through *Agrobacterium tumefaciens*-mediated transient transformation (agroinfiltration), we established a rapid system for assessing the requirement of candidate plant genes for Ve1-mediated immune signaling.

Key words VIGS, ATTA, Agroinfiltration, Hypersensitive response (HR), Ve1, Ave1, Effector, Tobacco, *Tobacco rattle virus*

1 Introduction

Virus-induced gene silencing (VIGS) is a technology for transient knock-down of target genes that is based on sequence-specific RNA degradation triggered by double stranded RNA (dsRNA). Recombinant virus carrying a partial sequence (>23 base pairs) of a host target gene is used to infect the plant. When the virus reproduces and spreads throughout the plant, the viral gene transcripts (mRNA) are targeted by the endogenous host silencing machinery as a natural defense mechanism against viral infection. Together with the viral gene transcripts, also the transcripts of the target gene are degraded, and thus the expression of the target gene is knocked down [1].

Although many viral vectors have been developed for VIGS in several host plant species, the *Tobacco rattle virus* (TRV)-based silencing vector has been most extensively used [1–3]. When compared with other viral vectors, TRV has a wide range of host plant

Paul Birch et al. (eds.), *Plant-Pathogen Interactions: Methods and Protocols*, Methods in Molecular Biology, vol. 1127, DOI 10.1007/978-1-62703-986-4_14, © Springer Science+Business Media New York 2014

species, infects large areas of adjacent cells, and induces only mild viral symptoms [1, 4]. Thus, many TRV-based vectors have been made to target candidate genes in tomato and other Solanaceous plant species, exploiting the generally high degree of sequence homology that exists between orthologous genes within this plant family [5–10].

Another tool that has been widely exploited for gene functional analysis is transient expression of candidate genes via the *Agrobacterium tumefaciens*-mediated transient transformation assay (agroinfiltration) [11, 12]. Interestingly, multiple genes can simultaneously be transferred into the same cells through agroinfiltration, and thus different proteins can be co-expressed [12]. Co-expression of plant immune receptors and the corresponding ligands, for instance particular effectors that are secreted by the pathogen to mediate disease establishment, often leads to a hypersensitive response (HR), which is a rapid and localized cell death of plant tissue at the site of recognition [13]. Therefore, the occurrence of HR can be used as a proxy for successful immune signaling. The combination of VIGS and agroinfiltration has been used extensively to study the requirement of candidate genes for immune signaling; the candidate gene can be silenced by VIGS, followed by co-expression of the immune receptor and its corresponding (pathogen-derived) ligand. The requirement of the candidate gene for immune signaling can be evaluated by evaluating the (extent of) occurrence of the HR, as a compromised HR indicates that the candidate gene is required for successful immune signaling.

The efficiency of VIGS is largely determined by the compatibility of the plant and the virus. When compared with VIGS on the model tobacco species *N. benthamiana*, TRV-based VIGS is notoriously difficult for most *N. tabacum* cultivars [8]. However, we recently screened a number of *N. tabacum* cultivars and discovered that *N. tabacum* cv. Samsun is amenable to TRV-based VIGS [8]. Here, we describe a protocol for TRV-based VIGS in combination with agroinfiltration in *N. tabacum* for the investigation of Ve1-mediated immune signaling.

2   Materials

2.1   *Plant Growth*

1. Seeds of *N. tabacum* cv. Samsun or *N. tabacum* cv. Samsun NN (*see* **Note 1**).

2. 9-cm Petri dishes.

3. Tissue papers.

4. Toothpicks.

5. 12-cm plastic pots with soil.

2.2 Constructs for Gene Silencing

1. TRV1, TRV2::PDS, TRV2::GFP, TRV2::EDS1 used for VIGS were described previously [4–6]. 35S::Ave1 and 35S::Ve1 for inducing Ve1-mediated immune signaling that culminates in HR have also been previously described [7, 8, 14].

2. To generate novel target gene silencing constructs of interest, empty TRV2 cloning vectors are available either as pYL156 for digestion-ligation cloning or as the Gateway-compatible pYL279 vector (*see* **Notes 2** and **3**).

3. If using pYL156, reagents for digestion and ligation are required.

4. If using pYL279, Gateway BP Clonase II enzyme mix, Gateway LR Clonase II enzyme mix, and pDONR207 vector (Invitrogen, Carlsbad, California) are required.

5. For co-expression of immune receptors gene and the corresponding ligands, the Gateway-compatible destination vectors pEarleyGate100 [15] and pSol2092 [8] are recommended.

2.3 Media, Buffers, and Solutions

1. LB medium (1 L): 10 g bacteriological peptone, 10 g NaCl, 5 g yeast extract (*see* **Note 4**).

2. YEB medium (1 L): 5 g beef extract, 5 g bacteriological peptone, 5 g sucrose, 1 g yeast extract, 2 mL 1 M MgSO$_4$.

3. Infiltration buffer (1 L, should freshly be made before use): 20 g sucrose, 5 g MS salts (Murashige and Skoog medium without vitamins), 10 mL 1 M MES, 1 mL 0.2 M acetosyringone.

4. Kanamycin (1,000× stock): 50 mg/mL in demineralized water (filter sterilize).

5. Rifampicin (1,000× stock): 25 mg/mL in DMSO (*see* **Notes 5** and **6**).

6. 1 M MgSO$_4$: 246 g/L (autoclave).

7. 1 M MES (2-[*N*-morpholino] ethanesulfonic acid): 195 g/L (autoclave).

8. 0.2 M acetosyringone: 39.3 mg/mL in DMSO (*see* **Notes 5** and **6**).

2.4 VIGS and Agroinfiltration

1. *Agrobacterium tumefaciens* strain GV3101 (*see* **Note 7**).

2. Sterile 50 mL tubes.

3. Small syringe needle (BD Plastipak, Madrid, Spain).

4. Needleless 1 mL syringes (BD Plastipak, Madrid, Spain).

5. Latex gloves (VWR, Leuven, Belgium).

3 Methods

3.1 Plant Growth

1. Place the *N. tabacum* seeds in a Petri dish with wet tissue paper, and cover with the lid of the Petri dish to maintain high humidity. Incubate seeds at room temperature for 3 days to stimulate germination.

2. Transfer the seedlings to soil using a toothpick, and cover the pots with a clear plastic dome to maintain high humidity for 2 days, after which the plastic dome can be removed.

3. Tobacco plants are grown in the greenhouse with the following settings: 21 °C/19 °C during 16/8-h day/night periods, respectively, at 70 % relative humidity and 100 W/m^2 supplemental light when the light intensity drops below 150 W/m^2. Alternately, plants can be grown in a controlled climate chamber at 22 °C/19 °C during 16/8-h day/night periods, respectively, with 70 % relative humidity.

4. Two- to three-week-old plants are optimal to be used for VIGS.

3.2 Preparation of A. tumefaciens

1. Transform *A. tumefaciens* with TRV1 and with TRV2::GFP, TRV2::PDS, TRV2::EDS1, or TRV2 carrying a fragment of the gene of interest; or streak from glycerol stock.

2. Incubate on LB plates containing kanamycin (50 μg/mL) and rifampicin (25 μg/mL) for 2 days at 28 °C (*see* **Note 8**).

3. Inoculate *A. tumefaciens* carrying TRV1 and *A. tumefaciens* carrying TRV2::GFP, TRV2::PDS, TRV2::EDS1, or TRV2 carrying a fragment of the gene of interest in 5 mL LB medium containing kanamycin (50 μg/mL) and rifampicin (25 μg/mL) (*see* **Note 8**).

4. Grow cultures overnight at 28 °C under continuous agitation (200 rpm).

5. Dilute 100 μL *A. tumefaciens* culture in 900 μL LB. Measure the OD$_{600}$ of the diluted cultures with a spectrophotometer. An OD$_{600}$ > 1 is optimal to proceed (*see* **Note 9**).

6. One day prior to VIGS, add 50 μL of the *A. tumefaciens* culture into 20 mL YEB medium containing 50 μg/mL kanamycin (20 μL from stock solution), 20 μM acetosyringone (2 μL from 0.2 M stock solution), 10 mM MES (200 μL from 1 M stock solution).

7. Grow overnight at 28 °C under continuous agitation (200 rpm).

8. Dilute 100 μL *A. tumefaciens* culture with 900 μL YEB and measure the OD$_{600}$ with a spectrophotometer. If the *A. tumefaciens* grew properly, the OD$_{600}$ should now be between 0.8 and 1.2 and the culture is ready for use.

3.3 VIGS

1. Plants at the two-leaf stage (approximately 2–3 weeks old) are optimal for VIGS. Nevertheless, plants at the four-leaf stage can still be used.

2. Overnight culture of *Agrobacterium* cells was harvested at OD_{600} of 0.8 to 1.2 by centrifugation ($3,363 \times g$ for 10 min) and resuspended in infiltration buffer to a final OD_{600} of 2.

3. Make 1:1 mixtures of the *A. tumefaciens* suspensions carrying TRV1 and the appropriate TRV2 constructs.

4. Incubate the mixture of *A. tumefaciens* suspensions for 1–6 h at room temperature.

5. Fill a 1 mL needleless syringe with the appropriate mixture of *A. tumefaciens* suspensions and remove air bubbles from the syringe.

6. Place the tip of the syringe against the lower side of a cotyledon of a tobacco seedling, and gently provide counter pressure at the upper side of cotyledon with a finger. Infiltrate the entire cotyledon with the *A. tumefaciens* suspension by putting gentle pressure on the syringe. Both cotyledons of a tobacco seedling should be infiltrated (Fig. 1) (*see* **Note 10**).

7. After infiltration, plants can be growth in a greenhouse or a climate room under the previously described conditions. It is important to note that the photoperiod has dramatic effects on the silencing efficiency, and the plants should be grown under a long day conditions (16 h photoperiod) after infiltration.

Fig. 1 A mixture of *Agrobacterium* suspension carrying TRV1 and appropriate TRV2 constructs is infiltrated into the lower side of a tobacco seedling cotyledon

3.4 Agroinfiltration

1. At 2–3 weeks after TRV inoculation, properly infected TRV-inoculated plants show viral symptoms and are smaller than non-inoculated plants. Clear photobleaching will be observed at 3–4 weeks after inoculation with TRV::PDS (Fig. 2).

2. As soon as photobleaching is observed in *PDS*-silenced plants, immune receptor-ligand combinations can be co-expressed in the true leaves of the plant, to test whether an HR can still be induced upon silencing of the target gene of interest.

3. As an example, we co-expressed tomato *Ve1* and *Verticillium dahliae Ave1* in *N. tabacum* inoculated with TRV::EDS1, which is required for Ve1-mediated defense signaling, while *N. tabacum* inoculated with TRV::GFP is used as control [7, 16, 17] (*see* **Note 11**).

4. *A. tumefaciens* cultures are prepared as described, harvested by centrifugation and resuspended in infiltration buffer to a final OD of 2.

5. The *A. tumefaciens* strain carrying 35S::Ve1 is mixed in a 1:1 ratio with *A. tumefaciens* carrying 35S::Ave1, incubated at room temperature for 1–6 h and subsequently infiltrated into the leaves of TRV-inoculated tobacco. A slight scratch made with a small syringe needle at the lower side of the leaf will facilitate infiltration.

6. Occurrence of the HR can be monitored between 2 and 5 days after infiltration (Fig. 3) (*see* **Note 12**).

N. tabacum cv. SR1   *N. tabacum* cv. Samsun

Fig. 2 TRV-based VIGS of the *phytoene desaturase* (*PDS*) gene leads to efficient photobleaching in leaves of *N. tabacum* cv. Samsun, but not in *N. tabacum* cv. SR1. Both cultivars were inoculated with recombinant TRV carrying a fragment of the *PDS* gene, and the pictures were taken at 21 days after infiltration

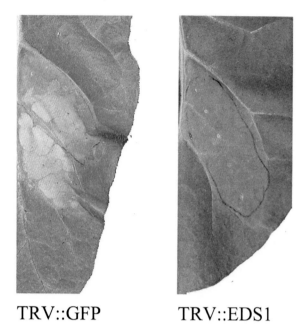

TRV::GFP   TRV::EDS1

Fig. 3 TRV-based silencing of the *EDS1* gene results in compromised HR upon induction with Ve1 and Ave1 in *N. tabacum* cv. Samsun. Plants were inoculated with recombinant TRV targeting *EDS1* (TRV::EDS1) or recombinant TRV targeting *GFP* as a control (TRV::GFP). Compromised HR was observed in *EDS1*-silenced plants at 5 days after infiltration

4 Notes

1. To the best of our knowledge, *N. tabacum* cv. Samsun and *N. tabacum* cv. Samsun NN are the only *N. tabacum* cultivars that display significant VIGS efficiency [8]. The VIGS efficiency is similar in both cultivars.

2. Both pYL156 and pYL279 display similar gene silencing efficiencies.

3. pYL156 and pYL279 are available from the Arabidopsis Biological Resource Center (ABRC, Ohio State University).

4. Bacteriological peptone can be replaced by bacteriological tryptone.

5. Dissolve in DMSO, do not filter sterilize.

6. Rifampicin and acetosyringone also can be dissolved in ethanol, not required to filter sterilize.

7. Other *A. tumefaciens* stains, such as AGL1, MOG101, or C58C1, are also frequently used for VIGS and agroinfiltration.

8. When using *A. tumefaciens* stain GV3101, 15 mg/L gentamycin can be added.

9. The *A. tumefaciens* culture can be stored at 4–10 °C for 2 months.

10. Some studies report infiltration of two true leaves; we found that this does not result in more efficient silencing.

11. Although an empty TRV2 vector is often used as control in VIGS of *N. benthamiana*, we found that inoculation of *N. tabacum* with this construct often result in strong viral symptoms. In contrast, TRV::GFP or TRV::GUS induces relatively mild viral symptoms in *N. tabacum*.

12. The timing of the occurrence of HR may vary considerably between different combination of immune receptors and their corresponding ligands. For Ve1 and Ave1, HR is typically visible at 2 days after infiltration [8, 14].

Acknowledgments

We acknowledge Bert Essenstam for excellent plant care. We are grateful to Yin Song and Yan Wang for their help with making the pictures.

References

1. Burch-Smith TM, Anderson JC, Martin GB, Dinesh-Kumar SP (2004) Applications and advantages of virus-induced gene silencing for gene function studies in plants. Plant J 39: 734–746

2. Burch-Smith TM, Schiff M, Liu YL, Dinesh-Kumar SP (2006) Efficient virus-induced gene silencing in Arabidopsis. Plant Physiol 142: 21–27

3. Ho FI, Chen YY, Lin YM, Cheng CP, Wang JF (2009) A tobacco rattle virus-induced gene silencing system for a soil-borne vascular pathogen *Ralstonia solanacearum*. Bot Stud 50:413–424

4. Liu YL, Schiff M, Dinesh-Kumar SP (2002) Virus-induced gene silencing in tomato. Plant J 31:777–786

5. Gabriëls SHEJ, Takken FLW, Vossen JH, de Jong CF, Liu Q, Turk SCHJ et al (2006) cDNA-AFLP combined with functional analysis reveals novel genes involved in the hypersensitive response. Mol Plant Microbe Interact 19:567–576

6. Gabriëls SHEJ, Vossen JH, Ekengren SK, van Ooijen G, Abd-El-Haliem AM, van den Berg GCM et al (2007) An NB-LRR protein required for HR signalling mediated by both extra- and intracellular resistance proteins. Plant J 50:14–28

7. Fradin EF, Zhang Z, Juarez Ayala JC, Castroverde CD, Nazar RN, Robb J et al (2009) Genetic dissection of *Verticillium* wilt resistance mediated by tomato Ve1. Plant Physiol 150:320–332

8. Zhang Z, Fradin EF, de Jonge R, van Esse HP, Smit P, Liu CM et al (2013) Optimized agroinfiltration and virus-induced gene silencing to study Ve1-mediated *Verticillium* resistance in tobacco. Mol Plant Microbe Interact 26:182–190

9. Vossen JH, Abd-El-Haliem AM, Fradin EF, van den Berg GCM, Ekengren SK, Meijer HJG et al (2010) Identification of tomato phosphatidylinositol-specific phospholipase-C (PI-PLC) family members and the role of PLC4 and PLC6 in HR and disease resistance. Plant J 62:224–239

10. Stulemeijer IJE, Stratmann JW, Joosten MHAJ (2007) Tomato mitogen-activated protein kinases LeMPK1, LeMPK2, and LeMPK3 are activated during the Cf-4/Avr4-induced hypersensitive response and have distinct phosphorylation specificities. Plant Physiol 144:1481–1494

11. Rossi L, Hohn B, Tinland B (1993) The Vird2 protein of Agrobacterium-Tumefaciens carries nuclear-localization signals Important for transfer of T-DNA to plants. Mol Gen Genet 239:345–353

12. Kapila J, DeRycke R, VanMontagu M, Angenon G (1997) An Agrobacterium-mediated transient gene expression system for intact leaves. Plant Sci 122:101–108

13. van der Hoorn RAL, Laurent F, Roth R, de Wit PJGM (2000) Agroinfiltration is a versatile tool that facilitates comparative analyses of Avr9/Cf-9-induced and Avr4/Cf-4-induced necrosis. Mol Plant Microbe Interact 13:439–446

14. de Jonge R, van Esse HP, Maruthachalam K, Bolton MD, Santhanam P, Saber MK et al (2012) Tomato immune receptor Ve1 recognizes effector of multiple fungal pathogens uncovered by genome and RNA sequencing. Proc Natl Acad Sci USA 109:5110–5115

15. Earley KW, Haag JR, Pontes O, Opper K, Juehne T, Song KM et al (2006) Gateway-compatible vectors for plant functional genomics and proteomics. Plant J 45: 616–629

16. Fradin EF, Abd-El-Haliem AM, Masini L, van den Berg GCM, Joosten MHAJ, Thomma BPHJ (2011) Interfamily transfer of tomato Ve1 mediates *Verticillium* resistance in Arabidopsis. Plant Physiol 156:2255–2265

17. Hu GS, de Hart AKA, Li YS, Ustach C, Handley V, Navarre R et al (2005) *EDS1* in tomato is required for resistance mediated by TIR-class *R* genes and the receptor-like *R* gene *Ve*. Plant J 42:376–391

Chapter 15

DIGE-ABPP by Click Chemistry: Pairwise Comparison of Serine Hydrolase Activities from the Apoplast of Infected Plants

Tram Ngoc Hong and Renier A.L. van der Hoorn

Abstract

Activity-based protein profiling (ABPP) is a targeted functional proteomics method that displays the active proteome by using small molecule probes that react covalently with the active sites of protein classes. Comparison of activity profiles from two different samples is not always easy, especially when using probes that generate too many signals. For accurate comparison of protein activities between two proteomes, we developed difference gel electrophoresis ABPP (DIGE-ABPP), which compares two fluorescently labeled proteomes in the same gel lane. This protocol describes the labeling of two proteomes with alkyne-labeled probes, followed by the coupling with two different fluorophores using "click chemistry," the separation of mixed proteomes on protein gels, and the quantification and comparison of the activity profiles. We applied DIGE-ABPP to investigate differential serine hydrolases activities in the apoplast of *Nicotiana benthamiana* challenged with *Pseudomonas syringae* p.v. *tomato* DC3000.

Key words Activity-based protein profiling, ABPP, Click chemistry, Fluorophosphonate, FP probe, Serine hydrolase, Apoplast, *Pseudomonas syringae*, DC3000, *Nicotiana benthamiana*

1 Introduction

Activity-based protein profiling (ABPP) is an advanced, targeted functional proteomics method that displays the active proteome using small molecules that label the active site of proteins [1]. The classical ABPP method often involves a comparison between two or more proteomes in different protein gel lanes. However, different gel lanes sometimes have different qualities, and dense profiles can make it difficult to distinguish signals with a minor difference in molecular weight or intensity. In this chapter, difference gel electrophoresis for ABPP (DIGE-ABPP) is described. This allows an accurate comparison of two activity profiles, by comparison within the same gel lane.

DIGE-ABPP contains a labeling reaction with minitagged activity-based probes, resulting in labeled proteins that carry minitags

Paul Birch et al. (eds.), *Plant-Pathogen Interactions: Methods and Protocols*, Methods in Molecular Biology, vol. 1127, DOI 10.1007/978-1-62703-986-4_15, © Springer Science+Business Media New York 2014

(alkynes or azides) [2]. In the coupling reaction, these minitags are coupled to a fluorescent reporter using a "click chemistry" reaction [3]. Click chemistry generates a covalent bond between an alkyne and an azide, catalyzed by the presence of copper (II) ions and Tris[(1-benzyl-1H-1,2,3-triazol-4-yl)methyl]amine (TBTA).

In practice, the two proteomes are first separately labeled with the same alkyne-probe under the same labeling conditions (Fig. 1a). After labeling, each proteome is clicked to a different cyanine-azide fluorophore, such as the cyanide-azides Cy3N3 or Cy5N3. These fluorophores have similar structure with a minor

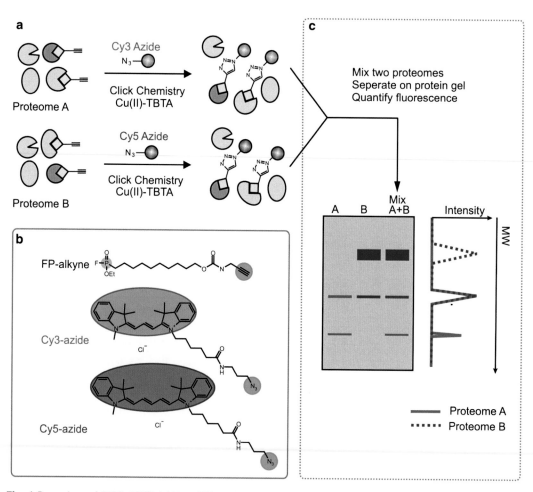

Fig. 1 Procedure of DIGE-ABPP. (**a**) Two different proteomes are labeled with FP≡. Click chemistry is used to couple fluorescent reporters to the alkyne-tags, resulting in two proteomes labeled with different fluorophores. (**b**) Structures of used chemicals. The reactive fluorophosphonate (FP) group that reacts with active site serine residues is indicated in *yellow*. Minitags for click chemistry and the fluorophores are indicated in *green* and *red/blue*, respectively. (**c**) The proteomes are mixed and separated on protein gels. The different fluorophores are detected using two different excitation lasers and emission filters. The fluorescent signals are quantified from the scanned gel

difference in molecular weight (575.19 Da versus 601.22 Da, respectively) (Fig. 1b) and non-overlapping spectral properties. Mixing of the two different labeled proteomes allows the labeled samples to undergo identical conditions of gel electrophoresis, but generate distinguishable profiles by scanning the fluorescence signals using different settings (Fig. 1c). In this experiment we use a fluorophosphonate probe carrying an alkyne minitag (FP≡) (Fig. 1b) and cyanide fluorophores carrying an azide minitag (CyN3, Lumiprobe). The same results can be achieved using the commercially available FP-azide (Thermo Scientific) and cyanine-alkyne (Lumiprobe).

The FP probe forms an irreversible covalent bond with the active site serine residue of serine hydrolases (SHs), which includes serine proteases, acyltransferases, lipases, and other esterases [4]. The catalytic serine is activated by a proton relay involving an acidic and a basic residue, usually aspartate and histidine. Plant genomes encode for over 200 SHs, and some SHs play important roles in metabolism, development, and immunity. Well-studied SHs in plant immunity include salicylic acid binding protein-2 (SABP2), enhanced disease susceptibility-1 (EDS1), pathogenesis-related protein-7 (PR7, a subtilase), and GDSL lipase-1 (GLIP1) [5–8]. We previously identified over 50 different SHs using ABPP from *Arabidopsis* proteomes [9], used ABPP to detect selective inhibition of plant SHs with agrochemicals [10], and identify SH proteins from infected plants [9, 11]. Here, we describe how to detect differential SH activities in the apoplast of *Nicotiana benthamiana* upon infection with *Pseudomonas syringae* p.v. *tomato* DC3000 (PtoDC3000), which triggers a non-host response because of its type-III effector hopQ1-1 [12].

2   Materials

2.1 Preparation of Infected Plants

1. *Pto*DC3000 frozen stock stored at −80 °C.

2. DMSO: dimethylsulfoxide (Sigma 41640).

3. Rif 1,000×: Rifampicin 50 mg/mL (Sigma-Aldrich, R8883). In a 1.5 mL Eppendorf tube, measure 50 mg rifampicin and dissolve into 1 mL DMSO. Cover the tube with aluminum foil and store at −20 °C.

4. NYG medium 1 L: 5 g/L of Bacto™ peptone (BD 211677), 3 g/L of Bacto™ yeast extract (BD 28862), 20 g/L of glycerol (CarlRoth 3783). In a 1 L beaker, stir 5 g peptone and 3 g yeast extract in 0.5 L MilliQ water using a magnetic stirrer. Add 20 mL of glycerol into the mixture, remove the mixer, and fill up to 1 L with MilliQ water. Autoclave the medium at 121 °C and 100 kPa for 15 min.

5. Five-week-old *Nicotiana benthamiana* plants with some fully expanded leaves. Grow *Nicotiana benthamiana* in a growth chamber with 60 % relative humidity and 16:8 light:dark cycles at 26 °C during the day and 22 °C during the night. Reserve three plants for each treatment and use two leaves from each plant.

6. Two 1 mL sterile syringes without needle.

7. Ethanol (70 %) for sterilization.

2.2 Isolation of Apoplastic Fluid

1. 800 mL Beaker and styrofoam lid: measure the size of beaker to cut the styrofoam, so that styrofoam lid fits tightly in the beaker, pressing against the wall.

2. Custom-made centrifuge tube, designed based on [13]. The tube consists of two parts. The upper part has a perforated bottom, which allows apoplastic fluid to be collected in the lower part by centrifugation.

3. Vacuum desiccator with pump.

4. Centrifuge with swing-out rotor for the custom-made centrifuge tubes (above).

2.3 Labeling of Apoplastic Fluid with FP-Alkyne Probe

1. 10× PBS: Phosphate-buffered saline, 10.6 mM KH_2PO_4 (Merck Millipore 1051080050), 1,552 mM NaCl (Merck Millipore 1064001000), 30 mM Na_2HPO_4 (Merck Millipore 1065660500), pH 7.4. In a 1 L beaker mix 1.44 g KH_2PO_4, 90.70 g NaCl and 4.26 g Na_2HPO_4. Add 800 mL water and dissolve using a magnetic stirrer. Then adjust to pH 7.4 using hydrochloric acid and make up to 1 L with MilliQ water. Filter sterilize the buffer using a 0.22 μm filter unit. Can be stored at room temperature (RT) indefinitely.

2. 1× PBS: In a 1 L bottle, mix 100 mL of 10× PBS with 900 mL of MilliQ water. Invert the bottle several times. Store the solution at RT.

3. DTT 1 mM: Dithiothreitol (Sigma 43815-1G). Dissolve 154 mg DTT powder in 1 mL MilliQ water by vortexing. Aliquot into 100 μL into a 500 μL Eppendorf tube and store at –20 °C.

4. FP alkyne 1 mM: The probe is dissolved in DMSO to a concentration of 1 mM and stored at –20 °C. Always open and handle the stocks at room temperature. This reduces the likelihood of water getting into the stock due to condensation.

2.4 Acetone Precipitation

1. Cold acetone: 99.9 % Acetone for HPLC (CarlRoth 7328.1). Store in a glass bottle at –20 °C until use.

2. 10 % SDS: Sodium dodecyl sulfate (CarlRoth 23262). In a chemical hood, dissolve 10 g SDS in 80 mL MilliQ water at 40 °C using a magnetic stirrer until the solution is clear. Then adjust the volume to 100 mL. The solution can be stored at RT.

3. PBS-SDS buffer: 1× PBS buffer containing 1 % SDS. Mix 10 mL of 10× PBS with 10 mL of 10 % SDS and add MilliQ water up to 100 mL.

2.5 Click Chemistry

1. Cyanine-azide: Cy3N3 and Cy5N3 (Lumiprobe 11030 and 13030, resp.). Dissolve the powder in DMSO to a final concentration of 100 µM and store at –20 °C.

2. 3.4 mM TBTA: Tris[(1-benzyl-1*H*-1,2,3-triazol-4-yl)methyl] amine 97 % (Sigma 678937). Dissolve 9 mg TBTA in 1 mL of DMSO/t-butanol 3:1. Add water up to 5 mL. Make 0.5 mL aliquots and store at –20 °C.

3. 100 mM TCEP: Tris(2-carboxyethyl)phosphine hydrochloride (Sigma C4706). Dissolve 28.7 mg of powder in 0.5 mL water and bring the volume up to 1 mL by adding water. Store at –20 °C.

4. 50 mM $CuSO_4$ (Sigma C1297). Dissolve 7.98 mg powder in 0.5 mL water and bring the volume up to 1 mL by adding water. Store at –20 °C.

5. 6× GLB: Gel Loading Buffer pH 6.8. Mix 6.06 g Tris with 400 mL water, adjust to pH 6.8 with concentrated HCl. Add to the mixture 3 mL glycerol, 1.4 g SDS, 0.93 g DTT (0.6 M final concentration) and 1.2 mg bromophenol blue. Heat the mixture at 65 °C until all SDS is dissolved. Bring the volume up to 10 mL by adding water. Store 0.5 mL aliquots at –20 °C.

2.6 Protein Gel Electrophoresis

1. 1 M Tris pH 6.8: Tris-(hydroxymethyl)-aminomethane (Roth 54293). In a 500 mL bottle, dissolve 60.57 g Tris in 400 mL MilliQ water. Adjust the pH to 6.8 with concentrated HCl. Fill with MilliQ water up to 500 mL. The solution can be stored at RT indefinitely.

2. 1.5 M Tris pH 8.8. Dissolve 90.85 g Tris in 400 mL MilliQ water. Adjust the pH to 8.8 with concentrated HCl. Then transfer the solution to a 500 mL measuring cylinder and add water up to 500 mL. Transfer the solution to a 500 mL bottle and autoclave. The solution can be stored at RT indefinitely.

3. 10 % APS: 400 mM Ammonium persulfate $(NH_4)_2S_2O_8$ (Sigma A3678). Add 1 g APS to a 15 mL tube and add MilliQ water up to 10 mL. Mix well until the chemical is completely dissolved. The solution can be stored at 4 °C for 1 week, or in aliquots at –20 °C for longer periods.

4. 30 % Acrylamide/Bis (Sigma A3574).

5. 15 % Resolving gel: for one gel mix the following components in a clean glass beaker: 1.6 mL water, 1.75 mL 1.5 M Tris pH 8.8, 70 µL 10 % SDS, 70 µL 10 % APS, 3.5 mL 30 % Acrylamide/Bis solution, and finally 28 µL TEMED. Pour the

gel between the gel plates and overlay the surface with Isopropanol. After 3 h, the resolving gel is polymerized. Pour off the Isopropanol and dry the gel surface with Whatman paper.

6. 6 % Stacking gel: For one gel mix the following components in a glass beaker on ice: 2 mL water, 378 μL 1.0 M Tris pH 6.8, 30 μL 10 % SDS, 30 μL 10 % APS, 6 mL 30 % Acrylamide/Bis solution, and finally 3 μL TEMED. Fill the gel cassette with stacking gel solution and add the comb. Incubate the gel at RT for 1 h. In case storage is needed wrap the cassette in wet tissue paper and store in a plastic bag at 4 °C for a few days.

7. 10× SDS running buffer: 248 mM Tris, 2 M glycine, 35 mM SDS. In a 1 L bottle, mix 30 g Tris, 144 g glycine, and 10 g SDS with 800 mL water with a magnetic stirrer until all the components are dissolved. Fill the bottle up to 1 L with water and store the solution at RT.

8. 1× SDS running buffer: 24.8 mM. In a 1 L bottle, mix 100 mL 10× SDS running buffer with 900 mL MilliQ water.

3 Methods

3.1 Preparation of Infected Plants

1. Streak out the *Pto*DC3000 on an LB medium plate containing 50 μg/mL rifampicin. Incubate bacteria at 28 °C for 24 h. Prepare 10 mL of liquid NYG medium containing 50 μg/mL rifampicin in a 50 mL falcon tube. Use a sterile pipette tip or toothpick to inoculate with *Pto*DC3000. Incubate overnight at 28 °C with shaking at 200 rpm. Spin down the bacteria for 5 min at $1,000 \times g$ and discard the medium. Resuspend the bacteria in 1 mL sterile water by vortexing. Measure OD at 600 nm. Dilute bacteria with water until $OD_{600} = 1$ (10^9 bacteria/mL). Dilute bacterial stock to 10^6 bacteria/mL in two steps by adding 100 μL bacteria to 10 mL sterile water.

2. Select three plants for each treatment, label them with "C" (control) or "I" (infected) and the date of infiltration. "C" plants are infiltrated with sterilized water, and "I" plants are infiltrated with bacteria.

3. Start infiltrating the "C" plants. Use a 1 mL syringe for infiltration. Take in sterilized water and remove air bubbles (*see* **Note 1**). Infiltrate water into the abaxial (lower) face of the tobacco leaf. Do not wound the leaves too much (*see* **Note 2**).

4. After finishing "C" plants, do the same for "I" plants using the bacterial culture.

5. Use a color marker to label the infiltrated leaves. Decontaminate the working area with 70 % ethanol.

6. Keep the plants in a growth chamber for 48 h.

3.2 Preparing Apoplastic Fluid Sample

This is a brief protocol based on Joosten [13].

1. Label two 800 mL beakers with "C" and "I."

2. Place 300 mL ice in the beaker and add water up to 500 mL.

3. Detach the infiltrated leaves and remove the petiole.

4. Submerge the leaves into the prepared beaker containing ice water. Cover the leaves with styrofoam and press down to keep the leaves under water. Add 100 mL ice to the top.

5. Place ice beaker into vacuum desiccator and apply vacuum for 10 min. A sufficient pressure is about 60 mbar (hPa) or 46 Torr (mmHg). Release vacuum slowly to allow water to enter the leaves (see **Note 3**).

6. Place the infiltrated leaves on paper towels, and quickly dry the surface with paper towels. Do not damage the leaves and proceed to the centrifugation step within 15 min.

7. Label the centrifugation tubes with "C" and "I." Stack the leaves, roll them up, and place them into the centrifugation tubes.

8. Centrifugate for 15 min, $3,000 \times g$ at 4 °C.

9. Collect the apoplastic fluid from the bottom of the tube and keep on ice. A gram of *Nicotiana benthamiana* leaf should result in about 0.5 mL apoplastic fluid.

10. Aliquot apoplastic fluids in Eppendorf tubes, and keep them on ice for not more than 8 h. Apoplast samples can be frozen with liquid nitrogen and stored at −80 °C for some months, but protein precipitation often occurs upon thawing.

3.3 Labeling of Apoplastic Fluid with FP-Alkyne Probe

1. Mix 500 µL apoplastic fluid with 2.5 µL of 1 M DTT (5 mM final), vortex well for 30 s, spin down at full speed for 15 s. Use the supernatant for labeling (see **Note 4**).

2. Label four 1.5 mL tubes with "CCy3," "ICy3" (red ink) and "CCy5," "ICy5" (blue ink). Add sample "C" to CCy3 and CCy5. Add sample "I" to the others. Add buffer and probe in the following amounts:

Stock	Volume (µL)	Final
1× PBS	20	0.1×
1 mM FP-alkyne	1	5 µM
Apoplastic fluid	179	
Total	200	

3. Vortex well for 10 s, spin down at full speed 5 s. Rotate the samples for 1 h at RT.

3.4 Acetone Precipitation

1. Add 1 mL of ice-cold 100 % acetone to each tube and vortex for 10 s (*see* **Note 5**).

2. Immediately centrifuge all tubes at full speed for 2 min at RT.

3. Discard the supernatant and use a narrow tip to remove most of the remaining solution without disturbing the protein pellet.

4. Let the protein pellet dry for 10 min at RT in an open tube (*see* **Note 6**).

5. Add 44 µL of PBS-SDS buffer to each tube. Close the cap and vortex protein at RT for 15 min to dissolve the pellet completely (*see* **Note 7**).

6. When the protein is completely dissolved in PBS-SDS buffer, denature the protein by heating at 90 °C for 10 min (*see* **Note 8**).

3.5 Click Chemistry

1. All chemicals are kept on ice. The samples are kept at RT.

2. Perform click reaction by adding components in the order shown below. Vortex the sample for 5 s after adding each chemical. Cy3N3 is added into red ink tubes (CCy3 and ICy3). Cy5N3 is added into the others.

Stock	Volume (µL)	Final
Sample in PBS–SDS buffer	44.0	
100 µM Cy3 (or Cy5)	2.5	5 µM
3.4 mM TBTA	1.5	100 µM
100 mM TCEP	1.0	2 mM
50 mM $CuSO_4$	1.0	1 mM
Total	50.0	

3. Vortex the samples well for 15 s. Spin down at full speed for 10 s. Rotate all samples in the dark (wrap in aluminum foil) at RT for 1 h (*see* **Note 9**).

4. Perform acetone precipitation as described in Subheading 3.3 (*see* **Note 10**). Dissolve the protein pellet in 50 µL 1× PBS buffer containing 1 % SDS. Add 10 µL of GLB to the solution. Samples are ready for separation in protein gels. The samples can be stored at –20 °C.

3.6 Protein Electrophoresis

1. Place a 15 % Bis–Tris gel in an electrophoresis unit (*see* **Note 11**). Fill both the inside and outside compartments with 1× SDS running buffer. Rinse the wells with SDS running buffer.

2. Mix 5 µL of CCy3 with 5 µL ICy5 in a new tube, label it "Mix1." Do the same for CCy5 and ICy3, labeled "Mix 2."

3. Use PageRuler protein ladder from Thermo Scientific as a molecular weight control (or a Spectra multicolor broad range

protein ladder). Load the samples with the following volumes and in the following order:

Lane	1	2	3	4	5	6	7
Sample	PageRuler	CCy3	Mix 1	ICy5	CCy5	Mix 2	ICy3
Volume (µL)	1	5	10	5	5	10	5

4. Perform protein electrophoresis at 200 V for 1 h until the blue dye front almost runs out from the bottom of the gel.

5. Remove the gel from the cassette and rinse gently under running tap water. Remove the blue dye area at the bottom of the gel. Keep the gel in a clean box containing water and immediately scan for fluorescence signals with the Typhoon Scanner.

6. Set up Typhoon Scanner FLA 9000 for scanning fluorescence multiple channels (DIGE). Reader setting is fluorescence, first layer use Cy3 method (532 nm LPG) and second layer use Cy5 method (635 nm LPR). Set PMT at 900 V and pixel size is 10 µm (see **Note 12**).

3.7   Create Multiplex Image and Quantification of Fluorescence Signal

1. The details for the process can be checked in the manual of the ImageQuantTL (IQTL) version 8.1 (GE Healthcare) [14].

2. By using multiple channels scanning setting with the typhoon scanner (**step 6** of Subheading 3.6), the program creates a folder with overlay gel under .ds file, and two separate gels under .gel file. Open overlay .ds file with IQTL, show overlay mode for both channels by clicking "overlay" bottom in the main menu. Set layer 1 in red for Cy3 gel, and layer 2 in blue for Cy5 gel. The overlay image can be exported as a .bmp file (Fig. 2a).

3. Select "edit mode" to create lanes, choose "automatic," and then click "create." The lanes can be adjusted with "edit multiple lanes" mode. Click "accept" when you have finished editing lanes. The background subtraction is optional. Under lane window, IQTL gives a line chart presenting counts against pixel position for each lane. Under measurement window, IQTL shows the band intensity value for each lane. Lane number four for Mix1 sample is quantified and presented in a line chart of the intensity against the apparent molecular weight (MW) (Fig. 2b).

4   Notes

1. Avoid air bubbles in the syringe, so that only liquid is infiltrated into the leaf. Pull the plunger to take in 2 mL of the liquid, turn the syringe with the opening up, flick the syringe to release the air bubbles and push out all the air slowly.

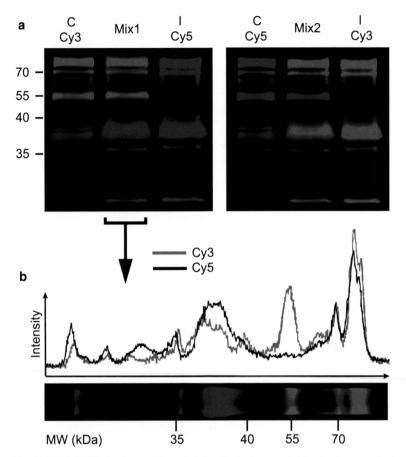

Fig. 2 DIGE-ABPP displays differential Ser hydrolase activities in the apoplast upon *Pseudomonas* infection. *N. benthamiana* plants were infiltrated with (I) and without (C) 10^6 PtoDC3000 bacteria/mL. Apoplastic fluids were harvested at 48 h post infiltration and labeled with FP-alkyne. Alkyne minitags were coupled to Cy3 and Cy5 fluorophores using click chemistry and labeled proteins were separated on protein gels, separately or mixed pairwise. The gel was scanned for fluorescence with different lasers and filters and an overlay image was created using ImageQuant TL and Adobe Photoshop (**a**). The Mix1 lane was quantified using ImageQuant TL and the intensity plotted against the apparent molecular weight (MW) (**b**)

2. Wear gloves during infiltration. Hold the leaf with one hand, placing a finger behind the leaf at the injected site. Inject water or bacteria culture from the underside of the leaf.

3. Check if all leaves are fully infiltrated with water, repeat the vacuum step if necessary.

4. DTT is a reducing agent that prevents protein aggregation and it is necessary for some proteins to maintain activity and stability.

5. The protein concentration in the apoplastic fluid is low. Protein precipitation increases the protein concentration and removes all unreacted probe and other chemicals from the solution.

6. Drying acetone pellet longer than 10 min makes it difficult to dissolve the protein later.

7. Run the tubes across an Eppendorf tube holder to mix vigorously. If the protein pellet is not over dried, they dissolve easily in PBS-SDS buffer. If a pellet still remains after centrifugation, the protein will need another 15 min rotating in the buffer. If the protein is over-dried, try to solubilize the pellet as much as you can for up to 1 h, and then move on to click chemistry.

8. Heating in the presence of 1 % SDS will expose the alkyne group of the FP probe for the click reaction.

9. Cyanine fluorophores are sensitive to light. Cover the shaker with aluminum foil to protect the Cy3 and Cy5 fluorophores.

10. Acetone precipitation is necessary to remove all the unbound Cy3-azide and Cy5-azide, which can cause smearing background signals in the fluorescence gel. Also, most copper ions will be removed in this step.

11. 12 or 15 % protein gels are commonly used, depending on the size of the expected proteins. Apoplast proteomes show clear differentials on a 15 % protein gel.

12. If the signal is saturated, the gel should be scanned at lower PMT voltage. You know the signal is saturated during scanning when signals turn red in the scanning image.

Acknowledgments

We would like to thank Dr. Sabrina Nickel and Prof. Dr. Markus Kaiser (University of Essen-Duisburg, Germany) for providing FP-alkyne. This work was financially supported by the Max Planck Society, the Deutsche Forschungsgemeinschaft (project HO 3983/7-1) and COST program CM1004.

References

1. Kolodziejek I, Van der Hoorn RAL (2010) Mining the active plant proteome in plant science and biotechnology. Curr Opin Biotechnol 21:225–233

2. Kaschani F, Verhelst SHL, Van Swieten PF, Verdoes M, Wong C-S, Wang Z, Kaiser M, Overkleeft HS, Bogyo M, Van der Hoorn RAL (2009) Minitags for small molecules: detecting targets of reactive small molecules in living plant tissues using 'click-chemistry'. Plant J 57:373–385

3. Speers AE, Cravatt BF (2004) Profiling enzyme activities *in vivo* using click chemistry. Chem Biol 11:535–546

4. Liu Y, Patricelli MP, Cravatt BF (1999) Activity-based protein profiling: the serine hydrolases. Proc Natl Acad Sci USA 96:14694–14699

5. Kumar D, Klessig DF (2003) High-affinity salicylic acid-binding protein 2 is required for plant innate immunity and has salicylic acid-stimulated lipase activity. Proc Natl Acad Sci USA 100:16101–16106

6. García AV, Blanvillain-Baufumé S, Huibers RP, Wiermer M, Li G, Gobatto E, Rietz S, Parker JE (2010) Balanced nuclear and cytoplasmic activities of EDS1 are required for a complete plant innate immune response. PLoS Pathog 6:e1000970

7. Tornero P, Conejero V, Vera P (1997) Identification of a new pathogen-induced member of subtilisin-like processing protease family from plants. J Biol Chem 272:14414–14419

8. Oh IS, Park AR, Bae MS, Kwong SJ, Kim YS, Lee JE, Kang NY, Lee S, Cheong H, Park OK (2007) Secretome analysis reveals an *Arabidopsis* lipase involved in defense against *Alternaria brassicicola*. Plant Cell 17: 2832–2847

9. Kaschani F, Gu C, Niessen S, Hoover H, Cravatt BF, Van der Hoorn RAL (2009) Diversity of serine hydrolase activities of non-challenged and Botrytis-infected *Arabidopsis thaliana*. Mol Cell Proteomics 8:1082–1093

10. Kaschani F, Nickel S, Pandey B, Cravatt BF, Kaiser M, Van der Hoorn RAL (2012) Selective inhibition of plant serine hydrolases by agrochemicals revealed by competitive ABPP. Bioorg Med Chem 20:597–600

11. Kaschani F, Gu C, Van der Hoorn RAL (2012) Activity-based protein profiling of infected plants. Methods Mol Biol 835:47–59

12. Wei CF, Kvitko BH, Shimizu R, Crabill JE, Alfano JR, Lin NC, Martin GB, Huang HC, Collmer A (2007) A *Pseudomonas syringae* pv. *tomato* DC3000 mutant lacking the type II effector HopQ1-1 is able to cause disease in the model plant *Nicotiana benthamiana*. Plant J 51:32–46

13. Joosten MH (2012) Isolation of apoplastic fluid from leaf tissue by the vacuum infiltration centrifugation technique. Methods Mol Biol 835:603–610

14. GE Healthcare Life Sciences handbook: imaging principles and methods for ImageQuantTL 8.1 (2012). www.gelifesciences.com article 29-0203-01

Chapter 16

A Simple and Fast Protocol for the Protein Complex Immunoprecipitation (Co-IP) of Effector: Host Protein Complexes

Jens Steinbrenner, Matthew Eldridge, Daniel F.A. Tomé, and Jim L. Beynon

Abstract

Plant pathogens are responsible for enormous damage in natural and cultured ecosystems. One strategy most pathogenic organisms follow is the secretion of effector proteins that manipulate the host immune system to suppress defense responses. There is considerable interest in finding host targets of pathogen effectors as this helps to shape our understanding of how those proteins work *in planta*. The presented protocol describes a protein complex immunoprecipitation method aimed at verifying protein–protein interactions derived from protein complementation assays like Yeast-two-Hybrid.

Key words Effector, Co-IP, *Hyaloperonospora arabidopsidis*, Oomycetes, Transient expression

1 Introduction

One major virulence strategy of phytopathogenic organisms is to deliver effector proteins into the plant cell that target and disable the host immune machinery and, therefore, enable successful colonization. Some effectors bind to DNA and manipulate host transcription [1, 2], whereas others interact with host proteins and may influence protein turnover, phosphorylation, dephosphorylation and exhibit transferase or ubiquitin ligase activities. The *Pseudomonas syringae* effector AvrPphB shows protease activity and cleaves the kinase PBS1 [3]. The endosome localized *P. syringae* effector HopM1 interacts with the *Arabidopsis thaliana* adenosine diphosphate (ADP) ribosylation factor guanine nucleotide exchange factor Arf-GEF protein AtMin7/BEN1 leading to its ubiquitin-dependent degradation via the 26S proteasome, thereby impairing the defense secretory pathway in plants [4, 5]. The *P. syringae* effector HopZ1a exhibits an acetyltransferase activity and interacts with plant tubulin causing depolymerization of the

Paul Birch et al. (eds.), *Plant-Pathogen Interactions: Methods and Protocols*, Methods in Molecular Biology, vol. 1127, DOI 10.1007/978-1-62703-986-4_16, © Springer Science+Business Media New York 2014

plant microtubule network which leads to the disruption of protein secretion [6]. The *P. syringae* effector AvrPtoB shows intrinsic E3 ligase activity and targets the Chitin Elicitor Receptor Kinase 1 (CERK1) for degradation [7, 8].

In the case of oomycetes the *Phytophthora infestans* (*P. infestans*) effector CRN8 shows *in planta* kinase activity but the target of this effector is unknown [9]. Other oomycete host protein targets have been reported, but the biochemical and molecular processes underlying these interactions in promoting virulence are not understood. The *P. infestans* effector Avr3a interacts and stabilizes host U-box E3 ligase CMPG1 possibly to prevent host-induced cells death during the biotrophic phase of the pathogen [10]. The *P. infestans* effector AVRblb2 prevents secretion of the host papain-like cysteine protease C14 [11]. For Avr2 an interaction with BSL1 (BSU Like 1) orthologs from the major hosts of *P. infestans*, tomato and potato, was shown by Protein Complex Immunoprecipitation (Co-IP) and Yeast-two-Hybrid (Y2H) analyses. This interaction is needed for R2-mediated recognition of Avr2 [12]. In a recent study, new host targets of effectors from the plant bacterial pathogen *P. syringae* and the oomycete *Hyaloperonospora arabidopsidis* (*Hpa*) were identified using a matrix-based yeast-two-hybrid assay [13]. This study showed that *P. syringae* and *Hpa* effectors converged onto 18 shared targets and in total both effector classes targeted 165 plant proteins. Although this plant pathogen interaction network gives new insights into the immune system of plants few of the interactions have been validated by independent secondary techniques such as Co-IP or Bimolecular Fluorescence Complementation (BiFC) [14]. Here, we describe a simple and effective protocol to validate experimentally proven and theoretically predicted protein–protein interactions from available datasets using the technique of Co-IP. As an example we prove the interaction between the *Hpa* effector HaRxL14 and an armadillo repeat only 2 (Aro2) protein which was reported earlier in a Y2H assay [13]. Armadillo repeat only proteins are involved in cytoskeleton binding and cell polarity [15]. HaRxL14 was described recently to enhance susceptibility of *P. syringae* DC3000 [16], and the *Arabidopsis thaliana* (Arabidopsis) accession Columbia-0 overexpressing this effector is more susceptible to the *Hpa* isolates Noks1 (Tome et al. this issue).

1.1 Strategy

Since the development of the epitope tagging technology at the end of the 1980s the need to raise specific antibodies (AB) against an antigen is no longer a prerequisite for immunoprecipitation [17]. Good monoclonal antibodies are now available for many epitope tags. These include the human influenza Hemagglutinin (HA), tag derived from a virus surface glycoprotein [18], or the c-Myc tag, a polypeptide derived from the human oncogenic transcription factor c-Myc [19]. For a more detailed overview we refer the reader to the review on this topic by Brizzard [20].

Epitope tags can be placed anywhere in the protein of interest but are usually added to the N-terminus or C-terminus of the protein. Monoclonal antibodies that recognize the tags can then be used to purify tagged proteins from a cell lysate.

1.2 The Choice of Vector

We use the pGWB and pEarleygate vector series for transient expression of epitope-tagged proteins in *N. benthamiana* [21, 22]. Other vectors containing full-length GFP or the C-terminal half of YFP can be used as well, since the C-terminal half of YFP is recognized by some GFP antibodies (*see* **Note 1**).

1.3 Controls

It is advisable to use several non-interacting control proteins. We use other effector proteins which show similar localization pattern as the effector of interest [23]. An example is shown for GFP fusions to the *Hpa* effectors HaRxL14 and HaRxL21. Both effectors show nuclear/cytoplasmic localization when transiently expressed in *N. benthamiana* (Fig. 1a).

It is also important to ensure that the effector and control proteins are expressed to similar levels. We perform test infiltrations with increasing amounts of *Agrobacterium tumefaciens* cells harboring the tagged HaRxL-effector constructs, run samples on an SDS-PAGE gel, and perform a Western blot. An example is shown in Fig. 1b.

Finally, an IP should be performed without the Bait protein to rule out the possibility that the Prey protein binds nonspecifically to the antibody-coupled matrix or antibody-coupled magnetic beads (Fig. 1c).

2 Materials

2.1 Cloning of Target Genes

1. Primers:

   AttB1_FWD:GGGGACAAGTTTGTACAAAAAAGCAGGCT

   AttB2_REV:GGGGACCACTTTGTACAAGAAAGCTGGGT

   M13_FWD: GTAAAACGACGGCCAGTC

   M13_REV: AACAGCTATGACCATG

2. KOD Hot Start Master Mix (Merck Biosciences Cat. No. 71842-3).

3. BP Clonase II (Invitrogen Cat. No. 11789020).

4. pDONR™Zeo Vector (Invitrogen Cat. No. 12535035).

5. Incubator at 37 °C.

6. Shaking Incubator at 37 °C.

7. One Shot® OmniMAX™ 2 T1 Phage-Resistant Cells (Invitrogen Cat. No. C8540-03).

8. PCR tubes.

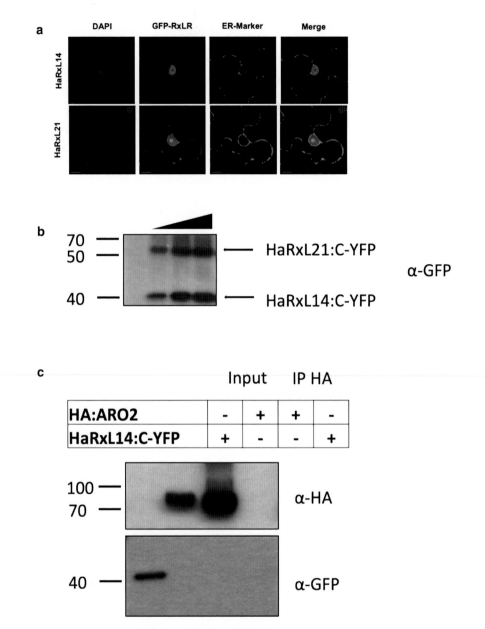

Fig. 1 (a) GFP:HaRxL14 and GFP:HaRxL21 show nuclear/cytoplasmic distribution in *N. benthamiana* cells when transiently expressed. Blue channel: DAPI stain, Green channel: HaRxLRs, Magenta channel: Endoplasmatic Reticulum (ER) marker [24]. (b) Western blot of protein extracts of *N. benthamiana* leaves that were co-infiltrated with increasing amounts of *Agrobacteria* cells harboring plasmids coding for the effector HaRxL14 and the control effector HaRxL21. Proteins were extracted 36 h after infiltration. Effectors were fused C-terminally to the C-terminal half of YFP of the pBIFP4 vector (BIFC-Vector). *Lane 1* non-infiltrated. *Lane 2* final OD_{600} 0.05 for each construct, *lane 3* OD_{600} 0.1 for each construct, *lane 4* OD_{600} 0.2 for each construct. Blot was probed with an anti-GFP antibody. Protein sizes on the left in kilo Daltons. (c) Western blot showing that HaRxL14:C-YFP does not bind nonspecifically to the μMACS HA magnetic beads. *N. benthamiana* leaves were infiltrated with *Agrobacterium* cells harboring plasmids coding for the effector HaRxL14:C-YFP or HA:Aro2 alone. Blots were probed with anti-HA and anti-GFP antibodies. HA:Aro2 binds to the μMACS HA magnetic beads and gets enriched whereas HaRxL14:C-YFP does not bind nonspecifically. Protein sizes on the left in kilo Daltons

9. Thermal Cycler.

10. QIAquick PCR Purification Kit (Qiagen, Cat. No. 28104).

11. QIAquick Gel Extraction Kit (Qiagen, Cat. No. 28704).

12. QIAprep Spin Miniprep Kit (Quiagen, Cat. No. 27104).

13. Zeocin (Invitrogen Cat. No. R25001).

14. SOC medium: Dissolve 20 g tryptone, 5 g yeast extract, 0.5 g NaCl, 2.5 mM KCl, 10 mM MgCl$_2$ in 900 ml of water, and adjust pH to 7.00 with NaOH. Autoclave at 121 °C for 15 min. After cooling the solution to room temperature add 100 ml of 200 mM Glucose and sterilize the final solution by passing it through a 0.2 µm filter.

15. Luria-Bertani (LB) low-salt medium: Dissolve 10 g Bacto Tryptone (Difco), 5 g yeast extract (Difco), 5 g NaCl in 1 L water, and adjust pH to 7.00 with NaOH. For plates add 15 g bacteriological agar (Sigma-Aldrich) per L before autoclaving. Autoclave at 121 °C for 15 min.

16. Nanodrop (Thermo Scientific).

17. 80 % (v/v) autoclaved glycerol.

2.2 Cloning into Destination Vector

1. LR Clonase II (Invitrogen Cat. No. 11791020).

2. Based on the selected cloning strategy, choose the appropriate destination vector (*see* **Note 3**).

3. Luria-Bertani (LB) medium: Dissolve 10 g Bacto Tryptone (Difco), 5 g yeast extract (Difco), 10 g NaCl in 1 L water, and adjust pH to 7.00 with NaOH. For plates add 15 g bacteriological agar (Sigma-Aldrich) per L before autoclaving. Autoclave at 121 °C for 15 min.

2.3 Preparation of Agrobacterium Competent Cells

1. *Agrobacterium tumefaciens* strain GV3101 or LB4404.

2. Incubator at 28 °C.

3. Shaking Incubator at 28 °C.

4. Spectrophotometer.

5. BD Falcon 50 ml conical tubes (Falcon).

6. YEB medium: Dissolve 5 g beef extract (Sigma), 1 g yeast extract (Difco), 5 g peptone (Sigma), 5 g sucrose, and 2 mM MgCl$_2$ in 1 L of water, and adjust the pH to 7.00 with NaOH. For plates add 15 g bacteriological agar (Sigma-Aldrich) per liter before autoclaving. Autoclave at 121 °C for 15 min.

2.4 Transient Expression in Nicotiana benthamiana

1. *N. benthamiana* plants grown at a 22/20 °C day/night temperature cycle and 16-h light/8-h-dark cycle in a growth chamber or in the glass house for 4–5 weeks.

2. Infiltration medium: 10 mM MgCl$_2$, 10 mM MES pH 5.7, 100 mM Acetosyringone (in DMSO Sigma Aldrich Cat. No. D134406).

3. Large plant propagator (Stewart, $35 \times 37 \times 21$ cm).

4. 50 ml conical screw cap tubes (Falcon).

2.5 Protein Complex Immunoprecipitation

1. Mortar and pestle.

2. Lysis Buffer: 50 mM Tris/HCl pH 8.0, 150 mM NaCl, 10 % Glycerol, 0.1–1 % Triton X-100 or Igepal CA-630 (NP-40) (*see* **Note 4**).

3. Protease inhibitor tablets: Roche complete ULTRA Tablets, Mini, EDTA-free (Roche Applied Science Cat. No. 05892791001) (1 tablet/10 ml).

4. 1.5 ml and 2 ml reaction tubes (Eppendorf).

5. 100 % Acetone (–20 °C).

6. μMACS HA Isolation Kit for 40 isolations (Miltenyi Biotec Cat. No. 130-091-122), μMACS GFP Isolation Kit (Miltenyi Biotec Cat. No. 130-091-125), or μMACS Myc Isolation Kit (Miltenyi Biotec Cat. No. 130-091-123).

7. μMACS Separator (Miltenyi Biotec Cat. No. 130-042-602).

8. μ Columns (Miltenyi Biotec Cat. No. 130-042-701).

9. MACS MultiStand (Miltenyi Biotec Cat. No. 130-042-303).

3 Methods

3.1 Primer Design

For a successful Gateway® reaction, recombination sites (AttB1 and AttB2) have to be introduced on both sides on the target sequence of interest (*see* **Note 2**). We add these sites to our PCR primer pairs. The strategy here is to fuse only AttB1/2 mini-sites to the primer to keep the primer costs low. In a two-step PCR the full-length AttB1/2 recombination sites are then added.

3.1.1 Forward Primer Design

1. Design gene-specific primers of 20–24 bp starting from the ATG start codon. If possible, the 3′-end of the primer should end with a G or C. For N-terminal fusion constructs the ATG start codon can be omitted and the recombination mini-site (AttB1 **AAAAAGCAGGCT**<u>**CC**</u>) is fused to the gene-specific primer. This results in a sequence: 5′-**AAAAAGCAGGCT**<u>**CC**</u>**NNNNNNNNNNNNNNNNNNNN**-3′ (N is the gene-specific sequence and the underlined bases are added to keep the construct in frame with the fusion construct).

2. For C-terminal fusion to the target sequence or for untagged constructs the ATG is maintained and a plant-specific Kozak sequence <u>CC</u>ACC is added [25]. The resulting primer is then:

   5′-**AAAAAGCAGGCT**<u>**CC**</u>**ACCATGNNNNNNNNNN NNNNNNNNNN**-3′.

3.1.2 Reverse Primer Design

1. In keeping with the forward primer design we choose a gene-specific primer sequence around 20–24 bp in length. For C-terminal fusion constructs we omit the STOP codon, for N-terminal fusion constructs the STOP codon is maintained. Note that an additional base needs to be included (underlined) for C-terminal fusion constructs; otherwise, a frame shift will be introduced. The resulting primer is then: 5′-AGAAAGCTG GGT<u>C</u>NNNNNNNNNNNNNNNNNNNNNN-3′ (N reverse complement sequence of the gene of interest).

3.1.3 Two-Step PCR and Cloning into pDONR™ Vector

In a two-step PCR the target sequence is initially amplified using the gene-specific primers with added AttB1/2 mini-sites. In the second step the full-length AttB1 and AttB2 primers are used to introduce the full AttB1/2-sites to the construct.

1. For a 20 μl PCR pipet the following components:

   10 μl of KOD Hot Start Master Mix (Merck Biosciences).

   0.6 μl of 10 μM gene-specific forward primer containing AttB1 mini-sites (final concentration 300 nM).

   0.6 μl of 10 μM gene-specific reverse primer containing AttB2 mini-sites (final concentration 300 nM).

   1 μl of template cDNA (50–500 ng/μl).

   7.8 μl of sterile autoclaved water.

2. Mix reaction well and spin down in a bench-top centrifuge. Run the PCR using the following conditions:

   95 °C for 2 min.

   10 cycles: 95 °C for 30 s (Denaturation step).

   55 °C for 45 s (Primer annealing step, temperature should be determined experimentally).

   70 °C for 20 s/kb (Elongation step).

   Final elongation step at 70 °C for 2–5 min.

   Hold at 10 °C.

   Remove unbound primers from reaction by using a QIAquick PCR Purification Kit and elute with 30 μl Elution Buffer (EB). Pipet EB directly onto the columns and incubate for 5 min to increase elution efficiency.

3. For the second PCR step using the full-length AttB1_FWD and AttB2_REV primer set up a 40 μl PCR reaction with following components:

   20 μl of KOD Hot Start Master Mix (Merck Biosciences).

   1.2 μl of 10 μM AttB1_FWD primer (final concentration 300 nM).

   1.2 μl of 10 μM AttB2_REV (final concentration 300 nM).

4 µl of purified PCR product from **step 3**.

13.6 µl of sterile autoclaved water.

4. Mix reaction well and spin down in a bench-top centrifuge. Run PCR using the following conditions:

95 °C for 2 min.

20 cycles: 95 °C for 30 s.

55 °C for 45 s.

70 °C for 20 s/kb.

Final elongation step at 70 °C for 2–5 min.

Hold at 10 °C.

5. Separate the PCR product on a TAE (Tris–acetate–EDTA) agarose gel and excise the appropriate band with a scalpel by placing it close to a UV light source. Be as quick as possible as long exposure to UV light damages DNA.

6. Extract DNA from gel slice using a QIAquick Gel Extraction Kit according to the manufacturer's instructions. Pipet 20–30 µl EB directly onto the columns and incubate for 5 min to increase elution efficiency.

7. Determine the DNA concentration by separating 1–2 µl on a TAE agarose gel and compare it to a DNA ladder or use a Nanodrop to determine DNA concentration.

3.2 BP and LR Reactions

1. Perform BP reaction according to manufacturer's instructions. We use the pDONR™Zeo Vector since none of the Gateway® Destination vectors harbors the Zeocin resistance cassette and therefore no linearization of the pDONR vector is necessary before the LR reaction. Incubate the reaction at least for 1 h at 25 °C, preferentially overnight.

2. Add 1 µl of the Proteinase K solution to each sample to stop the reaction. Vortex briefly. Incubate samples at 37 °C for 10 min.

3. Pipet 2 µl of BP reaction into 10 µl of One Shot® OmniMAX™ 2 T1 Phage-Resistant Cells.

Phage-Resistant Cells and perform transformation according to manufacturer's instructions.

4. Plate 20 µl and the rest of each transformation onto low-salt LB plates containing 25 µg/ml Zeocin.

5. Seal plates and incubate at 37 °C overnight.

6. Incubate one colony into 5 ml of low-salt LB medium containing 25 µg/ml Zeocin. Do this for at least three colonies and grow in a shaking incubator at 37 °C at 220 rpm overnight.

7. Prepare archival stocks by mixing 500 µl of bacterial culture with 500 µl 80 % (v/v) autoclaved glycerol in 2 ml reaction tubes. Mix well by vortexing and store at –80 °C.

8. Use the remainder of the liquid culture to isolate plasmids using QIAprep Spin Miniprep Kit according to manufacturer's instructions.

9. Determine the DNA concentration using the nanodrop.

10. Sequence plasmids using M13_FWD and M13_REV primers.

LR Reaction

1. Choose an appropriate Gateway destination vector (*see* **Note 3**) and the pDONR™ vector harboring the gene of interest.

2. Perform LR reaction according to manufacturer's instructions. Incubate the reaction at least for 1 h at 25 °C, preferentially overnight.

3. Add 1 µl of the Proteinase K solution to each sample to stop the reaction. Vortex briefly. Incubate samples at 37 °C for 10 min.

4. Pipet 2 µl of LR reaction into 10 µl of One Shot® OmniMAX™ 2 T1 Phage-Resistant Cells.
   Phage-Resistant Cells and perform transformation according to manufacturer's instructions.

5. Plate 20 µl and the rest of each transformation onto LB plates containing the appropriate antibiotic.

6. Seal plates and incubate at 37 °C overnight.

7. Incubate one colony into 5 ml of LB medium containing the appropriate antibiotic. Do this for at least three colonies and grow in a shaking incubator at 37 °C at 220 rpm overnight.

8. Prepare archival stocks by mixing 500 µl of bacterial culture with 500 µl 80 % (w/v) autoclaved glycerol in 2 ml reaction tubes. Mix well by vortexing and store at –80 °C.

9. Use the remainder of the liquid culture to isolate plasmids using a QIAprep Spin Miniprep Kit according to manufacturer's instructions.

10. Sequence plasmids using vector-specific primers to ensure that the constructs are correct and the epitope tag is in frame.

3.3 Preparation of Agobacterium Competent Cells

The preparation of competent *Agrobacterium* cells and the transformation procedure were adapted from [26].

1. Streak *Agrobacterium* strain on YEB Agar plates containing appropriate antibiotics and grow at 28 °C for 48 h.

2. Pick a single colony and inoculate 10 ml of YEB broth containing appropriate antibiotics. Grow bacteria at 28 °C in a shaking incubator at 220 rpm overnight.

3. Inoculate 200 ml of YEB broth containing appropriate antibiotics with the 10 ml overnight culture. Incubate flasks in orbital shaker for 2–6 h at 28 °C and 220 rpm till the culture reaches an OD_{600} of 0.6–0.8.

4. Chill the flask immediately on ice for 10 min and shake occasionally. Centrifuge cells for 20 min at $2,500 \times g$ at 4 °C.

5. Resuspend pellet in 50 ml ice cold TE-Buffer (10 mM Tris/HCl, 1 mM EDTA pH 8.0).

6. Centrifuge cells for 20 min at $2,500 \times g$ at 4 °C.

7. Resuspend pellet in 20 ml ice cold TE-Buffer (10 mM Tris/HCl, 1 mM EDTA pH 8.0).

8. Centrifuge cells for 20 min at $2,500 \times g$ at 4 °C.

9. Resuspend cells in 10 ml ice cold YEB medium and aliquot 100 µl in pre-cooled 1.5 ml reaction tubes. Snap freeze cells in liquid nitrogen. Larger volumes can be aliquoted if multiple transformations are planned at the same time.

10. Store cells at –80 °C (cells should be used within 6 months).

3.4 Agrobacterium Transformation

1. Thaw cells on ice.

2. Add 3–5 µg of plasmid DNA and mix cells by flicking the tube. In parallel prepare a control reaction by adding sterile water instead of DNA.

3. Incubate cells for 5 min on ice.

4. Snap freeze cells for 5 min in liquid nitrogen using a polystyrene float.

5. Heat shock cells at 37 °C for 5 min.

6. Chill cells for 2 min on ice and add 900 µl of YEB medium.

7. Incubate cells for 2–4 h at 28 °C on a shaker at 900 rpm.

8. Pellet cells by centrifugation in a bench-top centrifuge for 30 s and remove 900 µl of supernatant.

9. Resuspend the cells in the remaining liquid and plate 10 µl and the rest on YEB agar plates containing the appropriate antibiotics.

10. Grow plates for 48 h at 28 °C in an incubator. Colonies should have a diameter of around 1 mm. On the control plate no colonies should have grown.

11. Three to five colonies per construct should be streaked onto a new YEB agar plate containing the appropriate antibiotics.

12. Grow *Agrobacteria* for 24–48 h at 28 °C in an incubator.

13. Pick a small amount of *Agrobacteria* cells (no more than can fit in the very end of a standard 200 µl pipette tip) and resuspend in 100 µl of sterile water in a PCR tube.

14. Heat the resuspended cells for 5–10 min at 95 °C in a thermal cycler.

15. Spin cells >15,000×*g* in a microfuge to pellet debris and unlysed cells.

16. Use 1 μl of this in a 20 μl PCR with primers specific for the destination vector.

17. Run the PCR product on a TAE agarose gel and check for the expected size of the PCR product.

18. Grow clones that contain the right insert size at 28 °C in a shaking incubator in 5 ml liquid YEB medium containing the appropriate antibiotics at 220 rpm for 24 h.

19. Prepare archival stocks by mixing 500 μl of bacterial culture with 500 μl 80 % (w/v) autoclaved glycerol in 2 ml reaction tubes. Mix well by vortexing and store at –80 °C.

3.5 Transient Expression in Nicotiana benthamiana

This *Agrobacterium* transient expression protocol was adapted from [27, 28].

1. Grow *N. benthamiana* at a 22/20 °C day/night temperature cycle and 16-h light/8-h-dark cycle in a growth chamber or in the glass house. We generally infiltrate 4- to5-week-old plants that have six fully expanded leaves and that have not started to flower (Fig. 2).

2. Streak the untransformed *Agrobacteria* and the *Agrobacteria* harboring the constructs for bait, prey, and prey control from archival stocks on YEB agar plates containing appropriate

Fig. 2 Five-week-old *N. benthamiana* plant. *Arrows* indicate the leaves that should be infiltrated

antibiotics and grow them for 48 h at 28 °C in a plate incubator. Additionally, we grow *Agrobacteria* harboring a construct encoding the p19 protein of the tomato bushy stunt virus [29]. This is a silencing suppressor which improves the levels of transgene expression.

3. Inoculate one colony in 5 ml of YEB broth containing appropriate antibiotics and grow overnight in 50 ml conical screw cap tubes until stationary phase is reached. (OD_{600} should be between 2.5 and 3).

4. Spin 50 ml conical tubes for 20 min at $2,500 \times g$ at 20 °C and resuspend each pellet in 10 ml of infiltration medium without Acetosyringone.

5. Spin cells again for 20 min at $2,500 \times g$ and resuspend each pellet in 10 ml infiltration medium containing 100 μM Acetosyringone.

6. Adjust OD_{600} to 0.6 for *Agrobacteria* harboring constructs and no constructs with infiltration medium containing 100 μM Acetosyringone. The p19 carrying *Agrobacteria* are adjusted to OD_{600} 0.9.

7. Mix 3 ml of each strain 1:1:1 in the following combinations(Table 1):

   Incubate for 2–4 h in the dark. The final OD_{600} for each of the *Agrobacteria* harboring the epitope-tagged constructs is 0.2 and the p19 anti-silencing strain is at OD_{600} 0.3. In parallel, place *N. benthamiana* plants in a plant propagator (Stewart, $35 \times 37 \times 21$ cm) for 1–2 h under light to allow the stomata to open, which will greatly facilitate the infiltration.

8. Infiltrate the different construct combinations into the leaves as shown in Fig. 2. Using a 1 ml needleless syringe gently infiltrate the *Agrobacterium* suspension from the lower side of the leaf through the stomatal openings and support the syringe with a fingertip from the other side. Infiltrate the whole leaf. Use two leaves per construct on different plants.

9. Leave the plants for 36–48 h in the glasshouse or growth room.

Table 1

Example of required combinations of *Agrobacteria* harbouring constructs encoding for bait and prey proteins

1	Bait	Prey	p19
2	Bait	Prey_Control	p19
3	Empty *Agrobacteria*	Prey	p19
4	Bait	Empty *Agrobacteria*	p19

3.6 Protein Complex Immunoprecipitation (Co-IP)

1. Chill mortar and pestle on ice.

2. Add one protease inhibitor tablet to 10 ml lysis buffer and mix on shaker until completely dissolved. Approximately 5 ml of Lysis Buffer is needed for each leaf.

3. Harvest the two infiltrated *N. benthamiana* leaves used for each combination and determine the fresh weight. Roll one leaf and place it in a 15 ml conical tube and snap freeze it in liquid nitrogen. Store this sample at –80 °C as backup.

4. Place the remaining leaf in a mortar on ice.

5. Add 3 ml of lysis buffer per gram of plant material and homogenize using a mortar and pestle on ice.

6. Transfer 2 ml of the homogenate to a 2 ml reaction tube and store on ice.

7. Repeat **steps 3–5** for all construct combinations.

8. Spin all samples at $10,000 \times g$ for 10 min at 4 °C using a benchtop centrifuge.

9. Transfer 1.1 ml of the supernatant into a new pre-cooled 2 ml reaction tube.

10. Precipitate part of the lysate by pipetting 100 μl of the supernatant into 900 μl of cold (–20 °C) acetone in a new 2 ml reaction tube. A precipitate might be observed immediately. Store the samples for at least 1 h at –20 °C. This will be your IP-input sample.

11. Mix 50 μl of μMACS Anti-tag MicroBeads to 1 ml of lysate and mix by inverting the tube, avoid foaming. Incubate tubes for 30–60 min on ice.

12. Place four μColumns in the μMACS Separator and equilibrate columns by applying 200 μl of lysis buffer onto the column.

13. Apply 333 μl of the lysate containing the μMACS Anti-tag MicroBeads and let the lysate run through. Repeat this step three times until no sample is left. Avoid the formation of bubbles. Run all samples in parallel.

14. Wash the columns with 400 μl of lysis buffer.

15. Prepare 80 μl of Elution Buffer (SDS-Sample Buffer) for each column and incubate at 95 °C.

16. Wash the columns with 300 μl of lysis buffer.

17. Wash the columns twice with 200 μl of lysis buffer.

18. Finally wash the columns with 100 μl of 20 mM Tris–HCl (pH 7.5).

19. Apply 20 μl of preheated elution buffer onto each column and incubate for 5 min. We remove any droplets formed at the nozzle of the column.

20. Add 50 μl of Elution Buffer and collect flow through into a 1.5 ml reaction tube.

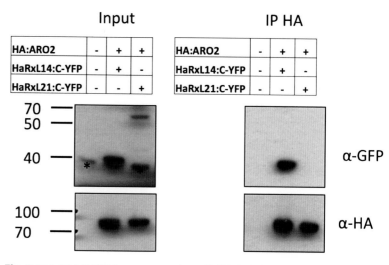

	Input				IP HA		
HA:ARO2	-	+	+	HA:ARO2	-	+	+
HaRxL14:C-YFP	-	+	-	HaRxL14:C-YFP	-	+	-
HaRxL21:C-YFP	-	-	+	HaRxL21:C-YFP	-	-	+

α-GFP

α-HA

Fig. 3 HaRxL14:C-YFP forms a complex with HA:Aro2 in *N. benthamiana*, but the negative control HaRxL21:C-YFP does not. HaRxL14:C-YFP or HaRxL21:C-YFP and Ha:Aro2 were transiently co-expressed in *N. benthamiana* and immunoprecipitated (IP) using μMACS HA magnetic beads. Input and IP samples were detected by Western blot using anti-HA and anti-GFP antibodies. Protein sizes on the left in kilo Daltons. *Asterisk* indicates an unspecific band

21. Store the eluted samples on ice.

22. Take the IP-Input samples from the –20 °C freezer and spin in a bench-top centrifuge at 21,000×*g* for 30 min at 4 °C.

23. Remove the supernatant and dry the pellets.

24. Add 100 μl of SDS-Sample Buffer to each tube and heat at 95 °C for 5 min.

25. Load 5 μl of each IP-Input sample and 5 μl of the eluted samples on two SDS-PAGE gels in duplicate and perform Western blot using antibodies against the epitope tags that were used (*see* **Note 5**). An example result is shown in Fig. 3. For SDS-PAGE preparation and protein transfer follow the protocol of [30].

4   Notes

1. The anti-GFP-HRP antibody (Miltenyi Biotec, Cat. No. 130-091-833) or the monoclonal anti-GFP antibody (Roche Applied Science Cat. No. 11814460001) recognizes the C-terminal half of the YFP protein. When BiFC constructs for the genes of interest are already available the vectors for the C-terminal half of the YFP protein can be used for transient expression in *N. benthamiana* and Co-IP.

2. In this section we will guide you through the process of cloning a cDNA into Gateway® pDONR™ vectors, which will then allow subsequent recombination into a wide range of destination vectors for the purpose of epitope tagging or observations of effector and target localization using fluorescence microscopy. We advise the reader to read additional literature about Gateway® cloning on the Invitrogen website (http://www.Invitrogen.com). cDNA clones from Arabidopsis gene transcripts can be obtained from the Arabidopsis Biological Resource Center (https://abrc.osu.edu) or from the TAIR website (http://www.arabidopsis.org). *Hpa* effector clones can be sourced from our laboratory. *P. syringae* effector clones can be cloned from isolated genomic DNA.

3. Since Gateway® cloning facilitates high-throughput cloning of target sequences, a wide choice of Gateway compatible expression vectors is available for a multitude of model organisms ranging from *Caenorhabditis elegans* to zebrafish [31, 32]. For plants Gateway® compatible binary vectors are available that allow *Agrobacterium*-mediated plant transformation for transient expression in *Nicotiana benthiamiana*, *Nicotiana tabacum* and the stable transformation of monocots and dicots [22, 33, 34] http://www.psb.ugent.be/gateway/.

4. A prerequisite for a successful Co-IP is to maintain the interacting proteins in their native conformation, which depends on an appropriate lysis buffer that maintains a slightly alkaline pH (7.4–8), inhibits protease activity, releases proteins from their compartments or membranes, and prevents unspecific ionic interactions. Widely used IP lysis buffers contain Tris/HCl buffer to maintain the pH. Non-ionic detergents such as Triton X-100 or Igepal CA-630 (formerly known as NP-40) in a range of 0.1–1 % are used to lyse organelles and release protein complexes from membranes. When extracting soluble, cytoplasmic protein complexes, detergent-free buffers can be used as mechanical disruption is sufficient. Low-salt extraction buffers with no or low concentrations (25–50 mM) of NaCl are used when weak interactions are expected. In standard buffers a NaCl concentration of 150 mM is used. Commercially available protease inhibitor cocktails are added to inhibit proteases released from organelles during the lysis process.

5. We use the antibodies at the following dilutions:

Anti-GFP-HRP antibody (PBS 0.1 % Tween20, 1 % BSA) Dilution 1:5,000 (Miltenyi Biotech).

Anti-HA antibody (TBS 0.1 % Tween20, 1 % BSA) Dilution 1:2,000 (Roche Applied Science).

Anti-Myc-HRP antibody (PBS 0.1 % Tween20, 1 % BSA) Dilution 1:7,500 (Miltenyi Biotech).

Acknowledgments

DFAT and JS are funded by the UK Biotechnology and Biological Sciences Research Council grant BB/G015066/1 to JLB. The authors thank Mrs. Rachel Clewes and Mrs. Christina Payne for their technical support. The authors thank Dr. François Parcy (University Grenoble, France) for the BIFC vectors pBIFP1-4.

References

1. Boch J, Bonas U (2010) Xanthomonas AvrBs3 family-type III effectors: discovery and function. Annu Rev Phytopathol 48:419–436

2. Kim JG et al (2008) XopD SUMO protease affects host transcription, promotes pathogen growth, and delays symptom development in xanthomonas-infected tomato leaves. Plant Cell 20:1915–1929

3. Shao F et al (2003) Cleavage of Arabidopsis PBS1 by a bacterial type III effector. Science 301:1230–1233

4. Nomura K et al (2006) A bacterial virulence protein suppresses host innate immunity to cause plant disease. Science 313:220–223

5. Nomura K et al (2011) Effector-triggered immunity blocks pathogen degradation of an immunity-associated vesicle traffic regulator in Arabidopsis. Proc Natl Acad Sci USA 108: 10774–10779

6. Lee AHY et al (2012) A bacterial acetyltransferase destroys plant microtubule networks and blocks secretion. PLoS Pathog 8:e1002523

7. Abramovitch RB et al (2006) Type III effector AvrPtoB requires intrinsic E3 ubiquitin ligase activity to suppress plant cell death and immunity. Proc Natl Acad Sci USA 103:2851–2856

8. Gimenez-Ibanez S et al (2009) AvrPtoB targets the LysM receptor kinase CERK1 to promote bacterial virulence on plants. Curr Biol 19:423–429

9. van Damme M et al (2012) The Irish potato famine pathogen Phytophthora infestans translocates the CRN8 kinase into host plant cells. PLoS Pathog 8:e1002875

10. Bos J et al (2010) Phytophthora infestans effector AVR3a is essential for virulence and manipulates plant immunity by stabilizing host E3 ligase CMPG1. Proc Natl Acad Sci USA 107:9909–9914

11. Bozkurt TO et al (2011) Phytophthora infestans effector AVRblb2 prevents secretion of a plant immune protease at the haustorial interface. Proc Natl Acad Sci USA 108: 20832–20837

12. Saunders DGO et al (2012) Host protein BSL1 associates with Phytophthora infestans RXLR effector AVR2 and the Solanum demissum Immune receptor R2 to mediate disease resistance. Plant Cell 24:3420–3434

13. Mukhtar MS et al (2011) Independently evolved virulence effectors converge onto hubs in a plant immune system network. Science 333:596–601

14. Kerppola TK (2008) Bimolecular fluorescence complementation (BiFC) analysis as a probe of protein interactions in living cells. Annu Rev Biophys 37:465–487

15. Gebert M, Dresselhaus T, Sprunck S (2008) F-actin organization and pollen tube tip growth in Arabidopsis are dependent on the gametophyte-specific Armadillo repeat protein ARO1. Plant Cell 20:2798–2814

16. Fabro G et al (2011) Multiple candidate effectors from the oomycete pathogen Hyaloperonospora arabidopsidis suppress host plant immunity. PLoS Pathog 7:e1002348

17. Braun P, Gingras A-C (2012) History of protein-protein interactions: from egg-white to complex networks. Proteomics 12: 1478–1498

18. Field J et al (1988) Purification of a RAS-responsive adenylyl cyclase complex from Saccharomyces cerevisiae by use of an epitope addition method. Mol Cell Biol 8:2159–2165

19. Evan GI et al (1985) Isolation of monoclonal antibodies specific for human c-myc proto-oncogene product. Mol Cell Biol 5: 3610–3616

20. Brizzard B (2008) Epitope tagging. Biotechniques 44:693–695

21. Nakagawa T et al (2007) Development of series of gateway binary vectors, pGWBs, for realizing efficient construction of fusion genes for plant transformation. J Biosci Bioeng 104:34–41

22. Earley KW et al (2006) Gateway-compatible vectors for plant functional genomics and proteomics. Plant J 45:616–629

23. Caillaud M-C et al (2012) Subcellular localization of the Hpa RxLR effector repertoire identifies a tonoplast-associated protein HaRxL17 that confers enhanced plant susceptibility. Plant J 69:252–265

24. Nelson BK, Cai X, Nebenführ A (2007) A multicolored set of in vivo organelle markers for co-localization studies in Arabidopsis and other plants. Plant J 51:1126–1136

25. Rangan L, Vogel C, Srivastava A (2008) Analysis of context sequence surrounding translation initiation site from complete genome of model plants. Mol Biotechnol 39:207–213

26. Weigel D, Glazebrook J (2006) Transformation of *Agrobacterium* using the freeze-thaw method. CSH Protoc 2006

27. Latijnhouwers M et al (2005) An Arabidopsis GRIP domain protein locates to the trans-Golgi and binds the small GTPase ARL1. Plant J 44:459–470

28. Boevink PC, Birch PRJ, Whisson SC (2011) Imaging fluorescently tagged Phytophthora effector proteins inside infected plant tissue. Methods Mol Biol 712:195–209

29. Voinnet O et al (2003) An enhanced transient expression system in plants based on suppression of gene silencing by the p19 protein of tomato bushy stunt virus. Plant J 33:949–956

30. Win J, Kamoun S, Jones AME (2011) Purification of effector-target protein complexes via transient expression in Nicotiana benthamiana. Methods Mol Biol 712:181–194

31. Zeiser E et al (2011) MosSCI and gateway compatible plasmid toolkit for constitutive and inducible expression of transgenes in the C. elegans germline. PLoS ONE 6:e20082

32. Villefranc JA, Amigo J, Lawson ND (2007) Gateway compatible vectors for analysis of gene function in the zebrafish. Dev Dyn 236:3077–3087

33. Karimi M, Inzé D, Depicker A (2002) GATEWAY vectors for *Agrobacterium*-mediated plant transformation. Trends Plant Sci 7:193–195

34. Nakagawa T et al (2007) Improved Gateway binary vectors: high-performance vectors for creation of fusion constructs in transgenic analysis of plants. Biosci Biotechnol Biochem 71:2095–2100

Chapter 17

An *Arabidopsis* and Tomato Mesophyll Protoplast System for Fast Identification of Early MAMP-Triggered Immunity-Suppressing Effectors

Malou Fraiture, Xiangzi Zheng, and Frédéric Brunner

Abstract

Transient expression in plant mesophyll protoplasts allows rapid characterisation of gene functions in vivo in a simplified and synchronized manner without bias due to the use of bacteria-based gene or protein delivery systems. It offers the possibility to test whether microbial effectors can subvert early events of plant immune signaling that are activated upon recognition of Microbe-Associated Molecular Patterns (MAMPs), the so-called MAMP-triggered immunity (MTI). Here, we describe the isolation and transfection with effector genes of *Arabidopsis thaliana* and *Solanum lycopersicum* mesophyll protoplasts, the use of a non-invasive luciferase reporter assay and a simple method to detect activated Mitogen-Activated Protein Kinases (MAPKs) to identify and study, in a medium-throughput manner, new effectors suppressing early signal transduction events of MTI.

Key words *Arabidopsis thaliana*, *Solanum lycopersicum*, Tomato, Mesophyll protoplast, Transient expression, Reporter assay, MAPK, MTI suppression, Effector

1 Introduction

Plants are constantly exposed to a multitude of pathogenic microorganisms, including bacteria, fungi and oomycetes. Many microbes that successfully invade plant tissue deploy an arsenal of effector proteins that they introduce into host cells to manipulate cellular activities to their benefit. A function of many effectors is the inhibition of the immune response that is initiated upon detection of invariant microbial structures, named Microbe-Associated Molecular Patterns (MAMPs), by corresponding Pattern-Recognition Receptors (PRRs) localized on the plant cell surface [1]. A prominent example for interference with MAMP-triggered immunity (MTI) is the blocking of signal transduction at the FLS2 receptor complex upon binding of bacterial flagellin by the *Pseudomonas syringae* effector AvrPto [2–6]. Analysis of effector

Paul Birch et al. (eds.), *Plant-Pathogen Interactions: Methods and Protocols*, Methods in Molecular Biology, vol. 1127, DOI 10.1007/978-1-62703-986-4_17, © Springer Science+Business Media New York 2014

function *in planta* is often performed upon protein delivery by the bacterial type III secretion system or gene transfer using *Agrobacterium* T-DNA. These techniques suffer experimental limitations, especially when the aim is to study the influence of effectors on early events recruiting Mitogen-Activated Protein Kinase (MAPK) cascades in MAMP-triggered immune responses. Transient expression of effectors in protoplasts offers a valuable alternative for studying their effect on MTI signaling, as the system does not require the use of microorganisms for effector delivery. Moreover, it permits the generation of highly synchronized responses and allows epistatic analysis. The Sheen laboratory extensively described the method of *Arabidopsis* protoplast isolation and transfection and made a major contribution to the development of protoplast-based reporter assays, monitoring signaling pathways of both inherent and adaptive processes that regulate growth, development and responses to external stimuli [7–9]. The methods and techniques in *Arabidopsis* protoplasts were also transferred to other plant species such as tomato; a demanding task usually requiring a specific experimental set-up [10].

Here, we describe how we have modified the existing protocols in *Arabidopsis* and tomato and adjusted them for use in a medium-throughput screen for MTI-suppressing functions of effectors from both prokaryotic and eukaryotic microorganisms. To set up this screen, candidate effector genes from the pathogen of interest are cloned without their signal peptide sequence into a plant vector harboring a CaMW 35S promoter for constitutive expression (*p35S-effector*). Most suitable are non-binary Gateway-based vectors as they permit rapid cloning by recombination and guarantee high protoplast transfection efficiency due to their small size. Effector plasmids are co-transfected into protoplasts with a *pFRK1-Luc* construct in which the firefly luciferase gene (*Luc*) is controlled by the MAMP-inducible promoter of *Arabidopsis FRK1* [5, 11]. This reporter construct is functional in both the *Arabidopsis* and the tomato protoplast system. Transfected protoplasts are then challenged with a genuine MAMP, such as flg22, a 22 amino acid immunogenic peptide derived from flagellin [3], and measurement of Luc activity serves as a read-out for the initiation of early MTI. The absence or significant inhibition of Luc activity induction compared to a Green Fluorescent Protein (GFP) control indicates that an effector acts as a suppressor of early immune signaling. A β-glucuronidase (GUS) activity assay, reflecting the constitutive expression of concomitantly transfected *pUBQ10*-GUS [5], allows the normalization of Luc activity and serves as an indicator for successful transfection. With our experimental set-up, 15 effectors can be tested simultaneously within less than 24 h. A significant improvement to measuring reporter gene activity is that we propose a non-invasive approach to monitor Luc activity in situ instead of collecting lysate from frozen cells. This method permits the recording of Luc activity kinetics over several hours. Figure 1

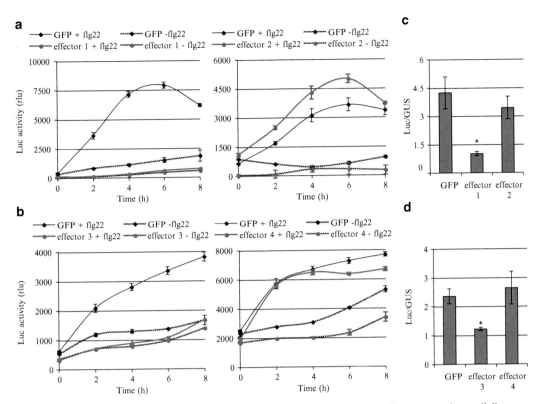

Fig. 1 Reporter assay to investigate the role of effectors in MTI inhibition. (**a**, **b**) Time-course Luc activity assay in *Arabidopsis* (**a**) and tomato (**b**) protoplasts. In these four independent representative experiments, effectors named 1–4 were tested for their ability to suppress the flg22-mediated expression of *Luc* under the control of the *FRK1* promoter. Protoplasts were co-transfected with *pFRK1-Luc*, *pUBQ10-GUS* and *p35S-effector* or *p35S-GFP* (*GFP* control) plasmids and incubated for 6 to 12 h. Cells were then challenged with flg22 or left untreated and Luc activity was monitored every 2 h for 8 h. *Error bars* represent SD from three technical replicates. rlu; relative light units. Effectors 1 and 3 strongly block Luc activity indicating that they function as suppressors of MTI. Effectors 2 and 4 have no influence on Luc activity induction, as the kinetics is similar to the GFP control. (**c**, **d**) Normalized Luc activity assay. The graphs depict the mean from three independent experiments 6 h after flg22 challenge in *Arabidopsis* (**c**) and 4 h after flg22 challenge in tomato (**d**) protoplasts expressing effector genes or *GFP*. Results are represented as ratio between the flg22-treated and non-treated sample and are normalized to the corresponding data sets from the GUS activity assay (Luc/GUS). *Error bars* show SEM. Statistical analysis was performed with the non-parametric Mann–Whitney test comparing effector gene-expressing samples to the *GFP* control. Data sets with $p < 0.05$ are marked by an *asterisk*. This representation of the reporter assay data highlights the strong and reproducible suppressing effects of effectors 1 and 3 in *Arabidopsis* and tomato, respectively

shows representative results of *pFRK1-Luc* activity upon flg22 treatment in the presence of either a putative early MTI-suppressing effector or an effector having no influence on reporter gene induction in both *Arabidopsis* and tomato protoplasts.

When an early MTI-suppressor is identified, the next step toward its functional characterisation is to investigate whether it blocks MAPK signaling triggered upon MAMP recognition either by acting upstream of the MAPK cascade or by directly targeting it. We describe an immunoblotting assay with *Arabidopsis* and

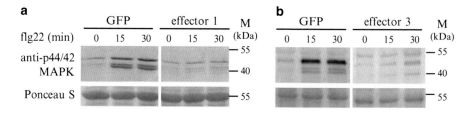

Fig. 2 MAPK activation in the presence of effectors. (**a**, **b**) Immunoblotting of phosphorylated MAPKs with anti-p44/42 MAPK antibody in *Arabidopsis* (**a**) and tomato (**b**) protoplasts transfected with either *GFP*, effector genes 1 or 3. Samples were collected 0, 15 or 30 min after flg22 treatment. Ponceau S Red staining served as a loading control. M; PAGE-Ruler. While MAPKs are phosphorylated upon flg22 challenge in the GFP control, effectors 1 and 3 strongly block their activation in *Arabidopsis* and tomato, respectively

tomato protoplasts using an anti-p44/42 MAPK antibody to visualize activated phosphorylated MAPKs upon MAMP challenge. This assay is easier and faster than a previously described in vitro kinase assay requiring immunoprecipitation of MAPKs and the use of radio-labeled (γ-^{32}P) ATP [7]. Figure 2 depicts representative immunoblots from both *Arabidopsis* and tomato protoplasts, showing the phosphorylation of MAPKs in *GFP*-expressing samples and, in comparison, a total inhibition of MAPK activation in the presence of an early MTI-suppressing effector.

It is important to note that the protoplast system presents some drawbacks. When effector genes are overexpressed, cells may collapse or ectopic effects may block Luc activity in a non-specific way. When a candidate effector displays *pFRK1-Luc* suppression, it is important to verify inhibition of endogenous *FRK1* up-regulation and other MAMP-induced genes by quantitative real-time PCR (qRT-PCR) to be sure that the immune signaling pathway is specifically targeted [7]. It is also worth noting that the protoplast system is not suitable for measuring all typical immune responses; for instance, cell wall-associated defense responses cannot be used as a read-out.

In summary, we have established a simple, fast and efficient protoplast-based system to pinpoint effectors subverting the earliest stages of the immune response in plants. Our system is suitable for the study of effectors from a wide range of pathogens, not only those that naturally infect *Arabidopsis* and/or tomato but also those with a different host range. For the latter, the use of an experimental model may reveal the repertoire of effectors that target ubiquitous components or cellular hubs of the plant immune system.

2 Materials

2.1 Plant Material

1. 4- to 5-week-old *Arabidopsis thaliana* plants (e.g. Col-0 ecotype).

2. 3- to 4-week-old *Solanum lycopersicum* plants (e.g. Moneymaker cultivar).

2.2 Stock Solutions for Protoplast Isolation

All solutions used for protoplasts should be prepared with Milli-Q-filtered H_2O (Millipore water purification systems). Stock solutions are stored at room temperature unless specified differently.

Stock solutions for the preparation of *Arabidopsis* protoplast enzyme solution, MMg, PEG, WI and W5 [9]:

1. 1 M $CaCl_2$.

2. 1 M $Ca(NO_3)_2$.

3. 0.1 M KCl.

4. 0.8 M sterile-filtered mannitol (*see* **Note 1**).

5. 0.2 M MES pH 5.7 (stored at –20 °C).

6. 0.15 M $MgCl_2$.

7. 5 M NaCl.

Stock solutions for the preparation of tomato protoplast enzyme solution [10, 12]:

1. Macro-stock: 1.5 g $NaH_2PO_4 \times H_2O$, 9.0 g $CaCl_2 \times 2\ H_2O$, 25 g KNO_3, 2.5 g NH_4NO_3, 1.34 g $(NH_4)_2SO_4$, 2.5 g $MgSO_4 \times 7\ H_2O$ for 1 l; autoclave for storage.

2. Micro-stock: 75 mg KI, 300 mg H_3BO_3, 1 g $MnSO_4 \times 7\ H_2O$, 200 mg $ZnSO_4 \times 7\ H_2O$, 25 mg $Na_2MoO_4 \times 2\ H_2O$, 2.5 mg $CuSO_4 \times 5\ H_2O$, 2.5 mg $CoCl_2 \times 6\ H_2O$ for 100 ml; filter sterilize and freeze at –20 °C.

3. Vitamin stock: 100 mg nicotinic acid, 100 mg pyridoxine–HCl, 1 g thiamine–HCl for 100 ml; filter sterilize and freeze at –20 °C.

4. FeNa-EDTA stock: 1 % (w/v) ethylenediaminetetraacetic acid (EDTA) ferric sodium salt (stored at 4 °C).

2.3 Arabidopsis Protoplast Isolation

1. Freshly prepared enzyme solution: 20 mM KCl, 0.4 M mannitol, 20 mM MES pH 5.7, 1.5 % (w/v) cellulase "Onozuka" R10 (Yakult Pharmaceutical Ind. Co., Ltd., Japan), 0.4 % (w/v) macerozyme R10 (Yakult Pharmaceutical Ind. Co., Ltd., Japan), 10 mM $CaCl_2$, 0.1 % (w/v) bovine serum albumin (BSA).

2. *Arabidopsis* W5 solution (may be stored at –20 °C): 125 mM $CaCl_2$, 5 mM KCl, 2 mM MES pH 5.7, 154 mM NaCl.

3. Nylon mesh (75 μm mesh size).

4. Desiccator.

5. Cellulose acetate membrane filter, 0.45 μm pore size.

6. 10 or 20 ml syringe.

7. Petri dish.

8. Forceps with bent flat tips.

9. Razor blades.

10. White sheet of paper.

11. Small funnel.

12. Bench-top centrifuge.

13. Haemocytometer (e.g. improved Neubauer chamber with a depth of 0.1 mm).

14. Light microscope.

15. 12 ml round-bottom polystyrene or polypropylene tubes with screw cap (e.g from Greiner, Germany).

2.4 Tomato Protoplast Isolation

1. K3 solution with 0.4 M sucrose: 10 ml macro-stock I, 0.1 ml micro-stock II, 0.1 ml vitamin stock, 0.5 ml FeNa-EDTA stock, 10 mg myo-inositol, 25 mg D-xylose, 13.7 g sucrose for 100 ml; adjust pH to 5.7 with 1 M KOH, filter sterilize and store at −20 °C.

2. Enzymes: 2 % (w/v) cellulose "Onozuka" R10 (Yakult Pharmaceutical Ind. Co., Ltd., Japan), 0.4 % (v/v) pectinase (Sigma).

3. Freshly prepared tomato W5 solution: 18.4 g $CaCl_2 \times 2\ H_2O$, 1 g glucose, 0.4 g KCl, 9 g NaCl for 1 l; adjust pH to 5.7.

4. Grated quartz.

5. Nylon mesh (100 μm mesh size).

6. 50 ml conical polystyrene or polypropylene tube with screw cap (e.g. Falcon).

7. **Items 5–15** listed in Subheading 2.3.

2.5 PEG-Mediated Protoplast Transfection

1. W5 solution (*see* Subheading 2.3 and 2.4 for *Arabidopsis* and tomato, respectively).

2. Freshly prepared MMg solution: 0.4 M mannitol, 4 mM MES pH 5.7, 15 mM $MgCl_2$.

3. Freshly prepared polyethylene glycol (PEG) solution: 0.1 M $CaCl_2$ (for *Arabidopsis*) or 0.1 M $Ca(NO_3)_2$ (for tomato), 0.2 M mannitol, 40 % (w/v) PEG4000 (Sigma-Aldrich, formerly Fluka) (*see* **Note 2**).

4. Freshly prepared WI solution: 20 mM KCl, 0.5 M mannitol, 4 mM MES pH 5.7.

5. Highly pure plasmid DNA (stored at −20 °C) (*see* **Note 3**): *pFRK1-Luc* [5], *pUBQ10-GUS* [5], *p35S-GFP* (Gateway-based *p2FGW7*; VIB, University of Ghent, Belgium) and *p35S*-driven effector gene constructs (Gateway-based *p2GW7* backbone; VIB, University of Ghent, Belgium). Effector genes must be devoid of their signal peptide sequence and can be with or without an N- or C-terminal fusion/tag such as *GFP*, *HA*, *myc* or *flag* (*see* **Note 4**).

6. 12 ml round-bottom polystyrene tubes with screw cap (Greiner, Germany).

7. Bench-top centrifuge.

2.6 Luciferase (Luc) Activity Assay

1. 20 mM D-luciferin acid free (stored at –20 °C) (*see* **Note 5**).

2. 50 μM flg22 peptide (stored at –20 °C) (*see* **Note 6**).

3. White flat-bottom 96-well-plate with lid.

4. Luminometer.

2.7 β-Glucuronidase (GUS) Activity Assay

1. 5× cell lysis buffer (stored at –20 °C): 125 mM Tris–HCl pH 7.8, 10 mM dithiothreitol (DTT), 10 mM 1,2-diaminocyclohexanetetraacetic acid (DCTA), 50 % (v/v) glycerol, 5 % (v/v) Triton X-100.

2. MUG solution (stored at –20 °C): 1 mM 4-methylumbelliferyl-β-D-glucuronide, 10 mM Tris–HCl pH 8.0, 2 mM $MgCl_2$.

3. Stop solution: 0.2 M Na_2CO_3.

4. 50 μM flg22 peptide (stored at –20 °C) (*see* **Note 6**).

5. 1 ml syringe with cannula.

6. 1.5 ml safe-lock tubes.

7. Liquid N_2.

8. Bench-top centrifuge.

9. Vortex.

10. White flat-bottom 96-well-plate with lid.

11. Fluorimeter with appropriate filter.

2.8 Cell Death Staining

1. 100 μg/ml propidium iodide.

2. 1.5 ml tubes.

3. Fluorescence microscope with appropriate filter.

2.9 Immunoblotting of Activated MAPKs

For preparation of material and solutions for polyacrylamide electrophoresis and protein blotting, refer to the relevant guides from Bio-Rad. Buffers can be prepared with demineralised H_2O and are stored at room temperature unless specified differently.

1. 50 μM flg22 peptide (stored at –20 °C) (*see* **Note 6**).

2. 1 ml syringe with cannula.

3. 1.5 ml safe-lock tubes.

4. Liquid N_2.

5. Bench-top centrifuge.

6. Vortex.

7. Heating block.

8. 13.5 % SDS mini-gel (0.1 cm thick, 10 or 15 slots) for SDS-PAGE.

9. Mini-Protean (Bio-Rad) vertical electrophoresis system.

10. Mini Trans-Blot (Bio-Rad) wet transfer system.

11. Nitrocellulose membrane.

12. Tris-glycine buffers for electrophoresis and transfer.

13. Laemmli protein loading buffer (stored at –20 °C).

14. PAGE-Ruler (Fermentas), prestained protein ladder (stored at –20 °C).

15. Ponceau S Red staining solution.

16. TBS-T: Tris Buffered Saline containing 0.1 % Tween 20.

17. Freshly prepared blocking solution: 5 % skimmed milk in TBS-T.

18. Primary antibody solution: anti-p44/42 MAPK (human Erk 1/2) antibody produced in rabbit (Cell Signaling Technology), diluted 1,000× in 5 % BSA TBS-T (fresh or stored at –20 °C for reuse).

19. Secondary antibody solution: anti-rabbit IgG produced in goat and coupled to alkaline phosphatase (AP) (Sigma), diluted 3,000× in TBS-T (fresh or stored at –20 °C for reuse).

20. AP buffer and reagents for colorimetric detection of AP activity.

3   Methods

3.1 Plant Growth Conditions

1. Cultivate *Arabidopsis thaliana* plants in a phytochamber or greenhouse under stable climate conditions: 8 h light at 22–24 °C/16 h dark at 20 °C, 40–60 % humidity, ~120 μE/m^2/s light intensity (*see* **Note** 7). Grow them on soil composed of a 3.5:1 mixture of GS/90 (Patzer, Germany) and vermiculite.

2. Keep *Solanum lycopersicum* in a phytochamber or greenhouse under controlled growth conditions: 16 h light at 24 °C/8 h dark at 22 °C, 40–45 % humidity, ~200 μE/m^2/s light intensity (*see* **Note** 7). Grow them on soil containing a 4.6:4.6:1 mixture of type P soil, type T soil (Patzer, Germany) and sand.

3.2 Arabidopsis Protoplast Isolation

1. To prepare enzyme solution, mix all the ingredients except CaCl$_2$ and BSA and heat the solution at 55 °C for 10 min (*see* **Note 8**). Chill the solution on ice and add the remaining components.

2. Syringe-filtrate the solution through a 0.45 μm filter into a Petri dish of adequate size.

3. Cut well-expanded leaves from 4- to 5- week-old plants and place them individually on a white sheet of paper. Cut only a few at a time.

4. Remove tip and stalk from the leaf and carefully cut the middle part into about 0.5 mm thick strips using a sharp razor blade (*see* **Note 9**).

5. Immediately submerge the leaf strips in the enzyme solution by dipping them with flat-tip forceps.

6. Vacuum-infiltrate leaf strips for 30 min in the dark using a desiccator covered with a black cloth.

7. Continue the digestion, without shaking, in the dark at room temperature for 3 h (*see* **Note 10**).

8. Release protoplasts by gentle swirling of the Petri dish. The solution should turn green and most of the leaf strips should become transparent.

9. Add an equal volume of W5 to the enzyme solution and swirl again.

10. Using a small funnel, filter the solution through a nylon mesh into one or several 12 ml round-bottom tubes.

11. Spin for 1 min at $200 \times g$ and remove the supernatant by pipetting.

12. Re-suspend the protoplasts in W5 by gently inverting the tubes. The total volume of W5 should be 2 ml. If several tubes are used, combine the protoplast suspensions into a single tube (*see* **Note 11**).

13. Take a small aliquot of protoplast suspension. Count protoplasts with a haemocytometer under a light microscope to determine their concentration (*see* **Note 12**).

14. Keep the protoplast solution on ice for 40 min to let protoplasts settle on the bottom of the tube.

15. Do a second wash with 2 ml W5 and keep protoplasts again on ice for at least 40 min.

16. Immediately before transfection, remove the supernatant by pipetting and add the required volume of MMg to obtain a final concentration of 2×10^5 protoplasts/ml. Gently mix by inverting the tube and keep at room temperature.

3.3 Tomato Protoplast Isolation

1. To prepare enzyme solution, thaw K3 solution containing 0.4 M sucrose at 55 °C for 30 min. Add cellulase and incubate again at 55 °C for 10 min. Chill the solution on ice, then add pectinase (*see* **Note 13**).

2. Syringe-filtrate the solution through a 0.45 μm filter into a Petri dish of adequate size.

3. Cut well-expanded leaves from 3- to 4- week-old plants and place them on a sheet of white paper with the bottom surface (lower epidermis) facing up.

4. Spread some grated quartz on the leaves and gently rub with a wet finger. Rinse with sterile water to remove the quartz.

5. Place the leaves on a new white paper. Remove leaf tip and stalk and carefully cut the middle part into about 2 mm thick strips using a sharp razor blade. Do **steps 3–5** for only a few leaves at a time (*see* **Note 9**).

6. Immediately float leaf strips on the enzyme solution with the rubbed epidermis facing down using a flat-tip forceps. Completely cover the surface of the enzyme solution with leaf sections.

7. Incubate the Petri dish, without shaking, in the dark at 26 °C for 3 h for enzymatic digestion (*see* **Note 10**).

8. Release protoplasts by gentle swirling of the Petri dish for several minutes. The solution should turn green and most of the leaf strips should become transparent.

9. Using a small funnel, carefully pour the enzyme-protoplast solution through a nylon mesh. Distribute the filtrate into several 12 ml round-bottom tubes (3 ml in each).

10. Carefully overlay the protoplast suspension in each tube with 1 ml W5 solution without mixing (*see* **Note 14**).

11. To form a sucrose gradient, spin at $200 \times g$ for 5 min, putting acceleration and brake settings of the centrifuge to zero.

12. Harvest around 1 ml viable protoplasts from each tube into a 50 ml conical tube (*see* **Notes 11** and **15**).

13. Suspend the protoplasts in 10 volumes W5. Gently mix by inverting the tube to dilute the remaining sucrose.

14. Spin at $100 \times g$ for 2 min and remove the supernatant by pipetting.

15. Re-suspend the protoplasts in 10 ml W5 by gently inverting the tube. If several tubes are used in a large experimental set-up, combine the protoplast suspensions into a single tube.

16. Take a small aliquot of protoplast suspension. Count protoplasts with a haemocytometer under a light microscope to determine their concentration (*see* **Note 12**).

17. Supplement the protoplast suspension with W5 to a volume of 35 ml.

18. Let the protoplasts recover for 1.5–2 h on ice in the dark. Maintain good aeration by inverting the tube every 15 min.

19. Immediately before transfection, pellet protoplasts at $100 \times g$ for 2 min and remove the supernatant by pipetting.

20. Add the required volume of MMg to obtain a final concentration of 7×10^5 protoplasts/ml. Gently mix by inverting the tube and keep at room temperature.

3.4 PEG-Mediated Protoplast Transfection

1. Prepare one 12 ml round-bottom tube for each transfection of a candidate effector construct. Pipette a total amount of 80 μg plasmid DNA into the bottom of the tubes. In each tube, the ratio of the co-transfected constructs should be 4.5:1:4.5 for *pFRK1-Luc*, *pUBQ10-GUS* and *p35-effector* or *p35S-GFP* as control (*see* **Note 16**).

2. Add 800 μl *Arabidopsis* or tomato protoplasts in MMg to the DNA.

3. Add 880 μl (1.1 protoplast volumes) PEG solution.

4. Mix by slowly inverting the tube (*see* **Note 17**).

5. Incubate at room temperature for 5–10 min.

6. To stop transfection, add 3.52 ml (4.4 protoplast volumes) W5 and mix gently.

7. Spin for 1 min at $200 \times g$ and remove as much supernatant as possible by pipetting.

8. Re-suspend the protoplasts in 800 μl (1 protoplast volume) WI.

9. Position the tubes horizontally and incubate the cells in the dark at room temperature for 6–16 h (*Arabidopsis*) or 6–10 h (tomato) to allow gene expression from the plasmids (*see* **Note 18**).

10. Carefully re-suspend the protoplasts. Use 600 μl for the Luc activity assay (leave in the tube), 100 μl for the GUS activity assay (transfer to two 1.5 ml safe-lock tubes, 50 μl each), 100 μl for propidium iodide staining of dead cells (transfer to a 1.5 ml tube).

3.5 Luc Activity Assay

1. Add 6 μl of 20 mM D-luciferin to the 600 μl protoplast suspension (200 nM final concentration) and gently invert.

2. Transfer protoplasts to a white flat-bottom 96-well plate, pipetting 100 μl into six wells in a row. The plate will be completely filled when 15 effectors plus the GFP control are tested.

3. Measure background luminescence in relative light units (rlu) using a luminometer. The detection time can be set to 1 s/well.

4. Incubate the plate for 30 min to 1 h in the dark at room temperature to allow protoplast recovery prior to MAMP challenge.

5. Treat the protoplasts from three of the wells with 1 μl of 50 μM flg22 peptide (500 nM final concentration) and leave the other three wells untreated. Gently tap the sides of the plate to mix.

6. Measure luminescence emitted by Luc activity every 2 h for 8 h using a luminometer (in rlu, 1 s/well detection time) (*see* **Note 19**). Between the measurements, incubate the plate covered by a lid in the dark at room temperature. A timesaving alternative is to run a program for automated Luc activity measurement (*see* **Note 20**).

3.6 GUS Activity Assay

1. Treat one of the two tubes containing 50 μl protoplast suspension with 0.5 μl of 50 μM flg22 (500 nM final concentration) and leave the other untreated (*see* **Note 21**). Gently tap the tube to mix.

2. Incubate *Arabidopsis* protoplasts for 6 h and tomato cells for 4 h in the dark at room temperature (*see* **Note 22**).

3. Spin the tubes for 10 s at maximum speed. Remove the supernatant with a 1 ml syringe.

4. Flash-freeze the protoplast pellet in liquid N_2. The tubes can be stored at −80 °C until the assay is performed.

5. Add 100 μl 1× cell lysis buffer prepared from the 5× stock solution to the pellet, mix by vortexing and briefly spin down at $10,000 \times g$.

6. Transfer the lysate to a white flat-bottom 96-well plate by pipetting three replicates of 10 μl. The plate will be completely filled when 15 effectors plus the GFP control are tested.

7. Add 90 μl MUG solution to the wells. Mix by tapping the sides of the plate.

8. Incubate the plate covered with a lid for 30 min at 37 °C.

9. Add 100 μl stop solution. Mix by tapping the sides of the plate.

10. Measure fluorescence emitted by GUS activity (production of 4-methylumbelliferone, 4-MU, $\lambda_{ex} = 365$ nm, $\lambda_{em} = 445$ nm) in relative light units (rlu) using a fluorimeter (*see* **Note 23**).

11. The values obtained in the GUS activity assay are used to normalize the data from the 6 h or 4 h time-point of the Luc activity assay for *Arabidopsis* or tomato, respectively (*see* **Note 24**).

3.7 Cell Death Rate

1. Continue incubating the tube containing 100 μl protoplasts for cell death staining for another 6 h in the dark at room temperature.

2. Add 1 μl of 100 μg/ml propidium iodide (1 μg/ml final concentration) to the protoplast suspension.

3. Determine the proportion of dead stained protoplasts under a fluorescence microscope ($\lambda_{ex} = 536$ nm, $\lambda_{em} = 617$ nm) (*see* **Note 25**).

3.8 Immunoblotting of Activated MAPKs

As immunoblotting is a commonly used technique in molecular biology, the protocol below is shortened and only focuses on the specific requirements for phosphorylated MAPK detection. Detailed information on polyacrylamide electrophoresis and protein blotting methods is available (*see* refs. 13, 14).

1. Prepare *Arabidopsis* or tomato protoplast samples (*see* Subheadings 3.2 and 3.3). Transfect 150 μl *Arabidopsis* protoplasts with 15 μg plasmid (*p35S-GFP* or *p35S-effector*) or 600 μl tomato protoplasts with 60 μg plasmid (*see* Subheading 3.4).

2. Incubate for 6–16 h (*Arabidopsis*) or 6–10 h (tomato) to allow gene expression from the plasmid.

3. Transfer into three 1.5 ml tubes (50 μl each for *Arabidopsis* or 200 μl each for tomato) and treat with 0.5 or 2 μl of 50 μM flg22 (500 nM final concentration). Gently tap the tubes to mix.

4. Immediately collect one of the samples (0 min time-point), the next after 15 min (15 min time-point) and the last again 15 min later (30 min time-point).

5. At each time-point, spin the tube for 10 s at $10,000 \times g$. Remove the supernatant with a 1 ml syringe.

6. Quickly flash-freeze the protoplast pellet in liquid N_2. The tubes can be stored at −80 °C until the assay is performed.

7. Put tubes on ice. Add 20 μl Laemmli buffer to the protoplast pellet and vortex for 10 s.

8. Heat at 90 °C for 5 min in a heating block to denature the protein samples. Mix again by vortexing and shortly spin down.

9. Load whole samples onto a 13.5 % SDS mini-gel using a Mini-Protean electrophoresis cell (Bio-Rad). Load 1.5 μl PAGE-Ruler (Fermentas) as molecular weight marker.

10. Perform SDS-PAGE. Stop electrophoresis when the 40 kDa band of the protein ladder has migrated approximately to the middle of the resolving gel.

11. Transfer proteins onto a nitrocellulose membrane using a Mini Trans-Blot cell (Bio-Rad) for wet protein transfer.

12. Stain the membrane with Ponceau S Red solution to visualize protein loads.

13. Rinse several times in H_2O to remove the Ponceau stain.

14. Block the membrane by incubating in 5 % skimmed milk TBS-T for 1 h at room temperature.

15. Rinse twice in TBS-T for 20 min at room temperature.

16. Incubate for 2 h at room temperature or overnight at 4 °C in primary antibody solution (*see* **Note 26**).

17. Wash three times in TBS-T for 10 min at room temperature.

18. Incubate for 1 h at room temperature in secondary antibody solution.

19. Wash three times in TBS-T for 10 min at room temperature.

20. Place in AP buffer and add reagents for colorimetric detection of AP activity.

21. Stop the reaction by washing the membrane in H_2O, once the signal for phosphorylated MAPKs in the GFP control is sufficiently strong.

4   Notes

1. The mannitol stock solution should be kept sterile, as there is a high microbial contamination risk. All working solutions containing mannitol should be prepared freshly on the day of protoplast preparation.

2. The quality of polyethylene glycol is very important. To dissolve PEG, heat the PEG solution for 15 min at 55 °C and mix from time to time.

3. For efficient transfection, it is crucial to prepare plasmid DNA using a commercial purification kit such as PureYield Plasmid Midipep System (Promega) or by conventional CsCl gradient purification. Accurately determine DNA concentration by OD_{260nm} measurement and adjust it to 1–2 μg/μl. DNA preparations that are too highly concentrated or that contain impurities may cause protoplast clumping in contact with PEG solution and consequently give very low transfection rates.

4. The presence of a tagged effector can be verified by immunoblotting of the transfected protoplast sample with corresponding anti-tag antibodies. The pellet from 50 μl protoplast suspension (10^4 protoplasts) is sufficient to obtain a detectable signal on an immunoblot.

5. To dissolve D-luciferin, carefully titrate with 1 M KOH. The solution will turn bright yellow when the powder is dissolved.

6. Chemically synthesized flg22, sequence from *Pseudomonas aeruginosa* as published [3]: QRLSTGSRINSAKDDAAGLQIA. It is also possible to use other MAMPs such as elf18 [15] or chitin [16].

7. Protect plants from any environmental stresses such as drought, flooding, heat or irregular light conditions during their growth. Even minor changes in growth conditions will affect protoplast yield and/or responsiveness.

8. Up to 30 leaves can be digested in 10 ml enzyme solution. In our laboratory, the yield is at least 5×10^5 protoplasts per ten leaves and can go up to 10^6. For a screen of 15 effector genes plus the *GFP* control—this corresponds to a whole 96-well plate for the Luc and GUS activity assays—about 55 leaves are necessary.

9. If possible, cut the leaves directly in the phytochamber or greenhouse where the plants are cultivated. Be very precise and avoid crushing the tissue. There should hardly be any green stains on the sheet of paper. Change razor blade when sharpness diminishes.

10. Protoplasts have the best survival rate and responsiveness when the leaf digestion (in the case of *Arabidopsis*) and all further experimental steps (both species) are performed in an air-conditioned room at 20–22 °C. This is especially important during warm weather conditions.

11. Protoplasts must be handled very carefully throughout the whole protocol. When adding a solution to the protoplasts, pipette it slowly along the tube wall. When pipetting protoplast suspension, cut the end of the pipet tip to slightly increase the size of the tip hole.

12. Count only round-shaped protoplasts containing chloroplasts and for which the plasma membrane is intact. Protoplast size is not relevant.

13. Up to 40 leaves can be digested in 20 ml enzyme solution. When a 12×12 cm rectangular Petri dish containing 20 ml enzyme solution is fully covered with leaf strips, this will yield approximately 4×10^6 protoplasts. For a screen of 15 effector genes plus the *GFP* control—this corresponds to a whole 96-well plate for the Luc and GUS activity assays—about 80 leaves are necessary.

14. W5 should be pipetted extremely slowly along the tube wall to overlay the protoplast suspension. The two-layered mixture must not be disturbed to ensure formation of a sucrose gradient after centrifugation.

15. Three layers form after sucrose gradient centrifugation. From the top to the bottom the gradient layers constitute: W5 solution, viable green protoplasts (to be collected), brownish green protoplasts and a pellet of dead cells and debris.

16. In general, 10 μg plasmid DNA is needed to transfect 100 μl protoplast suspension. Here, 80 μg plasmid DNA is used, as 800 μl protoplast suspension is required for each candidate effector to be tested for MTI suppression. If smaller or larger amounts of protoplasts are transfected, the DNA/protoplast ratio should stay the same, but the size of the round-bottom tube must be adapted.

17. To ensure that the handling of each tube is similar during transfection, place all tubes in a rack and invert them simultaneously. The suspension should stay homogeneously green. Protoplasts should not aggregate. Differences in transfection efficiency from one tube to another mainly depend on the quality of the DNA preparation and the size of the constructs.

18. If the incubation time is longer than 8 h, a background signal will appear in the Luc activity assay, which does not significantly disturb the outcome of the experiment. The background can be different from one sample to another, mostly depending on the transfection efficiency of the reporter construct (*see* Fig. 1). Avoid incubating *Arabidopsis* or tomato protoplasts longer than 16 or 10 h, respectively. They start dying after this time and will be less responsive to MAMP challenge.

19. In a reliable experiment, the luciferase activity induction of the flg22-treated samples compared to the non-treated samples of the *GFP* control is at least threefold after 6 h and twofold after 4 h for *Arabidopsis* and tomato, respectively (*see* Fig. 1). If the monitored rlu are very low for an effector data set compared to the GFP control and no Luc activity induction is observed, this does not always indicate that the effector is an MTI-suppressor. Artifacts are possible in the following cases: (a) the transformation efficiency is poor (it is optimally 70 % for *Arabidopsis* and 60 % for tomato and should be at least 40 %), (b) the introduced effector disturbs major cell functions or (c) the introduced effector causes cell death because it is present in non-physiological amounts. To identify a false-positive result, transfection efficiency can be determined by fluorescence microscopy when the tested effector is fused to a fluorescent protein such as GFP. The GUS assay described in Subheading 3.6 also indirectly indicates the transfection efficiency. To exclude enhanced protoplast dying in the presence of an effector, the cell death rate should be determined as described in Subheading 3.7. In addition to these tests, the induction of endogenous *FRK1* expression in flg22-treated transfected *Arabidopsis* protoplasts may be tested by qRT-PCR to validate the Luc activity assay data. A detailed protocol for RNA extraction from protoplasts, cDNA synthesis and qRT-PCR is available (*see* ref. 7). We suggest to use the following qRT-PCR primers: *FRK1* forward 5′-GATGGCGGACT TCGGGTT ATC-3′, *FRK1* reverse: 5′-CGAATAGTACTCG GGGTCAA GGTAA-3′. In tomato, another MAMP-induced gene may serve as marker gene for MTI suppression in qRT-PCR experiments.

20. When using automated measurement, the plate stays in the luminometer for at least 8 h. To avoid evaporation, it should be covered by optical or cling film. Note that the temperature inside the apparatus is slightly higher, which may affect protoplast responsiveness.

21. The comparison of flg22-challenged and untreated samples from the same transfection event can reveal a possible effect of flg22 on protoplast survival. We have rarely observed a notable difference between these samples when the protoplast incubation times after transfection were respected.

22. Sample collection for the GUS activity assay should ideally coincide with the 6 h or the 4 h measurement time-point of the Luc activity assay for *Arabidopsis* or tomato protoplasts, respectively. This time-point usually gives the best ratio of flg22-treated compared to non-treated samples for a data set. In the case of *Arabidopsis* protoplasts, it most often corresponds to the time-point where Luc activity peaks (*see* Fig. 1). We were not able to design a non-invasive GUS assay to perform GUS activity kinetics. Therefore, samples need to be collected at a specific time-point. It can be adapted to your experimental results.

23. Low rlu values for a given sample can be explained by the same reasons as mentioned in **Note 19**. If low values are observed in both assays for a data set, no conclusion can be drawn on the MTI suppression potential of the tested effector.

24. Data is normalized as follows: (value Luc +flg22/value GUS +flg22)/(value Luc −flg22/value GUS −flg22), using Luc and GUS values from the same time-point.

25. The cell death rate should not be above 30 % for *Arabidopsis* protoplasts and 60 % for tomato protoplasts. Note that the tomato protoplast system is very delicate. A substantial number of cells die during protoplast isolation, but this is compensated by high transfection efficiency. If cell death is much higher in one sample compared to the *GFP* control, the function or over-expression of the tested effector gene most probably kills the protoplasts.

26. Antibody incubation overnight will produce the best results, especially with a fresh antibody dilution. The primary antibody solution may be used again for up to three times.

Acknowledgements

We are thankful to all laboratory members who were involved in setting up the protoplast system and in constantly optimizing it, especially Dr. Anna K. Jehle and Christiane Jäntsch. We acknowledge Carolin Pfeifer and Birgit Löffelhardt for *Arabidopsis* plant maintenance and Gert Huber and his team from the plant growth facility for culturing tomato. This work was funded by the Deutsche Forschungsgemeinschaft (BR 3875/2-1 to F.B).

References

1. Dodds PN, Rathjen JP (2010) Plant immunity: towards an integrated view of plant–pathogen interactions. Nat Rev Genet 11:539–548

2. Chinchilla D, Bauer Z, Regenass M, Boller T, Felix G (2006) The *Arabidopsis* receptor kinase FLS2 binds flg22 and determines the specificity of flagellin perception. Plant Cell 18:465–476

3. Felix G, Duran JD, Volko S, Boller T (1999) Plants have a sensitive perception system for the most conserved domain of bacterial flagellin. Plant J 18:265–276

4. Gomez-Gomez L, Boller T (2000) FLS2: an LRR receptor-like kinase involved in the perception of the bacterial elicitor flagellin in *Arabidopsis*. Mol Cell 5:1003–1011

5. He P, Shan L, Lin NC, Martin GB, Kemmerling B, Nurnberger T, Sheen J (2006) Specific bacterial suppressors of MAMP signaling upstream of MAPKKK in *Arabidopsis* innate immunity. Cell 125:563–575

6. Xiang T, Zong N, Zou Y, Wu Y, Zhang J, Xing W, Li Y, Tang X, Zhu L, Chai J, Zhou JM (2008) *Pseudomonas syringae* effector AvrPto blocks innate immunity by targeting receptor kinases. Curr Biol 18:74–80

7. Niu Y, Sheen J (2012) Transient expression assays for quantifying signaling output. Methods Mol Biol 876:195–206

8. Sheen J (2001) Signal transduction in maize and *Arabidopsis* mesophyll protoplasts. Plant Physiol 127:1466–1475

9. Yoo SD, Cho YH, Sheen J (2007) *Arabidopsis* mesophyll protoplasts: a versatile cell system for transient gene expression analysis. Nat Protoc 2:1565–1572

10. Nguyen HP, Chakravarthy S, Velasquez AC, McLane HL, Zeng L, Nakayashiki H, Park DH, Collmer A, Martin GB (2010) Methods to study PAMP-triggered immunity using tomato and *Nicotiana benthamiana*. Mol Plant Microbe Interact 23:991–999

11. Asai T, Tena G, Plotnikova J, Willmann MR, Chiu WL, Gomez-Gomez L, Boller T, Ausubel FM, Sheen J (2002) MAP kinase signalling cascade in *Arabidopsis* innate immunity. Nature 415:977–983

12. Schutze K, Harter K, Chaban C (2009) Bimolecular fluorescence complementation (BiFC) to study protein–protein interactions in living plant cells. Methods Mol Biol 479: 189–202

13. Sambrook J, Russell DW (2006) SDS-polyacrylamide gel electrophoresis of proteins. CSH Protoc 2006. doi:10.1101/pdb.prot4540

14. Kurien BT, Scofield RH (2009) Introduction to protein blotting. Methods Mol Biol 536: 9–22

15. Kunze G, Zipfel C, Robatzek S, Niehaus K, Boller T, Felix G (2004) The N terminus of bacterial elongation factor Tu elicits innate immunity in *Arabidopsis* plants. Plant Cell 16:3496–3507

16. Shibuya N, Minami E (2001) Oligosaccharide signalling for defence responses in plant. Physiol Mol Plant Pathol 59:223–233

Chapter 18

Production of RXLR Effector Proteins for Structural Analysis by X-Ray Crystallography

Richard K. Hughes and Mark J. Banfield

Abstract

Structural analysis of RXLR effector proteins from oomycete plant pathogens is an emerging area of research. These studies are aimed at understanding the molecular basis of how these proteins manipulate plant cells to promote infection and also to help define how they can lead to activation of the plant innate immune system. Here, we describe a medium-throughput procedure for cloning and expression testing oomycete RXLR proteins in *Escherichia coli*. We also describe methods for purification of soluble protein and crystallization, with the aim of determining three-dimensional structures by X-ray crystallography. The procedures are generally applicable to any research program where the production of soluble recombinant protein in *E. coli* has proven difficult, or where there is a desire to evaluate *E. coli* thoroughly as a host before considering alternative hosts for heterologous expression.

Key words Effector, Oomycete, RXLR, Heterologous expression, pOPIN vectors, *E. coli*, Protein, Crystal, Diffraction, X-ray crystallography, Structure

1 Introduction

Determining the three-dimensional structures of proteins is an unparalleled approach for understanding molecular function. An experimentally defined structure can provide an atomic-level understanding of catalytic mechanisms or allow recognition surfaces to be mapped, for example, between proteins or between proteins and other ligands. Despite the absolute requirement for crystals, X-ray crystallography is still the most commonly employed method for structure determination. To be rewarded with the beauty of a three-dimensional structure of a protein of interest there are a number of significant hurdles that must all be overcome. Then the job of designing experiments to probe key biological questions based on the structure can begin. The steps that need to be followed in order to determine a structure can be broadly summarized as:

Paul Birch et al. (eds.), *Plant-Pathogen Interactions: Methods and Protocols*, Methods in Molecular Biology, vol. 1127, DOI 10.1007/978-1-62703-986-4_18, © Springer Science+Business Media New York 2014

1. Heterologous expression of the protein.

2. Purification of stable, homogeneous, monodisperse protein in milligram quantities.

3. Obtain well-ordered crystals that diffract X-rays to a resolution useful for analysis.

4. Solve the crystallographic phase problem and produce an experimental electron density map.

5. Complete the interpretation of the X-ray data and structural analysis.

The scope of this article limits us to detailed discussion of 1–3 above.

Expression of a chosen protein at sufficient levels and its purification in a stable, monodisperse form is, more often than not, a significant bottleneck in this process. Once conditions have been identified that allow production of stable, soluble, monodisperse protein in milligram quantities, crystallization trials are possible. If crystals are obtained that diffract X-rays to sufficient resolution, and the crystals do not show pathologies such as twinning, then structure solution should be achievable. In this article we concentrate on the processes of obtaining soluble protein of high quality for crystallization that have proven useful for obtaining structures of oomycete RXLR effector proteins.

Obtaining milligram quantities of a protein of interest almost always requires expression in a heterologous host as the protein is unlikely to be present in the host organism in sufficient quantities. Identification of a host that supports heterologous expression of the protein of interest in functional form is an important consideration. *Escherichia coli* is the host of choice for protein production in most laboratories [1]. However, *E. coli* will not always produce proteins that are correctly folded and functional. For example, proteins of eukaryotic origin may require chaperones or specific post-translational modifications not found in *E. coli*. Proteins with disulphide bonds may also be difficult to obtain in *E. coli* due to the oxidizing environment of the cytoplasm. Eukaryotic proteins may form inactive aggregates in *E. coli* and accumulate in inclusion bodies [1]. It is possible to re-fold such proteins following denaturation to obtain soluble, functional protein [2, 3], but frequently this is challenging. Despite these reservations, in terms of the number of structures that have been solved and deposited in the Protein Data Bank (http://www.rcsb.org/pdb/), *E. coli* is by far the most successful host for heterologous protein expression. It therefore remains common to exhaustively test protein expression in *E. coli* before considering alternative hosts, such as yeasts, insect cells, mammalian cells, other bacteria, xenopus oocytes or even plants. To this end, combinatorial library approaches for improving soluble protein expression in *E. coli* have been developed [4]. The details of protein expression outside *E. coli* are beyond the scope of this article.

To maximize the potential for yield of soluble, functional protein, many factors should be considered in the preliminary design of expression constructs [5]. These can include: optimizing the codon usage of the expressed sequence to match that of the expressing organism; elimination of cryptic splice sites and RNA destabilizing sequence elements for increased RNA stability; addition of RNA stabilizing sequence elements; and avoidance of stable RNA secondary structures. Many of these considerations are possible with the advent of cost-effective DNA synthesis technologies [6].

In the case of *E. coli*, the number of different expression vectors and host strains available can seem daunting [7]. Optimization of host strain–vector combinations, including the choice of inducible promoters, is known to be crucial for success in some cases, especially with expression of proteins toxic to *E. coli* [8]. Many expression vectors now encode fusion-tags, used to aid purification and/or enhance solubility (histidine, maltose binding protein, glutathione-S-transferase, thioredoxin, etc.) that can be localized to the N- or C-terminus of the protein [9]. For crystallization, an important consideration is the prediction and removal of putative disordered regions at the termini of expressed proteins [10]. For proteins with more than a single domain, expressing individual domains separately rather than trying to express the full-length protein may prove useful (a "divide-and-conquer" approach). It may also be necessary to test homologues or orthologues of the target protein of interest.

It is also important to give consideration to optimizing the liquid culture phase during protein expression. Maximizing plasmid stability in the culture, choice of culture density at which to induce protein expression, and the choice of media itself (e.g. LB, auto-induction, rich media [terrific broth]) are all important [11, 12]. Further, solubility of expressed proteins is often improved if the culture is grown at lower temperatures (8–20 °C) [13] or when additives such as sorbitol or other chemical chaperones are included in the growth medium [14].

Even when a large amount of protein of interest can be purified in a soluble and functional form it may not crystallize in standard crystallization screens, or may give unsuitable crystals. In such cases, re-designing the expression construct is frequently required. Various techniques can be employed to provide information that should improve the chances of producing good quality protein crystals. These include: Nuclear Magnetic Resonance (NMR—which can be used to identify regions of disorder in proteins that could not be predicted from sequence) or limited tryptic digest (to identify stable subdomains of a protein). The purified protein may also be relatively unstable and show signs of degradation in solution before crystals form. This can be monitored by comparing the behavior of the protein on SDS-PAGE gels immediately after purification and during storage. To help prevent degradation,

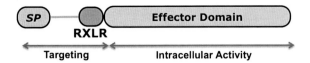

Fig. 1 Domain structure of RXLR effector proteins. RXLR proteins are modular and comprise a signal peptide (SP), followed by the characteristic RXLR sequence (*blue*). These sequences are involved in host cell targeting. The C-terminal effector domain encodes the activity in the host cell

proteins can be stored frozen in aliquots under liquid nitrogen. The chosen buffer in which the protein has been purified may also need to be optimized. Important parameters may include buffer pH, salt concentration, the presence of other additives like metal ions, metal ion chelators, reducing agents, etc., all of which can alter the structures and/or surface properties of proteins and affect crystallization. Dynamic light scattering (DLS) [15] and the Thermofluor assay [16] are useful tools to identify the buffer conditions for producing monodisperse and stable protein, respectively.

Whilst certainly not unique, the expression of RXLR effector proteins is complicated by a number of the factors detailed above. As the subject of this article, we focus on approaches used to obtain crystal structures of these proteins here. There are likely many means to the same end. In RXLR effectors, the region that contains the canonical RXLR motif (after the secretion signal and involved in the translocation of effectors into the host cell [17] (Fig. 1)) is frequently predicted to be disordered, but the disorder/order boundary transition is often not clear. The part of the protein that encodes biochemical function in the host cell (the "effector domain") may contain multiple sub-domains. Furthermore, pathogens can encode >>100 RXLR effectors [18, 19]. As many of these are of interest for understanding structure/function relationships, the expression pipeline must be streamlined for throughput. To rapidly screen for expressed proteins suitable for crystallization, we use the suite of recombination-based cloning expression vectors developed by the Oxford Protein Production Facility (OPPF) [20]. These pOPIN vectors allow cloning using In-Fusion technology (Clontech) and are available with a range of different tags and vector backbones (usually a histidine tag is included). They also allow for expression in *E. coli*, insect cells and mammalian cell culture without the need for re-cloning.

To screen for expression of constructs of interest established in pOPIN vectors, our procedural unit allows for the rapid production of 24 bacterial cultures (or multiples thereof). The system is highly flexible in terms of expression construct design (e.g. different fusion tags, dependent on the pOPIN vector chosen), the host strain/growth temperature and medium used for expression testing. It requires only simple laboratory equipment, consumables and multichannel pipettes; a robotic handling procedure can be used to streamline the process, if available. We use a system that permits the analysis of both total and soluble protein expression in cleared lysates. Our laboratory typically uses fully automated AKTA Xpress modules for purification of expressed soluble proteins to homogeneity from cleared lysates using a combination of Immobilized Metal-ion Affinity Chromatography (IMAC) and gel filtration. The latter technique also serves as a means of assessing the monodispersity and oligomeric state of the protein. Following purification, ultrafiltration devices are used to concentrate protein samples. Preliminary crystallization screens are purchased from suppliers and dispensed into crystallization plates using a liquid handling robot (multichannel pipettes can also be used). Nanoliter-scale crystallization experiments are set-up using a crystallization robot and these experiments are tracked for crystal formation using an imaging robot. Any crystals obtained in preliminary screens can be optimized (usually at the milliliter scale) or used directly for X-ray diffraction experiments.

In the sections below we detail the typical steps taken to obtain crystals of RXLR effector proteins in our laboratory. Much of the equipment and many of the consumables, or similar, would be available in most universities or research institutes.

2   Materials

2.1   Medium-Throughput Cloning into pOPIN Vectors

2.1.1   Equipment and Consumables

1. 37 °C incubator(s) with shaking capability and racks to accommodate deep-well blocks.
2. Benchtop centrifuge capable of $3,800 \times g$ with deep-well block holders.
3. 24-Deep-well block (Starlab, E2824-1000).
4. PCR machine, strips and caps.
5. Petri dishes (90 mm).
6. 96-Well PCR plate (non-skirted, Starlab E1403-0100) and plastic adhesive plate seal (Labtech Int., 4TI-0500).
7. Heating block at 42 °C with 96-well block for PCR plates and strips.
8. Agarose DNA gel electrophoresis equipment.
9. Syringes and syringe filters (0.2 μm).

2.1.2   Reagents

1. Fully digested (KpnI/HindIII—pOPINF, S3C or M; NcoI/PmeI—pOPINE) and gel purified pOPIN vectors (100 ng/μL).
2. PCR components and proofreading polymerase.
3. In-Fusion Cloning System Kit (Clontech, 50 reactions, 639646). In 2013, this product will be replaced with In-Fusion HD Cloning Plus Kit (50 reactions, 638910).

4. 12 PCR templates (ideally not ampicillin resistant) with cDNA encoding protein of interest.

5. 24 Primer pairs (pOPINF and pOPINE) or 12 primer pairs (pOPINF and pOPINS3C or pOPINM).

   Primers can be designed using the Opiner server at OPPF (http://www.oppf.rc-harwell.ac.uk).

6. 24 PCR products (pOPINF and pOPINE) or 12 PCR products (pOPINF and pOPINS3C or pOPINM).

7. Plasmid DNA miniprep and gel extraction kits.

8. Nanodrop spectrophotometer (ThermoFisher).

9. 1 % (w/v) agarose gels.

10. 24×50 μL aliquots of chemically competent DH5α and BL21(DE3) cells.

11. 100 mg/mL carbenicillin (water, filter sterilized).

12. 20 % (w/v) X-gal (in dimethylformamide, Fluka 40250).

13. 1 M IPTG (water, filter sterilized).

14. LB-carbenicillin (100 μg/mL, Formedium) agar containing isopropylthio-β-D-galactopyranoside (IPTG, 1 mM, Formedium) and X-gal (5-bromo-4-chloroindolyl-β-D-galactopyranoside, 0.02 % w/v, Melford).

15. 50 % (v/v) glycerol (sterile).

16. Ultrapure water (18 MΩ cm at 25 °C).

17. DNA sequencing reagents (or use service provider); T7 promoter (TAATACGACTCACTATAGGG) and pOPINRev (CACCACCTTCTGATAGGCAG) sequencing primers.

2.2 Medium-Throughput Protein Expression Testing

2.2.1 Equipment and Consumables

1. 24-Deep-well blocks (Starlab, E2824-1000).

2. 96-Deep-well blocks (Abgene, VWR, 732-4910).

3. 96-Well PCR plate (non-skirted) for SDS-PAGE samples (and foil seal).

4. 96-Well heating block at 42 and 95 °C for PCR plates and strips.

5. Gas permeable seals (Fisher Scientific, TUL980060V).

6. Expandable multichannel pipette (and 1.25 mL tips) (Matrix Science, ThermoFisher).

7. Standard (8 tip) multichannel pipette (5–50 μL).

8. Repeat pipetter (with 50 mL tip) and 100 mL beaker.

9. Incubator(s) with shaking and chilling capability + rack to accommodate deep-well blocks.

10. Bench top spectrophotometer and cuvettes for measuring absorbance at 600 nm.

11. Microtitre plate reader and 96-well clear microtitre plates (skirted, Qiagen 36985).

12. Sonicator (Sonics, Vibracell: Model CV33) with 24 head probe (630-0579).

13. Small ice box (to cool deep-well block during sonication).

14. Shaker for 96-deep-well blocks.

15. Liquid handling robot (e.g. Tecan, Freedom Evo) with optional microplate shaker (Inheco, Variomag Teleshaker, 3800047).

16. Benchtop centrifuge capable of $4,800 \times g$ with deep-well block holders.

17. Gel multicasting chambers (2) for 20 gels (Biorad, 165-4110) and 20 sets of glass plates: 1.0 mm spacer plates (Biorad, 165-3311) and small plates (Biorad, 165-3308).

18. SDS-gel tanks for running 12× SDS-PAGE gels (4 gels/tank, 3 tanks) (Mini Protean Tetra, Biorad).

19. SDS-PAGE running buffer (5×): 940 g glycine, 151 g Tris, 250 mL 20 % (w/v) SDS, water to 10-L.

20. Gel staining dishes (12) (Slaughter, 109).

2.2.2 Reagents

1. Ultrapure water (18 MΩ cm at 25 °C).

2. LB-carbenicillin (100 µg/mL) (300 mL).

3. Lysis buffer: 50 mM Tris–HCl pH 8.0, 0.3 M NaCl, 20 mM imidazole (Fluka, 56748), 5 % (v/v) glycerol, 50 mM glycine containing 0.25 % (w/v) polyethyleneimine (PEI) (*see* **Note 1**) + "Complete EDTA-free" protease inhibitors tablet (Roche, 1873580) (1 tablet per 50 mL lysis buffer).

4. 12×5 µL aliquots marker proteins (Life Technologies, SeeBlue+2, LC5925).

5. Instantblue (Expedeon, ISBIL).

6. 4× SDS loading buffer (50 mL): To a 50 mL Falcon tube add 20 mL glycerol, 10 mL 1 M Tris–HCl pH 6.8, 20 mL 20 % (w/v) SDS (Severn Biotech), 3.1 g dithiothreitol (Melford, MB1015) and 0.2 g bromophenol blue (Sigma, 114391). Shake gently for 10 min. Centrifuge at $3,800 \times g$ for 10 min and decant supernatant to remove undissolved bromophenol blue. Freeze in 0.5 mL aliquots at −80 °C. Store a few working aliquots at −20 °C.

7. *Optional*: Benzonase (Merck, ≥25 U/µL, Cat. No. 1.01694.0001); add 25 µL (625 U)/50 mL lysis buffer. For robotic platform: 25 % (v/v) ethanol (tip washing); 70 % (v/v) ethanol (tip sterilization); 2 % (v/v) Hycolin (Coventry Chemicals) (disinfectant for waste pot).

2.3 Scale-Up Expression of Soluble Proteins

2.3.1 Equipment and Consumables

1. Incubator(s) with shaking (and chilling capability if necessary).
2. Bench top spectrophotometer and cuvettes for measuring absorbance at 600 nm.
3. 1-L centrifuge tubes (Fisher, CFT-980-G).
4. Centrifuge and rotor for 1-L tubes ($5,670 \times g$).
5. 2-L conical flasks.
6. Falcon tubes (50 mL).

2.3.2 Reagents

1. LB ($2-6 \times$ 1-L in 2-L conical flasks and 1 or $2 \times$ 100 mL in 250 mL conical flasks), autoclaved.
2. 100 mg/mL carbenicillin (water, filter sterilized).
3. 1 M IPTG (water, filter sterilized).

2.4 Extraction and Purification of Soluble Proteins for Crystallization

2.4.1 Equipment and Consumables

1. AKTA Xpress system module(s) (GE Healthcare)—with sample loop extensions (15 mL max sample loading).
2. Hi-Trap FF Ni-IMAC columns (5 mL) (GE Healthcare, 17-5255-01).
3. Superdex 75 26/60 gel filtration columns (GE Healthcare, 17-5174-01).
4. 24-Deep-well blocks (for collecting protein fractions) (Starlab, E2824-1000).
5. Sonicator with single, large (10–12 mm) probe (suitable for lysis of 50 mL + volumes).
6. Falcon tubes (50 mL) or plastic beaker.
7. Centrifugal protein concentrators (Sartorius, Vivaspin 20, PES) and centrifuge ($7,500 \times g$).
8. Peristaltic pump (e.g. GE Healthcare, P1).
9. Spectrophotometer (e.g. Nanodrop).

2.4.2 Reagents

1. Lysis buffer: 50 mM Tris–HCl (pH 8.0), 0.5 M NaCl, 20 mM imidazole (Fluka, 56748), 5 % (v/v) glycerol, 50 mM glycine + 1 large "Complete EDTA-free" protease inhibitors tablet per 50 mL lysis buffer.
2. IMAC equilibration/binding buffer ("A1"): as lysis buffer, without protease inhibitors.
3. IMAC elution buffer ("B1"): as equilibration buffer with 500 mM imidazole.
4. Gel filtration buffer ("A4"): 20 mM HEPES pH 7.5, 0.15 M NaCl.
5. Stripping buffer: 38 mL 0.5 M NaH_2PO_4, 162 mL 0.5 M Na_2HPO_4, 146.1 g NaCl, 500 mL 0.5 M EDTA (pH 8.0), water to 5-L.

6. 0.1 M nickel chloride (Sigma, 223387).

7. Ultrapure water (18 MΩ cm at 25 °C).

8. 20 % (v/v) ethanol.

9. Purified HIS-tagged 3C protease (expression and purification details are available from the authors on request).

10. 1 M ß-mercaptoethanol (Sigma, M6250).

2.5 Setting Up Protein Crystallization Screens and Crystal Screening

2.5.1 Equipment and Consumables

1. Crystallization robot (Douglas Instruments, Oryx Nano).

2. MRC 2 drop 96-well plates (Molecular Dimensions, MD11-00).

3. Deep-well block seals (Molecular Dimensions, Sureseal, MD6-16).

4. MRC plate seals (Molecular Dimensions, ClearVue, MD6-01S).

5. Crystal imaging robot in room at 20 °C.

6. Binocular microscope.

2.5.2 Reagents

1. Ultrapure degassed water (18 MΩ cm at 25 °C).

2. Six different crystallization screens (1 mL dispensed from commercial preparations into 96-deep-well blocks using liquid handling robot), choose from: PEG (Qiagen, 130704), JCSG-*plus* (Molecular Dimensions, MDI-37), Ammonium sulfate (Qiagen, 130705), PACT (Molecular Dimensions, MDI-29), Structure Screen 1 (Molecular Dimensions, MDI-01) and 2 (Molecular Dimensions, MDI-02)—combined into a single 96-well block, Morpheus (Molecular Dimensions, MDI-46) and Midas (Molecular Dimensions, MDI-59).

2.6 Testing Protein Crystals for Diffraction Quality

2.6.1 Equipment and Consumables

1. X-ray diffractometer (or synchrotron).

2. Binocular microscope.

3. Cover slips and silicone grease.

4. Scalpel blade.

5. Clear tape to reseal well.

6. Crystal loops (various sizes: 0.1–1.0 mm).

7. Liquid nitrogen and Dewar flask.

2.6.2 Reagents

1. Reservoir solution (from plate with protein crystal).

2. Cryoprotectant solution(s). The exact composition of the cryoprotectant is dependent on the precipitant but can include organics, polyols, salts, sugars and polymers. Alternatively some oils, such as paratone, are suitable for cryoprotecting protein crystals.

2.7 Optimization of Protein Crystals for Improving Quality of X-Ray Diffraction

2.7.1 Equipment and Consumables

Hanging Drop

1. Crystallization room at 20 °C (to store plates).
2. Falcon tubes (15 mL) and 24 tube rack.
3. 24-Well plates with lids (Molecular Dimensions, MD3-11).
4. 22 mm cover "Crystalclene" slips (Molecular Dimensions, MD4-11).
5. Silicone grease and dispenser.
6. Fine forceps.
7. Pipette (0.1–2 μL).

Sitting Drop

1. Crystallization room at 20 °C (to store plates).
2. 48-Well plates (Molecular Dimensions, MRC MAXI, MD11-004) and seals (Molecular Dimensions, ClearVue, MD6-01S).
3. Standard multichannel (8 tips) pipette (50 μL).
4. Crystallization robot (Douglas Instruments, Oryx Nano).

Additive Screens and Seeding Experiments (*See* **Note 2**)

1. Everything listed in Subheadings "Hanging Drop" and/or "Sitting Drop".
2. Cat's whisker.
3. Acupuncture needle.
4. Super glue.

2.7.2 Reagents

Hanging Drop

1. Individual salts and solutions used in precipitant of interest from commercial 96-well screen.
2. 24 Homemade precipitant solutions (1–10 mL) (*see* **Note 3**).

Sitting Drop

1. Individual salts and solutions used in precipitant of interest from commercial 96-well screen.
2. 48 Homemade precipitant solutions (1–10 mL) (*see* **Note 3**).

Additive Screens and Seeding Experiments (*See* **Note 2**)

1. Additive Screen HT Kit (Hampton Research, HR2-138).
2. Seed Bead Kit (Hampton Research, HR2-320).

2.8 Solving the "Phase Problem"

2.8.1 Equipment and Consumables

Iodide Soaking of Native Crystals

1. Everything listed in Subheading 2.6.1.
2. Timer.

SAD Phasing Using Selenomethionine-Labeled Protein	For expression of selenomethionine-labeled protein, details are available on request from the authors.

1. Everything listed in Subheadings 2.3.1 (except LB liquid media), 2.4.1 and 2.6.1.

2. Sterile flasks (2-L).

Soaking of Native Crystals with Heavy Metal Derivatives (See Note 4)

1. Everything listed in Subheading 2.6.1.

2. Toxins cupboard.

3. Fine balance with extraction hood.

4. Face mask.

5. Timer.

2.8.2   Reagents

Iodide Soaking of Native Crystals

1. Everything listed in Subheading 2.6.2.

2. 1 M potassium (or sodium) iodide solution.

SAD Phasing Using Selenomethionine-Labeled Protein

1. Everything listed in Subheadings 2.3.2, 2.4.2, and 2.6.2.

2. B834 (DE3) cells transformed with expression construct.

3. Selenomethionine (Merck, 561505).

4. Minimal media (M9).

Soaking of Native Crystals with Heavy Metal Derivatives (See Note 4)

1. Everything listed in Subheading 2.6.2.

2. Heavy metal derivatives. Database of compounds is available at http://www.sbg.bio.ic.ac.uk/had/.

---

3   Methods

3.1   Medium-Throughput Cloning into pOPIN Vectors

This procedure is designed for cloning the sequences of 12 different target proteins in two different pOPIN vectors (pOPINF and either pOPINS3C, pOPINM or pOPINE) giving a total of 24 constructs for expression testing.

1. Prepare double-digested pOPIN vectors and 12 or 24 PCR products and purify them from agarose gels using a gel extraction kit. Elute DNA from spin columns in sterile water and store at a concentration of about 100 ng/μL.

2. Set up 24 In-Fusion reactions in 3 × 8 PCR strips on ice: 100 ng pOPIN vector, 100 ng PCR product, 1 μL 5× In-Fusion enzyme premix + sterile water to 5 μL. Incubate in PCR machine at 42 °C for 30 min. Store on ice or freeze at −20 °C.

3. Transform 2 μL each In-Fusion reaction into 24 × 50 μL chemically competent DH5a cells (in PCR tubes). Store remainder of In-Fusion reactions at −20 °C. Heat shock cells for 30 s at

42 °C. Transfer cells to 24-deep-well block and add 450 μL LB. Add gas permeable seal. Recover cells by shaking at 200 rpm at 37 °C for 1 h.

4. Plate 100 μL cells on LB-carbenicillin (100 μg/mL) agar plates containing IPTG (1 mM) and X-gal (0.02 % w/v). Incubate overnight at 37 °C. Select a white colony from each of the 24 transformations. Less than 10 % of colonies should be blue.

5. Use repeat pipetter to transfer 4 mL LB-carbenicillin (100 μg/mL) to each well of a 24-deep-well block and inoculate with white colonies. Add gas permeable seal and grow overnight at 37 °C with shaking at 200 rpm.

6. Add 400 μL sterile 50 % (v/v) glycerol to 24 × labeled 1.5 mL Eppendorf tubes. Use an expandable multichannel pipette and 1.5 mL Eppendorf tubes in a suitably spaced tube rack. Add 600 μL overnight culture to each tube, mix and flash-freeze samples in liquid nitrogen. Store at –80 °C.

7. Harvest remainder of cells by centrifugation ($3,800 \times g$, 10 min). Discard supernatants and blot upside down on a blue towel. Store plate with cell pellets at –20 °C, or proceed immediately to the next step.

8. Prepare 24 plasmid DNA preps using DNA miniprep kit. Elute DNA from spin column with 30 μL sterile water. Submit all samples for DNA sequencing using T7 promoter primer (and pOPINRev primer, if necessary).

9. Transform 24 × 50 μL aliquots of chemically competent BL21(DE3) as in **step 3** with 5 μL of each DNA preparation prepared in **step 8**. Store 24-deep-well blocks at 4 °C, but confirm sequences before proceeding to protein expression testing. Any *E. coli* expression strain can be used instead of BL21(DE3) as appropriate.

3.2 Medium-Throughput Protein Expression Testing

This procedure will test for protein expression from the 24 different constructs prepared above. Expression is induced at 37 °C for 4 h or at 18 °C overnight in a single host strain, in a single growth medium. The procedure will yield a total of 96 samples (48 whole cell and 48 soluble fractions) for analysis on 12× SDS-PAGE gels.

1. Use repeat pipetter to transfer 4 mL LB-carbenicillin (100 μg/mL) to a 24-deep-well block. Inoculate with single BL21(DE3) colonies on plates (Subheading 3.1, **step 9**). Add gas permeable seal and grow overnight at 37 °C with shaking at 200 rpm.

2. Make glycerol stocks of the constructs in BL21(DE3) if not available (repeat Subheading 3.1, **step 6**).

3. Use repeat pipetter to transfer 4 mL LB-carbenicillin (100 μg/mL) to 2 × 24-well blocks. One plate will be induced at 37 °C, the other at 18 °C overnight. Inoculate each well with 200 μL overnight culture from the 24-deep-well block in **step 1** (*see* **Note 5**).

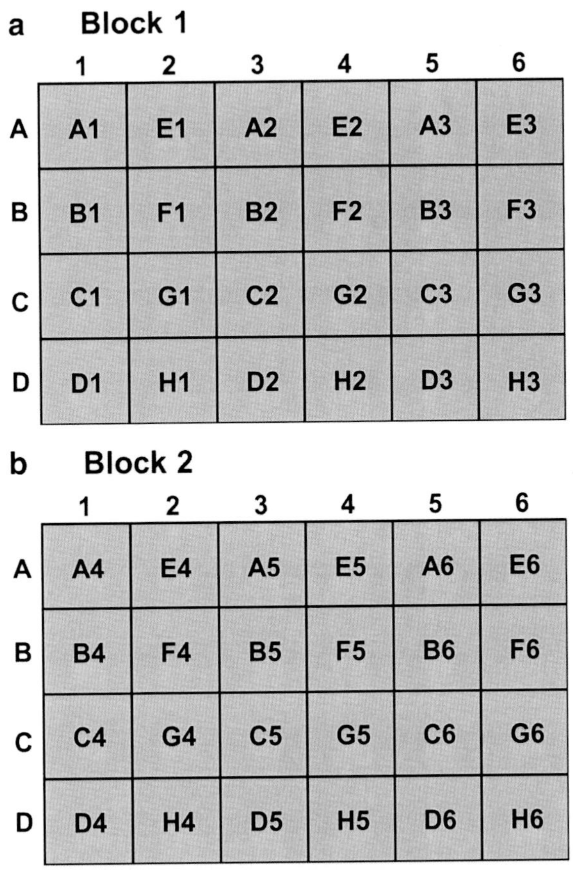

a   **Block 1**

	1	2	3	4	5	6
A	A1	E1	A2	E2	A3	E3
B	B1	F1	B2	F2	B3	F3
C	C1	G1	C2	G2	C3	G3
D	D1	H1	D2	H2	D3	H3

b   **Block 2**

	1	2	3	4	5	6
A	A4	E4	A5	E5	A6	E6
B	B4	F4	B5	F5	B6	F6
C	C4	G4	C5	G5	C6	G6
D	D4	H4	D5	H5	D6	H6

Fig. 2 Deep-well blocks used for protein expression testing. Two deep-well blocks are shown, as an example, for testing protein expression of 24 different constructs at 37 °C for 4 h (Block 1) or at 18 °C overnight (Block 2). Alternative testing regimes for blocks 1 and 2 can easily be adopted. The description in each well refers to the destination location for each sample in the 96-deep-well block after sonication (*see* Fig. 3a)

4. Grow cells with shaking at 200 rpm for 3 h (A_{600} 0.2–1.0). Remove seals and transfer 100 µL from each well of both blocks to 2×96-well clear microtitre plates. Leave 24-deep-well blocks at room temperature while recording A_{600} values using a plate reader.

5. At appropriate A_{600} add 40 µL 0.1 M IPTG to each well of both 24-deep-well blocks using an expandable multichannel pipette and replace with new gas permeable seals. Incubate one plate at 37 °C for 4 h (Fig. 2a) and the other at 18 °C overnight (Fig. 2b), both with shaking at 200 rpm.

6. Harvest cells as in Subheading 3.1, **step 7**. Cover blocks with plastic seals and store at −20 °C overnight or at −80 °C for >1 h.

7. From this point a robotic procedure can be followed to guide the user through both stages of cell lysis (using a Teleshaker). Alternatively, cell lysis can be carried out separately and the

robot can be used for just the preparation of samples for SDS-PAGE analysis. If a liquid handling robot is not available, the SDS-PAGE samples can be prepared manually using a multichannel pipette, if special care is taken when removing the supernatants.

8. Resuspend cells in lysis buffer (*see* Subheading 2.2.2, **item 3** and **Note 6**). If not using the Teleshaker, shake plate on a deep-well block shaker at room temperature for 10–15 min. With both shakers, additional pipetting up and down may be necessary to homogenize the cell suspensions.

9. Sonicate the lysed extracts using a 24-well probe on ice (40 % amplitude max., 1 s on, 3 s off, 1–2 min). Transfer 0.5–1 mL of each extract to a 96-deep-well block (pre-cooled on ice) using an expandable multichannel pipette. The samples should be loaded in columns (four samples at a time) in the following order: A1-H1, A2-H2, A3-H3—Plate 1; A4-H4, A5-H5 and A6-H6—Plate 2 (Fig. 3a).

10. Prepare 96-well PCR plate with 10 μL 4× SDS-PAGE loading buffer in every well.

11. Either using an appropriate script on a liquid handling robot (e.g. Tecan, Freedom Evo) or multichannel pipettes, prepare a PCR plate containing 96 samples: 48 whole cell and 48 soluble (prepared before and after centrifugation of block at $4,800 \times g$ for 40 min) (Fig. 3b). Store deep-well block containing centrifuged lysates at 4 °C. Cover plate containing SDS-PAGE samples with foil seal and heat at 95 °C for 3 min using the block provided. Allow to cool. Test all whole cell samples for viscosity problems using a multichannel pipette prior to loading SDS-PAGE gels (*see* **Note 7**).

12. Prepare 20× SDS-PAGE gels using 2× multicasting chambers and multichannel compatible 9-well combs (*see* **Note 8**).

13. Load 12×9-well SDS-PAGE gels with a standard (8 tip) multichannel pipette (gels 1–6, whole cell fraction; gels 7–12, soluble fraction) (Fig. 3b).

14. Stain gels with Instantblue. Inspect gels for expressed and soluble protein by comparing: Gels 1 and 7; 2 and 8; 3 and 10; 4 and 11; 5 and 12 (Fig. 4).

3.3 Scale-Up Expression of Soluble Proteins

1. Inoculate 50–100 mL LB-carbenicillin (100 μg/mL) in 250 mL conical flasks using glycerol stock prepared in Subheading 3.2, **step 2**. Grow overnight at 200 rpm with shaking at 37 °C. Alternatively, inoculate a fresh transformant from an agar plate.

2. Inoculate 2–6×1-L LB-carbenicillin (100 μg/mL) in 2-L conical flasks with 20 mL overnight culture. Grow at 200 rpm with shaking at 37 °C until the A_{600} recorded in Subheading 3.2, **step 4**, or for a maximum of 3 h.

a **Sonicated Lysates**

	1	2	3	4	5	6	7	8	9	10	11	12
A	1A1	1A3	1A5	2A1	2A3	2A5						
B	1B1	1B3	1B5	2B1	2B3	2B5						
C	1C1	1C3	1C5	2C1	2C3	2C5						
D	1D1	1D3	1D5	2D1	2D3	2D5						
E	1A2	1A4	1A6	2A2	2A4	2A6						
F	1B2	1B4	1B6	2B2	2B4	2B6						
G	1C2	1C4	1C6	2C2	2C4	2C6						
H	1D2	1D4	1D6	2D2	2D4	2D6						

Gel Number

b **Whole Cell**   **Soluble**

	1	2	3	4	5	6	7	8	9	10	11	12
A	1A1	1A3	1A5	2A1	2A3	2A5	1A1	1A3	1A5	2A1	2A3	2A5
B	1B1	1B3	1B5	2B1	2B3	2B5	1B1	1B3	1B5	2B1	2B3	2B5
C	1C1	1C3	1C5	2C1	2C3	2C5	1C1	1C3	1C5	2C1	2C3	2C5
D	1D1	1D3	1D5	2D1	2D3	2D5	1D1	1D3	1D5	2D1	2D3	2D5
E	1A2	1A4	1A6	2A2	2A4	2A6	1A2	1A4	1A6	2A2	2A4	2A6
F	1B2	1B4	1B6	2B2	2B4	2B6	1B2	1B4	1B6	2B2	2B4	2B6
G	1C2	1C4	1C6	2C2	2C4	2C6	1C2	1C4	1C6	2C2	2C4	2C6
H	1D2	1D4	1D6	2D2	2D4	2D6	1D2	1D4	1D6	2D2	2D4	2D6

Fig. 3 Layout of 48 samples in 96-deep-well block for sonication and SDS-PAGE analysis. (**a**) The layout of the 96-deep-well block containing 48 sonicated samples derived from both 24-deep-well blocks is shown. (**b**) The layout of the 96-well PCR plate for loading 96 samples on 12 gels with a multichannel pipette for SDS-PAGE analysis. The description in each well refers to the location of each sample in the original 24-deep-well block before transfer to the 96-deep-well block after sonication (e.g., 1A1, refers to Block 1, position A1; 2D6, refers to Block 2, position D6)

3. For constructs that express soluble protein at 37 °C, induce cultures by adding 1 mL 1 M IPTG/flask and shake at 37 °C, 200 rpm for 4 h. For constructs that require induction overnight at 18 °C to express soluble protein, transfer the cultures to an incubator at 18 °C and allow them to cool for 15 min before inducing them (*see* **Note 9**).

4. Harvest the cells at 5,670 × *g* for 7 min in 1-L centrifuge tubes. Store cells in centrifuge tubes at −20 °C or −80 °C, after transferring cell paste into a 50 mL Falcon tube.

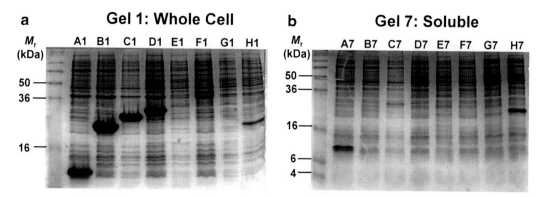

a **Gel 1: Whole Cell**   **b** **Gel 7: Soluble**

Fig. 4 SDS-PAGE analysis of protein expression. A typical example from a screen to compare the protein expression levels and solubility is shown. Gels 1 and 7 correspond to (**a**) whole cell and (**b**) soluble extracts of 8 expression tests. Sample *lanes 1* and *8* on each gel reveals that these constructs produce expressed and soluble protein; sample *lanes 2, 3, 4* and *6* on each gel are representative of constructs expressing protein, but it is not soluble (or barely soluble, *lane 3*); *lanes 5* and *7* on each gel show these constructs do not express any protein. Marker proteins were Seeblue+2. Gels were stained with Instantblue

3.4 Extraction and Purification of Soluble Proteins for Crystallization

1. Thaw frozen cell pellets (*see* **Note 10**) from Subheading 3.3, **step 4**. Hold on ice. Crude protein extracts from this point onwards should be kept on ice, or at 4 °C.

2. Resuspend cells in lysis buffer (*see* Subheading 2.4.2, **item 1** and **Note 11**) until a homogeneous mixture is obtained.

3. Lyse the cells by sonication (*see* **Note 12**) using a large (10–12 mm) probe. Use 50 mL Falcon tubes or larger plastic beakers completely surrounded by ice.

4. Centrifuge the cell lysate in Oakridge centrifuge tubes at $40,910 \times g$ for 15 min at 4 °C. Transfer supernatants to fresh Oakridge tubes and recentrifuge for 5 min. Decant and combine the supernatants in a 50 mL Falcon tube (or larger container). Dilute 3 μL sample with 27 μL A1 buffer and add 10 μL 4× SDS-loading buffer. Store at −20 °C for SDS-PAGE analysis (Fig. 5). Purify rest of the sample using AKTA Xpress system.

5. To use AKTA Xpress system: Insert sample inlet tubing (S1) into Falcon tube containing 40 mL IMAC equilibration buffer (A1, *see* Subheading 2.4.2, **item 2**); for multiple samples, repeat for inlets S2-S4. Attach 24-deep-well block for collecting protein fractions. Run program designed for 2-step purification: Affinity chromatography on a 5 mL Hi-Trap FF Ni-IMAC column followed by gel filtration chromatography on, for example, a Superdex 75 26/60 column. Note checklist for column positions and volumes of three purification buffers (A1, B1 [Subheading 2.4.2, **item 3**] and A4 [Subheading 2.4.2, **item 4**]) and any post-run solutions.

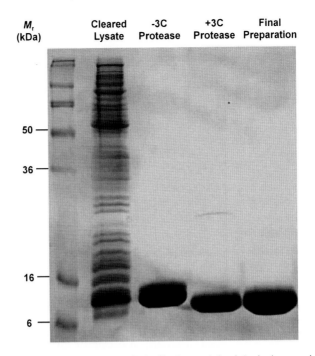

Fig. 5 SDS-PAGE analysis of purified effector protein. A typical example of the purification of an effector protein is shown. The expressed protein was purified from the cleared lysate by IMAC and gel filtration to prepare the "−3C" sample. The fusion protein was then incubated with 3C protease (12 μg/mg protein) overnight at 4 °C to remove the histidine tag ("+3C"). The final preparation was obtained after further purification by IMAC (collecting the flow-through) and gel filtration, to remove both the uncleaved effector protein and the HIS-tagged 3C protease, and concentrated to 10 mg/mL. Marker proteins were Seeblue+2. Gels were stained with Instantblue

6. Following the run, pool or analyze fractions (8 mL) on SDS-PAGE gels and concentrate monodisperse fractions to 2–3 mg/mL using ultrafiltration devices.

7. Dilute the protein solution and measure the A_{280} using a Nanodrop or other suitable spectrophotometer. Calculate the concentration and total amount of protein purified using the extinction coefficient predicted by, for example, ProtParam http://web.expasy.org/protparam/.

8. To a 30 μL sample ("−3C") add 10 μL 4× SDS-loading buffer. Store at −20 °C for SDS-PAGE analysis (Fig. 5).

9. N-terminally tagged pOPIN vectors contain a 3C protease site for tag removal. If this is required, add ß-mercaptoethanol to the protein solution to a final concentration of 5 mM from a freshly prepared 1 M stock (35 μL/0.5 mL A4 buffer).

10. Digest the remaining protein with 3C protease (12 μg/mg fusion protein) overnight at 4 °C to remove the tags (*see* **Note 13**). Remove a sample ("+3C") for SDS-PAGE analysis as above (Fig. 4).

11. On the bench, equilibrate a 5 mL FF Ni-IMAC column with A1 buffer (50 mL) using a peristaltic pump. With a syringe, apply the sample to the column and collect the unbound fraction. Wash the column with a further 25 mL A1 buffer using the pump and combine with the unbound fraction. Concentrate the protein to 5–8 mL using an ultrafiltration device.

12. Run the AKTA Xpress program for gel filtration with A4 buffer. Pool fractions (8 mL) and concentrate protein to at least 10 mg/mL.

13. Set up crystallization screens (*see* **Note 14**).

3.5 Setting Up Protein Crystallization Screens and Crystal Screening

An initial strategy is to set up 6×96-well screens with 0.6 µL drops consisting of 50 % protein and 50 % precipitant. Ideally, set up 2 drops at 1× and 0.5× the concentration at which the protein has been stored, e.g. 10 and 5 mg/mL. For a single drop you will require a total of 6×33 µL (approx. 200 µL) or 2 mg protein (at 10 mg/mL); for two drops you will require a total of 9×33 µL (approx. 300 µL) or 3 mg protein (at 10 mg/mL).

1. Prepare appropriate amount of either protein stored in the fridge, or thawed frozen protein, for the intended screens.

2. Spin protein at $17,000 \times g$ in a microcentrifuge at 4 °C for 15 min. Transfer supernatant to fresh tube. Check protein concentration. For setting up a second drop at 0.5× protein concentration, dilute 100 µL protein with an equal volume of A4 buffer.

3. Using a liquid handling robot (or multichannel pipettes), transfer 50 µL precipitants from each well of a 96-deep-well block containing the crystallization precipitants from commercial screens to the corresponding reservoirs of an MRC 2 drop 96-well plate. Place new seal on 96-deep-well block.

4. Transfer 33 µL protein to a PCR tube; for setting up two drops transfer 33 µL 0.5× diluted protein to a second PCR tube. Use the Oryx Nano robot, other crystallization robotics or multichannel pipettes to set up sitting drops.

5. Seal MRC plate and transfer to a controlled environment room. Use a crystal imaging robot if available.

6. Manually inspect crystallization trays (or images taken by the imaging robot) daily for 5 days and every week for 3 weeks. Record any conditions containing protein crystals.

7. Look up details of the precipitant in any well of interest using the lists provided with commercial screens. If multiple crystals are obtained, look for crystals with different morphologies (e.g., plates versus cubes).

8. Check if false positives have been seen before in this well. It may suggest the crystals are derived from salts in the precipitant and are not protein.

9. If possible, test any crystals to see if they diffract X-rays (see below) and assess diffraction quality. Confirm whether crystals obtained are salt or protein.

10. If crystals are protein and diffract reasonably well, set up optimization screens.

3.6 Testing Protein Crystals for Diffraction Quality

In-house X-ray diffractometer set-ups are ideal for screening initial crystals for diffraction quality. However, optimization screens are frequently required to obtain sufficient numbers of larger crystals required to determine the structure. Normally, the best crystals should be sent to a synchrotron radiation source for the collection of the very best X-ray dataset. Synchrotron radiation may also be required for solving the structure.

1. Examine crystal and estimate size of loop required to pick it up. Place drop of cryoprotectant solution onto a cover slip on the stage of the binocular microscope.

2. Working quickly, cut into the seal of 96- or 48-well plates covering the well with the crystal using a scalpel blade (or remove and invert cover slip over 24-well hanging drop plate).

3. Lift out crystal onto loop into cryoprotectant. Then loop-out the crystal and either flash-freeze it in liquid nitrogen or place it directly into position in a cryostream mounted on a diffractometer.

4. Collect X-ray diffraction data.

3.7 Optimization of Protein Crystals for Improving Quality of X-Ray Diffraction

A sensible strategy is to set up screens using 24-well plates (hanging drop) or 48-well plates (sitting drop) using home-prepared precipitant solutions based on those that gave crystals (*see* **Note 15**). Ideally use larger 2–4 µL drops (typically 50 % protein and 50 % precipitant). For hanging drop experiments the entire process will most likely be carried out manually. For sitting drops in 48-well plates, the Oryx Nano robot (or equivalent, or multichannel pipettes) can be used to set up the drops. First, define a broad range of conditions that span either side of the condition that gave crystals in the 96-well screen. In a 24-well plate, try 4× different buffers from pH 4–9, and 6 different (increasing) precipitant concentrations. Reduced increment steps can be used in 48-well format. For 24- and 48-well plates with single 2 µL drops (50 % precipitant) and protein at 10 mg/mL, you will need 0.24 or 0.48 mg protein, respectively.

1. Prepare 24×5–10 mL or 48×1 mL samples of each precipitant.

2. Transfer 1 mL or 0.2 mL of each precipitant to the reservoir of a 24-well or 48-well plate, respectively. Store remainder of precipitant solutions.

3. For 48-well plate (sitting drop) experiments, robotics or multichannel pipettes can be used to set up drops.

4. For 24-well plate (hanging drop) experiments: Line the perimeter of each well with silicone grease using a syringe pump dispenser or manually with a syringe. Lay out 24× cover slips in plate (6×4) format and pipette 1–2 μL protein onto a single row of six cover slips. Then, with a fresh tip, add the same volume of precipitant from the reservoir to the protein drop on the cover slip and slowly pipette up and down to mix the drop. Repeat this procedure for the other five cover slips in the first row. Invert the cover slips over the corresponding row of wells in the 24-well plate and gently press the cover slips in place using fine forceps to seal the edges completely. Repeat this procedure for the other three rows of the 24-well plate.

5. Transfer plate to crystallization room at 20 °C and monitor growth of crystals daily.

6. Review results frequently and test any crystals for diffraction. If necessary, set up other screens that have more finely tuned crystallization conditions.

7. Identify crystals giving the best diffraction.

3.8 Solving the "Phase Problem"

The phase-problem of crystallography refers to the fact that a diffraction experiment only records the intensities of diffracted X-rays and not their phase. Both sets of information are required for re-construction of electron density maps. There are many methods that can be used to solve the phasing problem in crystallography, which are outside the scope of this article (e.g., [21]). We have successfully used: a combined single anomalous dispersion (SAD)/single isomorphous replacement plus anomalous differences (SIRAS) approach using the anomalous signal from iodide soaking of native crystals; co-crystallization with ammonium bromide and anomalous scattering from bound bromide ions; and producing crystals from selenomethionine-labeled protein for collection of Se-SAD data, to solve crystal structures of bacterial and oomycete plant pathogen effectors [22–24]. Where an existing structure of a protein with similar sequence is available, molecular replacement is also an option. A range of well-developed computer programs, many with user-friendly interfaces, are available that will process X-ray data and, in straightforward cases, may solve the structure and produce an initial protein structure automatically.

A structure is considered "solved" when an interpretable electron density map is obtained, into which an initial protein model that satisfies the electron density can be placed. At this stage it is highly likely a structure determination can be completed in a matter of days or weeks.

4   Notes

1. This is prepared by dilution from a 12.5 % (w/v) PEI stock (1 in 4 dilution of a 50 % [w/v] PEI commercial stock, MW 60,000, Acros 178572500) that has been readjusted to pH 8.0.

2. If crystals grown in the optimization screens diffract poorly, then including additives or crystal "seeds" in the screen may improve the diffraction quality of crystals. Additives will manipulate sample–sample and sample–solvent interactions and alter sample solubility. Seeding experiments may encourage the growth of larger, single crystals.

3. If crystals cannot be reproduced in optimization screens using homemade solutions, the precipitant used in the 96-well plate commercial screens can be obtained from the supplier in larger volumes.

4. These compounds can be used in situations where attempts with iodide soaks have failed.

5. For strains with the pLysS or pLysE plasmid, or other strains where there is an additional antibiotic load, or which grow more slowly (judge by colony size), inoculate with 500 μL overnight culture. The aim is to obtain growth to an A_{600} at least 0.2 after 3 h. Where growth is poor in BL21(DE3) due possibly to the leaky expression of toxic proteins, induction can be attempted after 3 h regardless of A_{600}.

6. The volume of lysis buffer used will depend on the size of the cell pellet and the A_{600} of the cells when you induced the cultures with IPTG. Aim for a "slightly cloudy" mixture and about the same level of cloudiness in each well. You will need at least 2 mL to sonicate in a 24-deep-well block. If ANY of the cell pellets in the 24-deep-well block is very small, resuspend these pellets in 0.5 mL, then transfer 0.5 mL of ALL the extracts in the 24-deep-well block to a 96-deep-well block using the expandable multichannel pipette BEFORE sonication using the 24-well probe.

7. If this is a problem, then dilute difficult samples with 1× SDS-PAGE loading buffer until mobile. An alternative strategy would be to include Benzonase in the lysis buffer. This can almost always be avoided if the DNA has been sheared by effective sonication and the protein extracts were not too concentrated. If necessary, resuspend pellets in the centrifuged 96-deep-well block (Subheading 3.2, **step 11** above), then add 12.5 U/mL Benzonase for 10 min.

8. Excess gels can be individually wrapped in cling film and covered with wet blue towel in a sealed sandwich box at 4 °C.

9. Before extraction and purification it is worthwhile ensuring that the protein has expressed after scale-up. For this, take a 1–3 mL sample of the cells before and after induction. Spin the cultures at $2,400 \times g$ in a microcentrifuge, discard the supernatants and store the cell pellets at –20 °C for expression testing. Resuspend the cell pellets in Bugbuster reagent (in 50 mM Tris–HCl, pH 8.0) containing Benzonase (0.5 µL/mL). Leave for 10 min at room temperature. Transfer 30 µL to a 1.5 mL Eppendorf tube and add 10 µL 4× SDS loading buffer—this is the whole cell fraction. Spin remainder of the extract at $13,000 \times g$ (or max speed) in a microcentrifuge for 5 min at 4 °C. Transfer 30 µL supernatant and treat as above—this is the soluble fraction. Run both samples on SDS-PAGE gels.

10. If possible, prepare enough protein to set up 6 crystallization screens and have a few aliquots of protein leftover for setting up optimization screens (assuming crystals are obtained). This will reduce variation between batches of protein.

11. The volume of lysis buffer used for resuspension will depend on cell mass. Cells from 2 to 4-L cultures induced at 37 °C for 4 h can typically be resuspended in a total volume of 50 mL or less. Cells left to grow to saturation overnight at 18 °C, or cultures over 4-L should be resuspended in a minimum of 100 mL lysis buffer.

12. The length of time for sonication is variable. Ensure that no small cell clumps remain at the bottom of the tube/beaker. The final lysate should be fairly mobile and of low viscosity. Dilute with A1 buffer, if necessary, before loading on the IMAC column.

13. We routinely remove the histidine tag before setting up crystallization screens. In smaller proteins (<10 kDa) the flexible, unstructured tag may prevent crystal formation or adversely affects the quality of crystals.

14. The freshly purified protein can be used immediately for setting up crystallization screens. However, it will likely be necessary to store the protein preparation. A good practice is to flash-freeze the protein in 50 µL aliquots and store at –80 °C. If plenty is available, also store some protein at 4 °C. Depending on the project it may be appropriate to use the remaining protein for other studies.

15. Cover slips used in 24-well plates (hanging drop) can accommodate up to four drops. Two sitting drops can typically be set up in 48-well format.

References

1. Mukhopadhyay A (1997) Inclusion bodies and purification of proteins in biologically active forms. In: Scheper TH (ed) Advances in mechanical engineering/biotechnology, vol 56. Springer, Berlin

2. Swietnicki W (2006) Folding aggregated proteins into functionally active forms. Curr Opin Biotechnol 17:367–372

3. Nguyen LH, Jensen DB, Burgess RR (1993) Overproduction and purification of σ32, the *Escherichia coli* heat shock transcription factor. Protein Expr Purif 4:425–433

4. Hart DJ, Tarendeau F (2006) Combinatorial library approaches for improving soluble protein expression in *Escherichia coli*. Acta Crystallogr D Biol Crystallogr 62:19–26

5. Hartley JL (2006) Cloning technologies for protein expression and purification. Curr Opin Biotechnol 17:359–366

6. Glaser V (1999) Custom DNA synthesis—a crowded marketplace spurs competition in quality and price. Gen Eng News 19:10

7. pET System Manual, 11th Edition (2006) Novagen Technical Bulletin TB055. Merck (USA)

8. Esposito D, Chatterjee DK (2006) Enhancement of soluble protein expression through the use of fusion tags. Curr Opin Biotechnol 17:353–358

9. Park SJ, Lee SK, Lee BJ (2002) Effect of tandem rare codon substitution and vector–host combinations on the expression of the EBV gp110 C-terminal domain in *Escherichia coli*. Protein Expr Purif 24:470–480

10. Yang ZR, Thomson R, McNeil P, Esnouf RM (2005) RONN: the bio-basis function neural network technique applied to the detection of natively disordered regions in proteins. Bioinformatics 21:3369–3376

11. Hortsch R, Weuster-Botz D (2011) Growth and recombinant protein expression with *Escherichia coli* in different batch cultivation media. Appl Microbiol Biotechnol 90:69–76

12. Xu JS, Li WJ, Wu J, Zhang Y, Zhu Z, Liu JJ, Hu ZY (2006) Stability of plasmid and expression of a recombinant gonadotropin-releasing hormone (GnRH) vaccine in *Escherichia coli*. Appl Microbiol Biotechnol 73:780–788

13. Schein CH, Noteborn M (1988) Formation of soluble recombinant proteins in *Escherichia coli* is favoured by lower growth temperature. Bio/Technology 6:291–294

14. Prasad S, Khadatare PB, Roy I (2011) Effect of chemical chaperones in improving the solubility of recombinant proteins in *Escherichia coli*. Appl Environ Microbiol 77:4603–4609

15. FerreDAmare A, Burley SK (1997) Dynamic light scattering in evaluating crystallizability of macromolecules. Methods Enzymol 276:157–166

16. Nettleship JE, Brown J, Groves MR, Geerlof A (2008) Methods for protein characterization by mass spectrometry, thermal shift (thermofluor) assay, and multi-angle or static light scattering. Methods Mol Biol 426:219–318

17. Whisson SC, Boevink PC, Moleleki L, Avrova AO, Morales JG, Gilroy EM, Armstrong MR, Grouffaud S, van West P, Chapman S, Hein I, Toth IK, Pritchard L, Birch PR (2007) A translocation signal for delivery of oomycete effector proteins into host plant cells. Nature 450:115–118

18. Schornack S, Huitema E, Cano LM, Bozkurt TO, Oliva R, Van Damme M, Schwizer S, Raffaele S, Chaparro-Garcia A, Farrer R, Segretin ME, Bos J, Haas BJ, Zody MC, Nusbaum C, Win J, Thines M, Kamoun S (2009) Ten things to know about oomycete effectors. Mol Plant Pathol 10:795–803

19. Bozkurt TO, Schornack S, Banfield MJ, Kamoun S (2012) Oomycetes, effectors, and all that jazz. Curr Opin Plant Biol 15:483–492

20. Berrow NS, Alderton D, Sainsbury S, Nettleship J, Assenberg R, Rahman N, Stuart DI, Owens RJ (2007) A versatile ligation-independent cloning method suitable for high-throughput expression screening applications. Nucleic Acids Res 35:e45

21. Hauptman H (1997) Phasing methods for protein crystallography. Curr Opin Struct Biol 7:672–680

22. Boutemy LS, King SRF, Win J, Hughes RK, Clarke TA, Blumenschein MA, Kamoun S, Banfield MJ (2011) Structures of *Phytophthora* RXLR effector proteins: a conserved but adaptable fold underpins functional diversity. J Biol Chem 286:35834–35842

23. Sohn KH, Hughes RK, Piquerez SJ, Jones JDG, Banfield MJ (2012) Distinct regions of the *Pseudomonas syringae* coiled-coil effector AvrRps4 are required for activation of plant immunity. Proc Natl Acad Sci U S A 109:16371–16376

24. Wirthmueller L, Jones JD, Banfield MJ (2011) Crystallization and preliminary X-ray analysis of the RXLR-type effector RXLR3 from the oomycete pathogen *Hyaloperonospora arabidopsis*. Acta Crystallogr F Struct Biol Cryst Commun 67:1417–1420

Part IV

Methods to Identify Resistance Genes and Avirulence Genes

Chapter 19

The *Do's* and *Don'ts* of Effectoromics

Juan Du and Vivianne G.A.A. Vleeshouwers

Abstract

Effectoromics, a high-throughput functional genomics approach that uses effectors to probe plant germplasm to detect *R* genes, has proven a potent contribution to modern resistance breeding. Advantages of effectoromics are summarized in four aspects: (1) accelerating *R* gene identification; (2) distinguishing functional redundancy; (3) detecting recognition specificity and (4) assisting in *R* gene deployment. In this manuscript, we provide suggestions as well as some reminders for applying effectoromics in the breeding process. The two routine functional assays that are widely used, agroinfiltration and agroinfection, are presented. We briefly explain their advantages and disadvantages and provide protocols for applying them in the model system *Nicotiana benthamiana* as well as in potato (*Solanum tuberosum*).

Key words Effectoromics, Effectors, Resistance breeding, Agroinfiltration, Agroinfection, *Nicotiana benthamiana*, Potato

1 Introduction

Effectoromics has recently emerged as a powerful tool to identify resistance (*R*) genes in crop plants and matching avirulence (*Avr*) genes of pathogens [1–4]. This high-throughput functional genomics approach uses effectors to probe plant germplasm for specific recognition of R proteins. The required inflow of predicted effectors from plant pathogen genome sequences is becoming increasingly easy with the rapid development of sequencing technology. Still, after analyzing the huge amount of sequences by bioinformatics for putative functional genes, efficient biological assays are essential. Effectoromics perfectly bridges this gap, as we have shown for the late blight pathogen *Phytophthora infestans* and its interactions with wild potato species. Currently, researchers and breeders from various fields are adopting this approach for their own specific pathosystem.

For effectoromics in *Solanaceous* plants, two *Agrobacterium-based* functional assays, namely agroinfiltration and PVX (*Potato Virus X*) agroinfection, are routinely used to transiently express

Paul Birch et al. (eds.), *Plant-Pathogen Interactions: Methods and Protocols*, Methods in Molecular Biology, vol. 1127, DOI 10.1007/978-1-62703-986-4_19, © Springer Science+Business Media New York 2014

Table 1
Comparison of characteristics of agroinfiltration and PVX agroinfection

	Agroinfiltration	PVX agroinfection
Sensitivity	Sensitive	Very sensitive
Efficiency	Medium throughput	High throughput
Nonspecific responses	Sometimes suffers from plant responses to *Agrobacterium*	Sometimes suffers from plant responses to PVX
Reliability	Reliable	Sometimes risky to miss responses due to extreme resistance
Preparation of inoculation cultures	Relatively laborious	Quick
Phenotyping results	Short incubation time (3–5 days)	Long incubation time (up to 2 weeks)

effectors in plant cells [5–7]. Agroinfiltration is performed by infiltrating a suspension of recombinant *Agrobacterium* into a plant leaf, which will transfer the desired gene into plant cells. PVX agroinfection is also based on *Agrobacterium*, in this case as a binary PVX expression system. Agroinfiltration and PVX agroinfection both have advantages and disadvantages (Table 1). Briefly, the agroinfiltration assay (Box 1) is very suitable for functional analysis of single genes, such as resistance (*R*) or avirulence (*Avr*) genes, as well as for reconstructing the R-AVR relationships by delivering two transgenes into the same cell. However, nonspecific defense to *Agrobacterium* is regularly observed, especially in some potato genotypes. PVX agroinfection (Box 2) is more sensitive, more high-throughput and less sensitive to nonspecific defense responses to *Agrobacterium*. However, in this case, nonspecific defense to PVX can occur and there is a risk to miss responses due to virus-induced extreme resistance (ER). Generally, agroinfiltration experiments are more laborious in preparing the inoculation culture of *Agrobacterium*, while PVX agroinfection experiments take longer until symptoms can be scored.

As indicated above, the main issue for agroinfiltration and PVX agroinfection is the chance that nonspecific defense responses are raised against the infection agents, *Agrobacterium* or PVX, respectively, in diverse plant materials (Fig. 1a). To solve this issue, we usually search for related plant genotypes that carry the same resistance gene but do not suffer from this unpractical response to the infection agents (Fig. 1b). One way to do this is to identify resistant offspring plants that are amenable to the expression assays, and subject those genotypes to subsequent effector studies (Fig. 1c, d).

Box 1
Agroinfiltration Protocol

1. Around 4–5-week-old seed-grown *N. benthamiana* or potato from in vitro tissue culture can be used for agroinfiltration. Choose young, healthy and fully developed leaves for infiltrations.

2. Inoculate 20 μl glycerol stock of the desired *Agrobacterium* strains into 10 ml YEB medium supplemented with 1 μl acetosyringone (3′–5′ dimethoxy-4′-hydroxy acetophenone, 200 mM stock, 39.3 mg/ml DSMO), 100 μl MES buffer (2-(*N*-morpholino)-ethane sulfonic acid, 1 M stock, 195 g/L) and the appropriate antibiotics. Incubate cells for 1–2 days at 28 °C at 200 rpm to an OD_{600} of approximately 1.0.

3. Harvest cells by centrifugation at $3,400 \times g$ for 10 min, pour off the supernatant and resuspend the pellet in freshly made MMA medium (20 g sucrose, 5 g MS salts, 1.95 g MES, pH adjusted to 5.6 with NaOH, and 1 ml acetosyringone/L) to an OD_{600} of 0.3. For co-infiltration, mix the culture in a 1:1 ratio. Then incubate cells at room temperature for 1–6 h.

4. Place *Agrobacterium* suspensions into a 1 ml needleless syringe. Carefully inject the suspension from the syringe to the leaf.

5. Responses can be macroscopically scored about 3 days after infiltrations. Results can be quantified by assessing cell death percentages.

Box 2
PVX agroinfection Protocol

1. Around 2–3-week-old seed-grown *N. benthamiana* or potato from in vitro tissue culture can be used for PVX agroinfection. For large-scale tests, slightly older (4–5 weeks) plants can be used.

2. Inoculate 20 μl glycerol stock of the desired *Agrobacterium* strains into 3 ml YEB medium supplemented with the appropriate antibiotics. Incubate cells for 1–2 days at 28 °C and 200 rpm to an OD_{600} of approximately 1.0.

3. Pipet about 300 μl of each *Agrobacterium* strain and spread them onto LB solid agar medium plates supplemented with the appropriate antibiotics. Incubate cells at 28 °C for 1–2 days.

4. Dip a wooden toothpick in the culture of the recombinant *Agrobacterium* strain and pierce the leaves.

5. To make a quantitative scoring possible, make multiple inoculations sites for each strain. Use three leaves per plant to serve as triplicates. Inoculate at least three plants for each strain.

6. Symptoms can be scored about 2 weeks after inoculation. For high-throughput screens, summarize the qualitative responses (yes/no) for each inoculation spot. Then calculate the percentage of responding sites and compare them with controls.

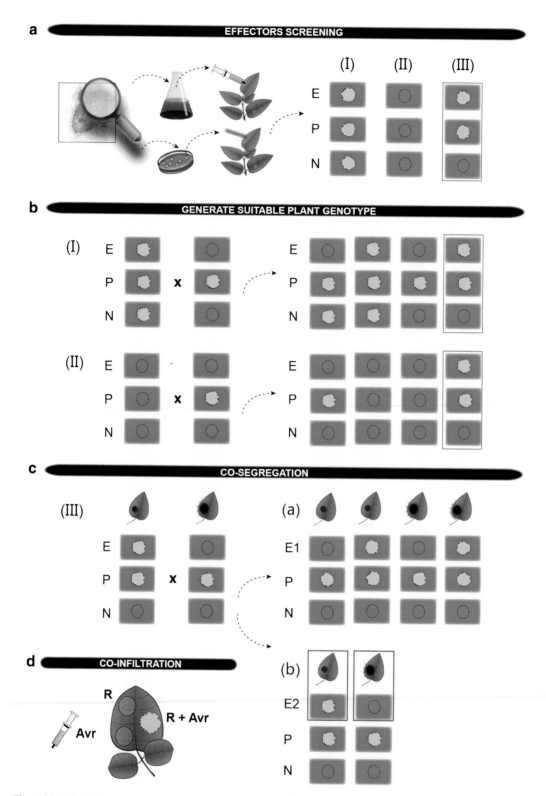

Fig. 1 Identification of *R* and *Avr* genes by effectoromics. RXLR effectors are retrieved from the *Phytophthora infestans* genome sequences and cloned into expression vectors. Constructs are then introduced in *Agrobacterium tumefaciens* for functional screening by agroinfiltration or *Potato Virus X* (PVX) agroinfection. Plant genotypes that are suitable for effectoromics screens (III) show response to the positive control (P) but

So far, effectoromics has shown to contribute to resistance breeding in four aspects, i.e. accelerating *R* gene identification, distinguishing functional redundancy, detecting recognition specificity and assisting in *R* gene deployment [4]. Distributed over these four aspects, we list seven "Do's" that may be helpful to design a resistance breeding program involving effectoromics in potato or other crops. In addition, we discuss three "Don'ts" that are important to consider when assessing the results of effectoromics experiments.

2 Do's

2.1 Exploit Effectors to Accelerate R Gene Cloning

During *R* gene cloning processes, various candidate genes are often obtained. Complementation studies are needed to select the true functional *R* gene. In a classic complementation test, e.g. for potato, it takes up to 6 months to generate stable transgenic plants that can be inoculated with *P. infestans* and assessed for resistance. By contrast, if the matching *Avr* gene is available, the test can be performed in several days; one needs simply to agro-co-infiltrate the candidate *R* gene with the matching *Avr* gene in leaves of *Nicotiana benthamiana* to test for *R* gene-specific cell death responses. In this way, the identification and characterization of *R* genes in potato can greatly be accelerated [1, 8–10].

2.2 Distinguish Functional Redundancy

2.2.1 Exploit Effectors for Circumventing Genetic Crossing Barriers

Since most cloned *R* genes have been defeated by the fast evolving pathogen *P. infestans*, potato breeders explore wild *Solanum* germplasm for new *R* genes. However, genetic crossing barriers, together with linkage drag, remain a problem in potato resistance breeding and can drastically delay introgression of *R* genes. By effectoromics approaches, this problem can be solved. Functional allele mining with *Avr* genes in large collections of germplasm can quickly lead to identification of functional *R* gene homologs in various species. Sexually more compatible species with a particular resistance specificity can then be selected for introgression. A good example is the well-known broad-spectrum potato resistance gene

Fig. 1 (continued) not the negative control (N), and response to effectors (E) can be studied (in *red box*). However, some potato genotypes (I) show nonspecific responses to negative controls, other genotypes (II) fail to show response to positive controls, and such genotypes are not suitable. (**b**) To generate more suitable genotypes, genetic crosses are made. Among the progeny, potato genotypes that show response to the effector as well as the positive control but not the negative control are selected (in *red boxes*). (**c**) Genetic studies for co-segregation of response to effectors with resistance to *Phytophthora infestans*. In (**a**), response to effector E1 does not co-segregate with resistance. In (**b**), response to one effector E2 co-segregates with resistance (in *blue boxes*), and E2 is a candidate *Avr* gene that matches the *R* gene in the tested plant. (**d**) Further validation by agro-co-infiltration of the candidate *Avr* gene with the matching *R* gene. Specific cell death occurs in leaf panels where *R* and *Avr* gene are simultaneously expressed

RB/Rpi-blb1 [11, 12], which originates from *Solanum bulbocastanum* that is not directly sexually compatible with cultivated potatoes. In the past, difficult somatic hybridisation experiments [11] and time-consuming bridge-crossing experiments [13] were used to introgress the *Rpi-blb1/RB* gene into *S. tuberosum*. Functional screens with *Avrblb1* in *Solanum* germplasm quickly led to identification of specific cell death responses in *Solanum stoloniferum*, which is directly crossable with cultivated potatoes. Based on *Rpi-blb1* homology and functional equivalence, the two *Rpi-blb1* homologues *Rpi-sto1* and *Rpi-pta1* were rapidly cloned from *S. stoloniferum* [1]. In addition to trans- or cisgenic approaches, traditional breeding strategies can now be more quickly used to introgress the *Rpi-blb1* specificity into potato.

In addition to identifying sexually compatible resistant species, breeders can avoid redundant breeding or cloning efforts by classifying germplasm or *R* genes based on their responses to effectors [1–3]. This is particularly important for *R* genes with broad-spectrum effects for which diagnostic pathogen races are not available. Therefore, effector-based resistance breeding also enables pyramiding of functionally complementary *R* genes.

2.2.2 Exploit Effectors for Dissecting Resistance Specificities

Although most cloned *R* genes have been defeated by local *P. infestans* populations, some cultivars and wild species still retain a certain resistance and plants remain healthy. In many cases, the resistance of those plants is based on pyramided *R* genes. Compared to traditional inoculations with *P. infestans* isolates, the effectors can more easily and more accurately dissect the activities of otherwise indistinguishable *R* genes into discrete recognition specificities. For example, resistance specificities were dissected in a segregating population of potato cultivar "Sarpo Mira", which has retained resistance in practice over many years. Responses to *P. infestans* RXLR effectors were matched with race-specific resistance responses to different *P. infestans* strains and "Sarpo Mira" was shown to contain at least three known *R* genes, *R3a*, *R3b* and *R4*, and two new genes, *Rpi-Smira1* and *Rpi-Smira2* [9]. In the same way, the two potato *R* gene differentials MaR8 and MaR9 [14] were dissected to have at least four (*R3a*, *R3b*, *R4* and *R8*) and seven (*R1*, *Rpi-abpt1*, *R3a*, *R3b*, *R4*, *R8*, *R9*) *R* genes, respectively [8]. Knowing which combinations of *R* genes are present in the resistant genotypes, breeders can introgress their favorite *R* genes into the current potato cultivars. The main prerequisite is obvious, namely that the matching *Avr* genes should be known.

2.3 Detect Specificity

2.3.1 Exploit Effectors for Detecting Weak Resistance Phenotypes

For many years, potato breeding has been focussed on the introgression of field resistance, which has been claimed to be more durable in various studies. However, the genetic basis of field resistance has remained unclear for two reasons. One is that the weak phenotypes of field resistance are too difficult to follow in the genetically complex potato breeding. The other reason is that the avirulence profiles of infecting *P. infestans* strains cannot always be accurately determined in routine detached leaf assays for *R* genes with weaker phenotypes [8]. In such cases, effectors can play an important role. For example, the so-called field resistance gene *Rpi-Smira2* in potato cultivar "Sarpo Mira" could not be distinguished in detached leaf assays with *P. infestans*, but was detected by response to *Avr-Smira2* [9]. Thus, field resistance phenotypes, which are often too weak to be reliably detected under laboratory conditions on detached leaves, can be accurately detected by effectors. It shows that effectors can act as functional markers and contribute to more efficient resistance breeding in potato.

2.3.2 Expand Effector Recognition Specificity of R Genes; Target for Broad-Spectrum Resistance

Expanding the effector recognition specificity of a given *R* gene to new virulent alleles can further improve breeding for durable resistance. As described above, *R* genes that can target all allelic forms of *Avr* genes could provide a full-spectrum resistance to pathogen isolates. A potential tool to accomplish this goal is the artificial evolution by random mutagenesis, as previously demonstrated for the PVX resistance gene *Rx* [15, 16]. Whenever the original *R* gene is present in the crop species, one can implement targeted mutagenesis (genome editing) by new technologies, such as zinc finger nuclease-based approaches [17, 18] or transcription activator-like (TAL) effectors that can be fused to DNA nucleases to target a precise site in a genome to create genetic variation [19, 20]. Genome editing could provide a non-transgenic resistant variety that does not carry extraneous pieces of DNA [21]. To design an efficient screening system to identify the mutated *R* gene candidates, basic knowledge of the pathogen effectors is a key requirement. Also, the right choice of the R-AVR pair will influence the new *R* gene durability in the field. For instance, as mentioned above, *R3a* but not *R4* would be a good target to manipulate and an engineered *R3a* that also recognizes $Avr3a^{EM}$ could provide a broader spectrum resistance than the original *R3a*. Indeed, expanded recognition specificity by a new *R* gene variant could be due to a single aa change in the R protein, as recently discovered for *R3a* [22].

2.4 Deployment

2.4.1 Exploit Effectors That Are Important for P. infestans; Target for Durable Resistance

Essential effectors of *P. infestans* are expected to be useful targets for potato resistance breeding. The genome and expression analyses of three *P. infestans* strains show that only a small subset of 45 RXLR genes is consistently induced *in planta* during the biotrophic infection stage [23]. Among those "core effectors", most known *Avr* genes occur, for example the extensively studied *Avr3a*. In *P. infestans* populations, two alleles of *Avr3a* have been identified that encode secreted proteins $AVR3a^{K80/I103}$ ($AVR3a^{KI}$) and $AVR3a^{E80/M103}$ ($AVR3a^{EM}$), which differ in two amino acids (aa) in their effector domains [24]. Only $AVR3a^{KI}$ can induce potato resistance to *P. infestans* by activating the resistance protein R3a. Interestingly, AVR3a was recently shown to be essential for full virulence of *P. infestans* [25]. These findings suggest that *R3a* can be an important target for durable resistance breeding when an *R3a* variant that targets both allelic forms of *Avr3a* can be identified from natural sources (*see* Subheading 2.2.1) or otherwise (*see* Subheading 2.3.2). In contrast, *R* genes such as *R4* are not considered as useful targets for durable resistance breeding because the matching *Avr4* gene encodes dispensable effectors [26].

In addition to breeding for resistance genes that recognize core RXLR effectors, another layer of resistance in the form of apoplastic receptors offers an alternative for durable resistance breeding. These apoplastic receptors can recognize apoplastic effectors that in some cases represent conserved pathogen-associated molecular patterns (PAMPs). Among the best-characterized oomycete PAMPs are elicitins, a conserved family of extracellular proteins that share a 98-amino-acid elicitin domain [27–30]. For oomycete species that cannot synthesize sterols, elicitins fulfill an important function as sterol scavengers. Therefore, we argue that receptors of PAMPs like elicitins can be good targets for breeding durable resistant potato.

2.4.2 Assist in R Gene Deployment in Agriculture

Monitoring effector allelic diversity in pathogen populations can improve the spatio-temporal deployment of *R* gene-based disease resistance. Functional profiling of *Avr* genes in local *P. infestans* populations can inform about the distribution of virulence alleles. This information can help breeders evaluate the potential of a given *R* gene. Moreover, it will help breeders to detect the emerging virulent races of *P. infestans* before they reach epidemic proportions. Normally, primary inoculum will increase during the following season along with genetic adaptation and selection and thus accelerate the emergence of highly aggressive clones. Once detected, breeders can choose appropriate cultivars to buffer the occurrence of the clone to manage the epidemics, or apply instant chemical control measures. With *P. infestans* genome sequences, next-generation sequencing technologies and sensitive tools based on, e.g., real-time PCR, it is now possible to rapidly profile the effector repertoires of emerging *P. infestans* genotypes. This effector profiling can assist in decision making for *R* gene deployment and chemical control measures in current and subsequent potato growing seasons.

3 Don'ts

3.1 Don't Give Up Too Fast

As described above, some plant genotypes can give nonspecific responses to agents like *Agrobacterium*, which is in fact also a pathogen on a wide range of plants (Fig. 1a). Also, some germplasm may not be amenable to *Agrobacterium*-mediated transient transformation. Especially in plant breeding, when working with a wide diversity of plant germplasm, such issues can occur at a certain frequency. Don't give up. One can easily generate alternative plant genotypes that do not suffer from these matters and respond to controls as they should. This procedure to generate and select suitable plant genotypes is presented in Fig. 1b.

3.2 Don't Conclude Too Fast on AVR Activity

Proteins that show cell death in transient assays do not necessarily have avirulence activity in the natural potato–*P. infestans* interaction. Other characteristics, such as level and timing of expression or localization to different sites in the host, also influence whether the proteins act as avirulence determinants [31]. For example, effector gene *PEX147-3* is a close relative of *Avr3a* that induces cell death with *R3a* by agro-co-infiltration [32], but is not expressed during infection [24] and not likely to represent a true *Avr* gene. Also, for example, some *crinklers* (necrosis-inducing proteins) induce cell death in a broad range of plants (and *Crn2* is therefore often used as a positive control) [33]. Some of these proteins target the host nucleus where the outcome of the plant–microbe interaction is determined, but avirulence activity has not been detected [31, 33, 34]. To draw conclusions about presumed AVR activity, one should perform follow-up experiments with independent assays. We often use genetic studies for co-segregation of response to the effector with resistance to *P. infestans* in a segregating population (Fig. 1c). In addition, co-infiltration of *R* genes and candidate *Avr* genes can provide complementary information (Fig. 1d). Complementation studies in *P. infestans* will provide the ultimate answer on avirulence activity [35].

3.3 Don't Conclude Too Fast on R Gene Identity

Once specific responses to an AVR protein are detected (Fig. 1a) and have been confirmed with independent assays (Fig. 1c, d), we can carefully conclude that the tested plant contains an *R* gene that matches the inoculated *Avr* gene. In functional allele mining studies, when an *Avr* gene is screened on diverse plant germplasm (*see* Subheading 2.2.1), identified specific resistance responses are most likely due to homologues of the known *R* gene [1]. However, it cannot be excluded that one effector interacts with different families of *R* genes, similarly as one *R* gene can interact with different effectors [36–38]. Therefore, independent experiments are required to confirm the identity of detected *R* genes.

4 Conclusions

Effectoromics provides innovative advantages that contribute to resistance breeding and *R* gene deployment. We summarize seven advantages in the *Do's*, and basically we classify them in four themes, i.e. acceleration, distinguishing functional redundancy, detecting specificity and deployment. *Don'ts* are less frequent than *Do's*. Apart from the prerequisite that sequence libraries should be available to apply effectoromics, the three presented *Don'ts* are just some practical notes to draw scientifically sound conclusions from phenotyping exercises. Ironically, these *Don'ts* mainly advocate patience during the experiments, whereas the effectoromics strategy in general focusses on enhancing the speed of *R* and *AVR* gene identity.

Acknowledgements

The work is partially supported by Wageningen University Fund (WUF), Wageningen UR Plant Breeding and China Scholarship Council Program for Graduate Students.

References

1. Vleeshouwers VGAA, Rietman H, Krenek P, Champouret N, Young C, Oh SK, Wang M, Bouwmeester K, Vosman B, Visser RGF, Jacobsen E, Govers F, Kamoun S, van der Vossen EAG (2008) Effector genomics accelerates discovery and functional profiling of potato disease resistance and *Phytophthora infestans* avirulence genes. PLoS One 3:1–10

2. Ellis JG, Rafiqi M, Gan P, Chakrabarti A, Dodds PN (2009) Recent progress in discovery and functional analysis of effector proteins of fungal and oomycete plant pathogens. Curr Opin Plant Biol 12:399–405

3. Oh SK, Young C, Lee M, Oliva R, Bozkurt TO, Cano LM, Win J, Bos JIB, Liu HY, van Damme M, Morgan W, Choi D, van Der Vossen EAG, Vleeshouwers VGAA, Kamoun S (2009) *In planta* expression screens of *Phytophthora infestans* RXLR effectors reveal diverse phenotypes, including activation of the *Solanum bulbocastanum* disease resistance protein Rpi-blb2. Plant Cell 21:2928–2947

4. Vleeshouwers VGAA, Raffaele S, Vossen JH, Champouret N, Oliva R, Segretin ME, Rietman H, Cano LM, Lokossou A, Kessel G, Pel MA, Kamoun S (2011) Understanding and exploiting late blight resistance in the age of effectors. Annu Rev Phytopathol 49:507–531

5. Kanneganti TD, Huitema E, Kamoun S (2007) *In planta* expression of oomycete and fungal genes. Methods Mol Biol 354:35–43

6. Vleeshouwers VGGA, Rietman H (2009) *In planta* expression systems. In: Lamour K, Kamoun S (eds) Oomycete genetics and genomics: diversity, interactions, and research tools. Wiley, New York, pp 455–475

7. Bhaskar PB, Venkateshwaran M, Wu L, Ané JM, Jiang J (2009) *Agrobacterium*-mediated transient gene expression and silencing: a rapid tool for functional gene assay in potato. PLoS One 4:1–8

8. Kim HJ, Lee HR, Jo KR, Mortazavian SMM, Huigen DJ, Evenhuis B, Kessel G, Visser RGF, Jacobsen E, Vossen JH (2012) Broad spectrum late blight resistance in potato differential set plants MaR8 and MaR9 is conferred by multiple stacked *R* genes. Theor Appl Genet 124:923–935

9. Rietman H, Bijsterbosch G, Cano LM, Lee HR, Vossen JH, Jacobsen E, Visser RGF, Kamoun S, Vleeshouwers VGAA (2012) Qualitative and quantitative late blight resistance in the potato cultivar Sarpo Mira is determined by the perception of five distinct RXLR effectors. Mol Plant Microbe Interact 25:910–919

10. Li G, Huang S, Guo X, Li Y, Yang Y, Guo Z, Kuang H, Rietman H, Bergervoet M, Vleeshouwers VGGA, van Der Vossen EAG, Qu D, Visser RGF, Jacobsen E, Vossen JH (2011) Cloning and characterization of *R3b*; members of the *R3* superfamily of late blight resistance genes show sequence and functional divergence. Mol Plant Microbe Interact 24:1132–1142

11. Song J, Bradeen JM, Naess SK, Raasch JA, Wielgus SM, Haberlach GT, Liu J, Kuang H,

Austin-Phillips S, Buell CR, Helgeson JP, Jiang J (2003) Gene *RB* cloned from *Solanum bulbocastanum* confers broad spectrum resistance to potato late blight. Proc Natl Acad Sci USA 100:9128–9133

12. van der Vossen E, Sikkema A, Hekkert BTL, Gros J, Stevens P, Muskens M, Wouters D, Pereira A, Stiekema W, Allefs S (2003) An ancient *R* gene from the wild potato species *Solanum bulbocastanum* confers broad-spectrum resistance to *Phytophthora infestans* in cultivated potato and tomato. Plant J 36: 867–882

13. Hermsen JGT, Ramanna MS (1973) Double-bridge hybrids of *Solanum bulbocastanum* and cultivars of *Solanum tuberosum*. Euphytica 22:457–466

14. Mastenbroek C (1953) Experiments on the inheritance of blight immunity in potatoes derived from *Solanum demissum* Lindl. Euphytica 2:197–206

15. Bendahmane A, Kanyuka K, Baulcombe DC (1999) The *Rx* gene from potato controls separate virus resistance and cell death responses. Plant Cell 11:781–791

16. Farnham G, Baulcombe DC (2006) Artificial evolution extends the spectrum of viruses that are targeted by a disease-resistance gene from potato. Proc Natl Acad Sci USA 103: 18828–18833

17. Shukla VK, Doyon Y, Miller JC, DeKelver RC, Moehle EA, Worden SE, Mitchell JC, Arnold NL, Gopalan S, Meng XD, Choi VM, Rock JM, Wu YY, Katibah GE, Zhifang G, McCaskill D, Simpson MA, Blakeslee B, Greenwalt SA, Butler HJ, Hinkley SJ, Zhang L, Rebar EJ, Gregory PD, Urnov FD (2009) Precise genome modification in the crop species *Zea mays* using zinc-finger nucleases. Nature 459: 437–443

18. Townsend JA, Wright DA, Winfrey RJ, Fu FL, Maeder ML, Joung JK, Voytas DF (2009) High-frequency modification of plant genes using engineered zinc-finger nucleases. Nature 459:442–446

19. Bogdanove AJ, Voytas DF (2011) TAL effectors: customizable proteins for DNA targeting. Science 333:1843–1846

20. Boch J, Scholze H, Schornack S, Landgraf A, Hahn S, Kay S, Lahaye T, Nickstadt A, Bonas U (2009) Breaking the code of DNA binding specificity of TAL-type III effectors. Science 326:1509–1512

21. Marton I, Zuker A, Shklarman E, Zeevi V, Tovkach A, Roffe S, Ovadis M, Tzfira T, Vainstein A (2010) Nontransgenic genome modification in plant cells. Plant Physiol 154: 1079–1087

22. Kamoun S, Segretin ME, Schornack S (2013) Late blight resistance genes. Patent WO/2013/009935

23. Cooke DEL, Cano LM, Raffaele S, Bain RA, Cooke LR, Etherington GJ, Deahl KL, Farrer RA, Gilroy EM, Goss EM, Grünwald NJ, Hein I, MacLean D, McNicol JW, Randall E, Oliva RF, Pel MA, Shaw DS, Squires JN, Taylor MC, Vleeshouwers VGAA, Birch PRJ, Lees AK, Kamoun S (2012) Genome analyses of an aggressive and invasive lineage of the Irish potato famine pathogen. PLoS Pathog 8:e1002940

24. Armstrong MR, Whisson SC, Pritchard L, Bos JIB, Venter E, Avrova AO, Rehmany AP, Bohme U, Brooks K, Cherevach I, Hamlin N, White B, Frasers A, Lord A, Quail MA, Churcher C, Hall N, Berriman M, Huang S, Kamoun S, Beynon JL, Birch PRJ (2005) An ancestral oomycete locus contains late blight avirulence gene *Avr3a*, encoding a protein that is recognized in the host cytoplasm. Proc Natl Acad Sci USA 102:7766–7771

25. Bos JIB, Armstrong MR, Gilroy EM, Boevink PC, Hein I, Taylor RM, Zhendong T, Engelhardt S, Vetukuri RR, Harrower B, Dixelius C, Bryan G, Sadanandam A, Whisson SC, Kamoun S, Birch PRJ (2010) *Phytophthora infestans* effector AVR3a is essential for virulence and manipulates plant immunity by stabilizing host E3 ligase CMPG1. Proc Natl Acad Sci USA 107:9909–9914

26. van Poppel PMJA, Guo J, van De Vondervoort PJI, Jung MWM, Birch PRJ, Whisson SC, Govers F (2008) The *Phytophthora infestans* avirulence gene *Avr4* encodes an RXLR-dEER effector. Mol Plant Microbe Interact 21: 1460–1470

27. Jiang RHY, Tyler BM, Whisson SC, Hardham AR, Govers F (2006) Ancient origin of elicitin gene clusters in *Phytophthora* genomes. Mol Biol Evol 23:338–351

28. Ponchet M, Panabieres F, Milat ML, Mikes V, Montillet JL, Suty L, Triantaphylides C, Tirilly Y, Blein JP (1999) Are elicitins cryptograms in plant–oomycete communications? Cell Mol Life Sci 56:1020–1047

29. Qutob D, Huitema E, Gijzen M, Kamoun S (2003) Variation in structure and activity among elicitins from *Phytophthora sojae*. Mol Plant Pathol 4:119–124

30. Vleeshouwers VGAA, Driesprong JD, Kamphuis LG, Torto-Alalibo T, Van'T Slot KAE, Govers F, Visser RGF, Jacobsen E,

Kamoun S (2006) Agroinfection-based high-throughput screening reveals specific recognition of INF elicitins in *Solanum*. Mol Plant Pathol 7:499–510

31. Schornack S, Huitema E, Cano LM, Bozkurt TO, Oliva R, Van Damme M, Schwizer S, Raffaele S, Chaparro-Garcia A, Farrer R, Segretin ME, Bos J, Haas BJ, Zody MC, Nusbaum C, Win J, Thines M, Kamoun S (2009) Ten things to know about oomycete effectors. Mol Plant Pathol 10:795–803

32. Bos JIB (2007) Function, structure and evolution of the RXLR effector *Avr3a* of *Phytophthora infestans*. PhD thesis, Plant Pathology Graduate Program

33. Torto TA, Li S, Styer A, Huitema E, Testa A, Gow NAR, van West P, Kamoun S (2003) EST mining and functional expression assays identify extracellular effector proteins from the plant pathogen *Phytophthora*. Genome Res 13:1675–1685

34. Huitema E, Bos JIB, Tian M, Win J, Waugh ME, Kamoun S (2004) Linking sequence to phenotype in *Phytophthora*–plant interactions. Trends Microbiol 12:193–200

35. Gilroy EM, Breen S, Whisson SC, Squires J, Hein I, Kaczmarek M, Turnbull D, Boevink PC, Lokossou A, Cano LM, Morales J, Avrova AO, Pritchard L, Randall E, Lees A, Govers F, van West P, Kamoun S, Vleeshouwers VGAA,

Cooke DEL, Birch PRJ (2011) Presence/absence, differential expression and sequence polymorphisms between *PiAVR2* and *PiAVR2-like* in *Phytophthora infestans* determine virulence on *R2* plants. New Phytol 191:763–776

36. Angel CA, Schoelz JE (2013) A survey of resistance to tomato bushy stunt virus in the genus *Nicotiana* reveals that the hypersensitive response is triggered by one of three different viral proteins. Mol Plant Microbe Interact 26:240–248

37. Lozano-Torres JL, Wilbers RHP, Gawronski P, Boshoven JC, Finkers-Tomczak A, Cordewener JHG, America AHP, Overmars HA, Klooster JW V 't, Baranowski L, Sobczak M, Ilyas M, van der Hoorn RAL, Schots A, de Wit PJGM, Bakker J, Goverse A, Smant G (2012) Dual disease resistance mediated by the immune receptor *Cf-2* in tomato requires a common virulence target of a fungus and a nematode. Proc Natl Acad Sci USA 109:10119–10124

38. Vos P, Simons G, Jesse T, Wijbrandi J, Heinen L, Hogers R, Frijters A, Groenendijk J, Diergaarde P, Reijans M, Fierens-Onstenk J, de Both M, Peleman J, Liharska T, Hontelez J, Zabeau M (1998) The tomato *Mi-1* gene confers resistance to both root-knot nematodes and potato aphids. Nat Biotechnol 16:1365–1369

Chapter 20

Protoplast Cell Death Assay to Study *Magnaporthe oryzae AVR* Gene Function in Rice

Hiroyuki Kanzaki, Kentaro Yoshida, Hiromasa Saitoh, Muluneh Tamiru, and Ryohei Terauchi

Abstract

We describe a protocol for transient gene expression in rice protoplasts and its application to the study of *Magnaporthe oryzae* avirulence (*AVR*) gene function. In this assay the gene encoding the firefly luciferase protein is transfected into rice protoplasts by electroporation together with the candidate *AVR* genes. The luminescence can then be used to assess the viability of rice protoplasts. The hypersensitive response (HR) caused by the interaction between *M. oryzae AVR* and rice *R* genes can subsequently be monitored by recording the decrease in luminescence from the transfected cells.

Key words Protoplast, Electroporation, Hypersensitive response (HR), Luciferase, Rice blast, Rice, Monocotyledons

1 Introduction

Monocotyledonous plants include economically important grain crops such as wheat, maize and rice. These crops are attacked by various pathogens whose control is crucial for food security worldwide. One of the most important pathogens of rice is rice blast, caused by the fungal pathogen *Magnaporthe oryzae*. Identification of new resistance sources is an important strategy for control of *M. oryzae* and other pathogens, and identification of *AVR* genes recognized by resistance genes provides an effective tool for resistance gene identification and characterization. In order to facilitate *AVR* gene identification, a tool for rapid characterization of pathogen effectors by transiently expressing them in their respective host plant cells is required. Agroinfiltration has been successfully used in dicotyledonous plant species for transient overexpression of transgenes, including pathogen effector genes. However, this approach is not suitable for monocots. As an alternative, assays using

Paul Birch et al. (eds.), *Plant-Pathogen Interactions: Methods and Protocols*, Methods in Molecular Biology, vol. 1127, DOI 10.1007/978-1-62703-986-4_20, © Springer Science+Business Media New York 2014

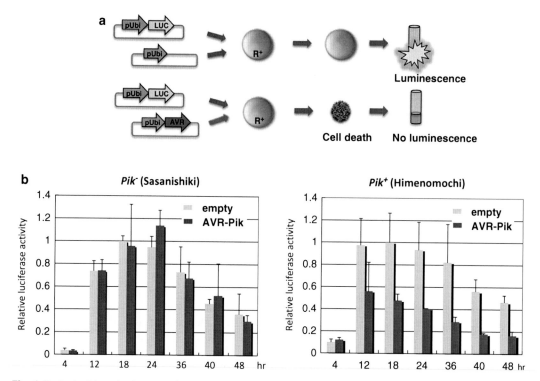

Fig. 1 Protoplast transient expression assay to study *AVR-R* interactions. (**a**) An illustration of the principle of the assay. A plasmid construct containing the firefly luciferase gene (*LUC*) driven by a strong maize ubiquitin promoter (pUbi) is transfected into rice protoplast cells to monitor the cell viability (*top*). If the LUC plasmid is co-transfected with a plasmid containing an *AVR*-gene driven by the same promoter, the resulting interaction between the *AVR* gene and a cognate rice *R*-gene triggers a hypersensitive response (HR), resulting in reduction in luciferase activity of the cells due to cell death (*bottom*). By measuring the reduction in luciferase luminescence, we can monitor the HR. (**b**) Examples showing application of the assay in rice. Protoplasts of a rice cultivar Sasanishiki lacking the *Pik R*-gene were transfected with either empty vector or *AVR-Pik* vector and luminescence from rice cells was measured at the given time (hours) after transfection. There is no difference in luminescence between empty- and *AVR-Pik*-transfected cells (*left*). By contrast, when protoplasts of Himenomochi cultivar, harboring the *Pik R*-gene, were transfected with the *AVR-Pik* plasmid, a significant reduction in luminescence was observed as compared to the cells transfected with the empty plasmid (*right*)

transient expression in protoplasts have frequently been employed in monocots, including rice (e.g. [1, 2]).

Here, we describe a protocol for transient expression in rice protoplasts [3] as specifically applied to the study of the interactions between *M. oryzae AVR* genes and their cognate rice (*Oryza sativa*) *R*-genes. We have previously cloned three *AVR* genes, *AVR-Pia*, *AVR-Pik* and *AVR-Pii* from *M. oryzae* [3] and the *Pia R* gene from rice [4]. In these studies, the interactions between the *M. oryzae AVR* and rice *R* genes were first tested using the protoplast transient expression assays described here (Fig. 1). Subsequent studies on a smaller number of functionally characterized candidates were performed by making stable transgenic fungi or plants. We have recently applied the protoplast assay to address the allele-specific recognition of *AVR-Pik* alleles by *Pik* alleles [5]. The protocol given here could potentially be useful for studying *AVR-R* interactions in other monocotyledonous species including wheat, barley and maize.

2   Materials

1. Buffer A: 0.6 M Mannitol, 1.5 % Cellulase RS (Yakult), 0.3 % Macerozyme R10 (Yakult), 10 mM MES, 1 mM $CaCl_2$, 5 mM β-Mercaptoethanol, 0.1 % BSA, 100 ppm Ampicillin (*see* **Note 1**), pH 5.7.

2. Buffer B: 0.6 M Mannitol, 20 mM KCl, 4 mM MES, adjusted to pH 5.7 with KOH.

3. Buffer C [Gamborg's B5 1 l]:

    2,500 mg KNO_3, 134 mg $(NH_4)2SO_4$, 150 mg $CaCl_2·2H_2O$.

    250 mg $MgSO_4·7H_2O$, 150 mg $NaH_2PO_4·2H_2O$, 3 mg H_3BO_3.

    10 mg $MnSO_4·4H_2O$, 2 mg $ZnSO_4·7H_2O$, 0.75 mg KI, 0.25 mg Na_2MoO_4.

    $2H_2O$, 0.025 mg $CuSO_4·5H_2O$, 0.025 mg $CoCl_2·6H_2O$.

    40 mg Fe-EDTA, 10 mg Thiamine·HCl, 1 mg Pyridoxine·HCl.

    1 mg Nicotinic acid, 100 mg Myo-inositol, 20 g Sucrose, 50 g Glucose.

    2 mg/L 2,4-D, pH 5.7, 100 ppm Ampicillin (*see* **Note 1**).

3   Methods

3.1 Plant Growth (Fig. 2)

Grow seedlings on MS medium at 26 °C for 2 weeks in the dark (*see* **Note 2**). 40–50 seedlings will give ~1×10^7 protoplasts, enough for making more than 40 samples for transfection.

3.2 Protoplast Isolation (Fig. 2)

All steps should be performed on a clean bench in a laminar flow hood.

1. Cut leaves and leaf sheaths of etiolated plants into 0.5 mm strips using a razor blade (*see* **Note 3**).

2. Place the cut leaves (strips) in Buffer A (25 plants/12.5 ml).

3. Apply vacuum infiltration for 10 min (*see* **Note 4**).

4. Gently shake samples at $0.02 \times g$ at 26 °C for 4 h or less. Confirm the release of protoplasts under a microscope (*see* **Note 5**).

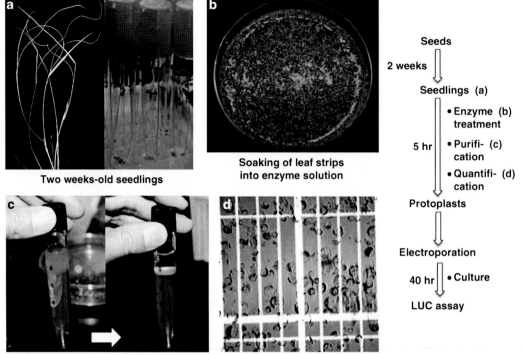

Fig. 2 Preparation of rice protoplasts for transfection assay. (**a**) Two-week-old seedlings of rice grown in darkness are used for protoplast extraction. Seedlings normally grown in soil (*left*) as well as seedlings grown in vitro (right) can be used for the assay. (**b**) Leaf blades together with leaf sheath are cut into 5 mm strips with a razor and treated with cell wall degrading enzymes. (**c**) Protoplasts are purified from other cell debris by Percoll-gradient centrifugation, and (**d**) intact protoplasts are quantified and used for the transfection assay by electroporation

5. Shake at $0.07 \times g$ for 5 min (*see* **Note 6**).

6. Transfer the cell suspension gently using a 10 ml pipette and filter using a 37 µm-pore nylon mesh (*see* **Note 7**). Cells will be transferred more easily if the top of the pipette tip is removed.

7. Centrifuge at $100 \times g$ for 10 min at room temperature (RT) (*see* **Note 8**).

8. Resuspend cell pellets in 2 ml of Buffer B by gentle mixing using a pasteur pipette (*see* **Note 9**).

9. Gently put 2 ml of protoplast cell suspension onto a Percoll-gradient. The gradient is formed by mixing 2.25 ml of Percoll with 5.25 ml of Buffer B. Mix this solution (30 % percoll solution) well by vortexing.

10. Centrifuge at $1,300 \times g$ for 10 min. No brake should be applied at the end of this centrifuging step.

11. Collect the protoplast layer (an yellow-colored band) into a new tube using a pasteur pipette (*see* **Note 10**).

12. Wash the protoplast suspension with 10 ml of Buffer B (*see* **Note 11**).

13. Centrifuge at $100 \times g$ for 10 min.

14. Remove the supernatant and discard.

15. Resuspend the cell pellet in Buffer B at a concentration of $1–2.5 \times 10^6$ cells/ml. Cell concentration is measured using a hemocytometer (*see* **Note 12**).

3.3 Transfection by Electroporation

1. Use 200 µl of protoplast suspension ($1–2.5 \times 10^6$ cells/ml) for each transfection with a total of 10 µg plasmids (comprising, e.g., 5 µg LUC plasmid and 5 µg AVR plasmid) (*see* **Note 13**).

2. Perform electroporation at 250 µFD, 300 V/0.4 cm, 100 Ω (*see* **Note 14**).

3. Place the electroporated cell suspension on ice for 10 min.

4. Add 300 µl of Buffer C containing Ampicillin (Final concentration 100 µg/ml) to each well of a 24-well plate (Costar® 24-Well Clear Flat Bottom Ultra Low Attachment Multiple Well Plates, Individually Wrapped, Sterile) and keep the plate at room temperature for 10 min.

5. Transfer 200 µl of electroporated cells to each well containing Buffer C (*see* **Note 15**).

3.4 Protoplast Culture and Reporter Assay

1. Incubate the electroporated cells at 26 °C in total darkness for a predetermined time period (4–48 h, depending on the experiment).

2. Centrifuge at $14,000 \times g$ for 6 s (twice) at 4 °C to collect the cells and remove the supernatant.

3. Add 100 µl 1× CCLR lysis buffer (Promega) to the cells and mix for 20 min by vortexing at 4 °C (*see* **Note 16**).

4. Put 100 µl of the lysate into a 96-well titer plate (black color) (e.g. PerkinElmer Black Opaque 96-well Microplate).

5. Add 100 µl of LUC substrate solution containing luciferin and ATP and mix with a pipette.

6. Measure luminescence with a luminometer (*see* **Note 17**).

4 Notes

1. Ampicillin is required to suppress growth of bacteria.

2. Culturing under full light conditions is likely to enhance leaf browning. Growing plants for more than 3 weeks leads to a decrease in protoplast yield.

3. Leaves and leaf sheaths should be cut into the smallest possible pieces. A larger surface area of samples increases protoplast yield. It is recommended to use sharp razor blades for preparing leaf samples in order to avoid crushing of the cells.

4. Vacuum treatment is required to infiltrate enzyme solution into the samples.

5. Following enzyme treatment, an yellow-colored haze should be visible at the bottom of the petri dishes as protoplasts are released from the tissues. If this is not visible, the enzyme soaking step should be performed for a longer time. Clear, round-shaped protoplasts should be visible when the sample is viewed under a microscope.

6. Stronger shaking for a short time enhances the release of protoplasts from tissues. If the release of protoplasts cannot be confirmed using a microscope, repeat this shaking step one more time.

7. Glass tubes 30 mm in diameter with the open ends covered with nylon mesh are used for the filtration of protoplasts. Gentle shaking of nylon mesh-attached tube in a petri dish enhances filtration of protoplasts. If it is difficult to filtrate, nylon mesh with larger pore sizes (e.g. 72 µm) should be used before using the 37 µm pore nylon mesh.

8. When transferring the protoplast suspension into a tube, gently place it onto the side of the tube held at an angle.

9. Use a pasteur pipette with short taper tube (3–4 cm tip) for transferring protoplast-containing solution.

10. Yellow-colored bands contain the intact protoplasts. Degraded protoplasts or tissue residues are left as pellets following centrifugation.

11. This step is performed to remove the remains of the Percoll solution. Small residual quantities of Percoll do not affect the electroporation and transient expression assay.

12. Lower or higher concentrations of protoplast suspension may affect the transfection efficiency.

13. Make the total amount of the plasmid to be used for the electroporation uniform (e.g. 10 µg) using empty plasmids. Unequal amounts of plasmids affect the number of plasmid DNA molecules delivered into plant cells, thereby introducing a bias in the activity of luciferase detected.

14. Pulse time for electroporation should be normally 19–24 ms.

15. To enhance the luciferase activity, nurse culturing of protoplasts is recommended. For this, put 1 ml of electroporated protoplasts suspension in a Millicell (e.g. Millicell®-CM 0.4 µm, 30 mm diameter, individually wrapped, sterile,

Millipore), which is then put inside a 60 mm diameter petri dish containing 4 ml of Buffer C and 100 mg of suspension cells (nurse cells) in log growth phase.

16. Instead of vortexing for 20 min, sonication on ice may accelerate the rupture of the plant cells and enhance the luminescence.

17. Downloaded from the Promega website: (http://www.promega.jp/resources/protocols/technical-bulletins/0/luciferase-assay-system-protocol/).

Acknowledgements

This work was supported by the Ministry of Agriculture, Forestry, and Fisheries of Japan (Genomics for Agricultural Innovation PMI-0010) and the Program for Promotion of Basic Research Activities for Innovative Biosciences (PROBRAIN), Japan, and Grant-in-aid for Scientific Research from the Ministry of Education, Cultures, Sports and Technology, Japan to HS and RT (Grant-in-Aid for Scientific Research on Innovative Areas 23113009).

References

1. Bart R, Chern M, Park C-J, Bartley L, Ronald PC (2006) A novel system for gene silencing using siRNAs in rice leaf and stem-derived protoplasts. Plant Methods 2:13

2. Chen S, Tao L, Zeng L, Vega-Sanchez ME, Umemura K, Wang G-L (2006) A highly efficient transient protoplast system for analyzing defence gene expression and protein-protein interactions in rice. Mol Plant Pathol 7:417–427

3. Yoshida K, Saitoh H, Fujisawa S, Kanzaki H, Matsumura H, Yoshida K, Tosa Y, Chuma I, Takano Y, Win J, Kamoun S, Terauchi R (2009) Association genetics reveals three novel avirulence genes from rice blast fungal pathogen *Magnaporthe oryzae*. Plant Cell 21:1573–1591

4. Okuyama Y, Kanzaki H, Abe A, Yoshida K, Tamiru M, Saitoh H, Fujibe T, Matsumura H, Shenton M, Galam DC, Undan J, Ito A, Sone T, Terauchi R (2011) A multifaceted genomics approach allows the isolation of the rice *Pia-blast* resistance gene consisting of two adjacent NBS-LRR protein genes. Plant J 66:467–479

5. Kanzaki H, Yoshida K, Saitoh H, Fujisaki K, Hirabuchi A, Alaux L, Fournier E, Tharreau D, Terauchi R (2012) Arms race co-evolution of *Magnaporthe oryzae AVR-Pik* and rice *Pik* genes driven by their physical interactions. Plant J 72:894–907

Chapter 21

A Bacterial Type III Secretion-Based Delivery System for Functional Assays of Fungal Effectors in Cereals

Narayana M. Upadhyaya, Jeffery G. Ellis, and Peter N. Dodds

Abstract

Large numbers of candidate effectors are being identified by genome sequencing of fungal pathogens and *in planta* expression studies. These effectors are both a boon and a curse for pathogens as they modulate the host cellular environment or suppress defense response to allow fungal growth as well as become targets of plant resistance (R) proteins. Recognition of a fungal effector by a plant R protein triggers a hypersensitive reaction (HR) leading to death of plant cells in and around the infection site, thus preventing further proliferation of the pathogen. Such HR induction has been used as an indicator of effector activity in functional assays of candidate effectors in dicots based on *Agrobacterium*-mediated transient expression. However, the *Agrobacterium* assay is not functional in cereal leaves. We therefore have adapted an alternative assay based on effector protein delivery using the type III secretion system (T3SS) of a non-pathogenic *Pseudomonas* spp. for use in wheat and other cereals. Here, we describe protocols for delivery of effector proteins into wheat and barley cells using the AvrRpm1 T3SS signal in the engineered non-pathogenic *Pseudomonas fluorescens* strain Effector-to-Host Analyzer (EtHAn). For ease of making expression clones we have generated the GATEWAY cloning compatible vectors. A calmodulin-dependent adenylate cyclase (Cya) reporter protein can be used as an effective marker for fusion protein delivery into wheat and barley by this system.

Key words Type-III secretion system, *Pseudomonas fluorescens*, Wheat, Barley, Wheat stem rust

1 Introduction

During infection, fungal pathogens deliver effector proteins into host cells which are important for establishing disease, but may also be recognized by host immune receptors [1]. Thus understanding the roles of these proteins in both infection and host resistance is critical for managing plant diseases. Recent advances in whole genome sequencing of plant fungal pathogens and concurrent advances in bioinformatic tools have enabled identification of large numbers of candidate effectors from diverse fungal and fungus-like pathogens, mostly detected as small secreted proteins expressed specifically in infection structures [2–4]. These include many important cereal pathogens such as the wheat stem rust

Paul Birch et al. (eds.), *Plant-Pathogen Interactions: Methods and Protocols*, Methods in Molecular Biology, vol. 1127, DOI 10.1007/978-1-62703-986-4_21, © Springer Science+Business Media New York 2014

fungus *Puccinia graminis* f. sp. *tritici* [5]. However, there is a lack of high-throughput functional assay in cereals that can be used to investigate the role and, in particular, the avirulence function of these effectors. *Agrobacterium*-mediated transient expression has been a powerful tool for effector analysis in dicot plant hosts, including those of rust pathogens [6, 7], but it is not applicable to cereals, although there is a recent report of successful transient expression in rice leaves [8]. An alternative approach to over-express effectors involves transient expression via biolistic transformation [9], which has been employed in rice with *AvrPi-ta/Pi-ta* [10] and barley [11]. In this case Avr function and corresponding induction of the HR can be detected by an indirect output, i.e. loss of detection of a reporter protein (GUS or GFP) whose expression is eclipsed by the induction of cell death. However, this is a labor-intensive method that is not easily adapted to high-throughput screening. A promising approach for high-throughput screening is to deliver effector proteins into host cells using the type III secretion system (T3SS) from a bacterial pathogen. For instance, Sohn et al. [12] showed that several oomycete effector proteins could be delivered into Arabidopsis cells by *Pseudomonas syringae* pv *tomato* (DC3000) when fused to the N-terminal secretion-translocation signals of the well-characterized bacterial effectors AvrRpm1 or AvrRps4. Similar systems have been used successfully for functional studies of oomycete effectors delivered as fusions [3, 13, 14]. Also, Sharma and co-workers [15] showed that *Burkholderia glumae* could deliver effectors from the fungal pathogen *Magnaporthe oryzae* into rice cells. A calmodulin-dependent adenylate cyclase assay has previously been used to demonstrate the T3SS-dependent translocation of effector proteins in different dicot pathosystems—*Xanthomonas* AvrBs2 in pepper [16], *Pseudomonas* AvrPto in tomato and tobacco [17], and *Erwinia* DspA/E in tobacco [18]. The principle of this reporter system is that Cya is inactive in prokaryotes but becomes active in the presence of calmodulin, which is exclusively present in plants and other eukaryotes. This Cya activity results in the production of cAMP which can be quantified by an enzyme immunometric assay (EIA).

Here we describe an efficient effector delivery system for wheat and barley based on T3SS delivery by an engineered non-pathogenic *Pseudomonas fluorescens* Pf0-1 strain Effector-to-Host Analyzer (EtHAn) [19]. This strain has the *P. syringae* pv. *syringae* 61 *hrp/hrc* cluster stably integrated into the chromosome, thereby making this non-pathogenic bacterium T3SS competent. We have used this reporter system to demonstrate protein delivery to wheat and barley by leaf infiltration [20]. In this system we use the AvrRpm1 T3SS secretion signal, which we have modified to remove the N-terminal myristoylation/palmitoylation sites to allow delivery of soluble cytosolic proteins.

2   Materials

2.1   General

1. Wheat and barley seedlings. Plant seedlings (wheat, barley) raised in small (5 cm) pots with vermiculite/perlite (50/50 v/v) mixture containing 2 g/L osmocote (*see* **Note 1**).

2. *P. fluorescens* Pf0-1 strain EtHAn [19].

3. Destination vectors pNR526 (or pEDV-AvrRpm1-Des1N) and its derivatives pNR526-G2AC3A (Fig. 1a).

4. Topo entry clones of Cya and candidate effectors produced using pENTR/D Topo vector (Invitrogen).

5. LB liquid/solid media and SOC liquid media (use established recipes to prepare these media).

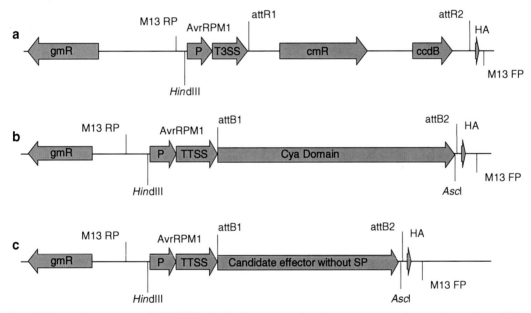

Fig. 1 Schematic diagram of GATEWAY destination vector, Cya Reporter expression vector, and candidate effector expression vector for T3SS-based delivery by *Pseudomonas fluorescence*. (**a**) Destination vector pNR526 (or pEDV-AvrRpm1-Des1N) has a *Pseudomonas syringae* T3SS delivery signal (promoter and part of the coding region of *AvrRpm1*) and an HA tag flanking the Reading frame A *att*R1-*ccdB*-cm^R-*att*R2 cassette. We have made a derivative of pNR526 with the myristoylation and/or palmitoylation sites at the N-terminal of AvrRpm1 [17] mutated (G2 to A and C3 to A) to abolish plasma membrane targeting of the delivered protein. This is designated pNR526_G2AC3A (pEDV-AvrRpm1_G2AC3A-Des). (**b**) Expression vectors pNR527 (pEDV-AvrRpm1::Cya) and pNR594 (pEDV-AvrRpm1-G2AC3A::Cya). (**c**) Candidate effector expression vectors. Entry clones used in making these expression vectors have the coding region without the predicted signal peptide. *GmR* gentamicin resistance gene; *M13 FP* M13 forward primer binding site; *M13 RP* M13 Reverse primer binding site; *P* Promoter; *T3SS* secretion signals; *CmR* chloramphenicol resistance gene; *ccdB E. coli* killer gene; *HA* single HA epitope tag; *attR1,attR2,attB1,attB2* bacteriophage lambda-specific recombination sites (GATEWAY Technology); *Hind*II, *Asc*I restriction enzyme sites

6. Sucrose peptone (or tryptone) agar, PSA/TSA: sucrose 20 g/L, peptone/tryptone 5 g/L, K_2HPO_4 0.5 g/L, $MgSO_4 \cdot 7H_2O$ 0.25 g/L, Agar 15 g/L, pH 7.3.

7. Minimal T3SS induction medium : 50 mM potassium phosphate buffer, 7.6 mM $(NH_4)_2SO_4$, 1.7 mM $MgCl_2$, 1.7 mM NaCl, 10 mM fructose, pH 5.7–5.8 (see **Note 2**).

8. Antibiotic stocks (×1,000)—gentamicin 20 mg/mL, chloramphenicol 30 mg/mL.

9. 10 mM $MgCl_2$, 10 mM $MgSO_4$, 0.1 M $CaCl_2$, DMSO.

10. 50 mL Falcon tubes, 50 mL Corex tubes, 1.5 and 2 mL Eppendorf tubes, 250 mL conical flasks.

11. Refrigerated bench top centrifuge.

12. Aquasol.

13. Glasshouse (18–22 °C) and growth room (24 °C constant temperature and 16 h light) facility.

14. Incubators (33, 29 and 20 °C) with shaking facility and water bath (42 °C).

15. 3,3′-Diaminobenzidine tetrahydrochloride hydrate (DAB) powder (Sigma-Aldrich, St. Louis, MO).

16. 1 mL syringes.

17. Liquid Nitrogen, ice bath, −80 °C storage facility.

2.2 Cya Reporter Assay

1. Cya expression clones produced with GATEWAY cloning system by LR Clonase reaction of entry clones and destination vector pNR526 or its derivatives (Fig. 1b).

2. 1.1 M $HClO_4$ (dilute 157 μL of 70 % stock with 843 μL water).

3. 6 M K_2CO_3 (824 mg/mL in water).

4. 10 mM $MgCl_2$.

5. Liquid Nitrogen.

6. Stirafoam cooler box.

7. Paper freezer box.

8. Eppi-grinder setup.

9. Cyclic AMP EIA Kit. Cat No 581001 (Ace™, Cayman Chemical MI USA).

10. Quick Start™ Bradford Dye Reagent Cat No. 500-0205 (Bio-Rad).

11. 1.5 M Tris–HCl, pH 8.8.

12. Flat bottom micro-titer plate.

13. Spectrophotometer (plate reader).

14. BSA 10 mg/mL stock.

3   Methods

3.1   Production of Expression Vectors by GATEWAY System

1. PCR amplify candidate effectors according to the method described in the pENTR™ Directional TOPO® cloning Kit (Invitrogen) (*see* **Note 3**).

2. Construct all GATEWAY entry vectors of candidate effectors and reporters (e.g. Cya) in pENTR™/D-TOPO® according to manufacturer's (Invitrogen) instructions (*see* **Note 4**).

3. Generate expression vectors of Cya reporter (Fig. 1b) or candidate effectors (Fig. 1c) by performing LR Clonase reactions (according to manufacturer's instructions) with respective entry clones and the AvrRpm1 T3SS destination vector (*see* **Note 5**).

4. Analyze resulting expression clones by gene/destination vector-specific PCR and/or restriction analyses (*see* **Note 6**) and sequencing to confirm correct integration (*see* **Note 7**).

5. Introduce expression vectors into the engineered *P. fluorescence* strain EtHAn by heat-shock transformation (Subheading 3.2) or tri-parental mating with helper plasmid strain (pRK2013/HB101) (*see* **Note 8**).

3.2   Heat-Shock Transformation of Pf EtHAn

Heat-shock competent *P. fluorescence* cells can be prepared by a method modified from the standard protocol described for *E. coli* [21]. Following modifications are adopted from a laboratory protocol developed by Stephan Heeb, The University of Nottingham.

1. Dilute 300 μL of a fresh overnight culture (in LB) into 30 mL LB and incubate with shaking at 33 °C until $OD_{600} = 1.8$–2.0.

2. Centrifuge at $4,500 \times g$ for 6 min at 4 °C in a Corex tube and resuspend cells in 7.5 ml of chilled 0.1 M calcium chloride gently. Leave on ice for 30 min.

3. Centrifuge at $4,500 \times g$ for 6 min at 4 °C and resuspend cells in 1.5 ml of chilled 0.1 M calcium chloride.

4. Add 50 μL of DMSO, mix gently by swirling, and store the suspension on ice for 15 min.

5. Add an additional 50 μL of DMSO to the suspension. Mix gently by swirling, and then return the suspension to an ice bath.

6. Dispense 100 μL aliquots of the suspensions into chilled, sterile Eppendorf tubes. Snap-freeze in liquid nitrogen and store the tubes at –70 °C.

7. When needed thaw the cells in an ice bath and leave for 10 min.

8. Mix 500 ng of plasmid DNA with 100 μL of competent cells and leave on ice for 30 min (*see* **Note 9**).

9. Heat pulse at 42–43 °C for 2 min and quench on ice.

10. Add 250 µL of pre-warmed SOC and incubate shaking at 33 °C for 3–4 h.

11. Plate on selective (gentamicin and chloramphenicol) media.

12. Analyze transformants by PCR and/or plasmid restriction analysis.

3.3 Preparation of Pf EtHAn Cells Under Non-inducing (T3SS) Conditions (See Note 10)

1. Streak Pf EtHAn containing effector/reporter expression vectors from glycerol stock to an LB plate with antibiotic (gentamicin and chloramphenicol) selection and grow 36 h at 29 °C (see **Note 11**).

2. Inoculate one loopful of culture into 50 mL LB with selection (gentamicin and chloramphenicol) in a 250 mL flask and incubate at 29 °C shaking (200 rpm) for 24 h.

3. Cool the culture on ice, transfer to a 50 mL Falcon tube and harvest cells by centrifugation $5,000 \times g$ at 4 °C for 10 min.

4. Remove all the supernatant and resuspend in 10 mM $MgCl_2$ to $OD_{600} = 0.4$ ($\sim 4 \times 10^8$ cfu/mL) or 2.0 as required (see **Note 12**).

3.4 Preparation of Pf EtHAn Cells Under T3SS Inducing Conditions (See Note 13)

1. Plate glycerol stock cultures onto the LB media with antibiotic selection (chloramphenicol and gentamicin) and grow at 29 °C for 36 h.

2. Inoculate a loopful of the cultures to 5 mL LB (with antibiotics) and grow O/N at 29 °C with shaking (200 rpm).

3. Add this O/N culture to 50 mL LB (with antibiotics) and further incubate at 29 °C for 4–6 h (200 rpm) to $OD_{600} > 0.8$.

4. Cool on ice, transfer to a 50 mL Falcon tubes and harvest cells by centrifugation ($5,000 \times g$, 4 °C, 10 min).

5. Resuspend culture pellets in 25 mL of cold 10 mM $MgSO_4$ and centrifuge to harvest cells.

6. Repeat **step 5** once.

7. Resuspend in a minimal medium with antibiotic (gentamicin 15 mg/L) and fructose (10 mM) to a OD_{600} 0.8 (see **Note 14**).

8. Incubate shaking (200 rpm) at 20 °C O/N.

9. Chill and harvest cells by centrifugation ($5,000 \times g$ at 4 °C for 10 min).

10. Resuspend pellets in 10 mM $MgCl_2$ to the required OD and use for infiltration (see **Note 12**).

3.5 Wheat Infiltration Assay

1. Prepare the EtHAn (carrying the expression constructs) cultures as described in Subheading 3.3 (non-induced) or as described in Subheading 3.4 (T3SS-induced culture) and adjust the cell density to required OD (see **Note 15**).

2. Transfer wheat seedlings from the glasshouse to the growth room and label the treatments.

3. Pre-warm the culture to growth room temperature.

4. Infiltrate first and second leaf blades in 2–3 spots (depending on the length of the leaf) using a needleless syringe (*see* **Note 16**).

5. Transfer the pots (in a tray with water) to a transparent box and cover with the lid and keep O/N before placing outside in the growth room.

6. Check for the induction of HR-like symptoms after 24–48 h (*see* **Note 17**) and up to 7 days.

7. Perform DAB staining (if required) as described in Subheading 3.6.

3.6 DAB (3,3′-Diaminobenzidine) Assay for Hydrogen Peroxide Accumulation

An early sign of the hypersensitive response is the production of reactive oxygen species including hydrogen peroxide (H_2O_2). This can be detected by staining with Diaminobenzidine (DAB) as described previously [22].

1. Dissolve 0.1 g of 3,3′-Diaminobenzidine tetrahydrochloride hydrate (DAB) powder (Sigma-Aldrich, St. Louis, MO) 100 mL of water and adjust pH to ~5.00 (*see* **Note 18**) with NaOH.

2. Immerse very fresh wheat leaf cuttings (harvested 24–48 h post infiltration in DAB solution in 2 mL Eppendorf tubes).

3. Incubate under light at 24 °C in a growth room for 8–12 h.

4. Transfer the samples to 10 mL screw-cap Falcon tubes with 5–6 mL of 96 % EtOH, close (loosely) with the screw cap and place in boiling water bath for 10 min.

5. Cool the tubes and replace the 96 % EtOH with 70 % EtOH, keep shaking O/N to clear the tissue of chlorophyll and then preserve for photography.

3.7 Cya Reporter Assay

1. Infiltrate *Pf* EtHAn containing the Cya expression construct (pNR527) into leaf blades (first and second fully expanded leaf, three spots per leaf) of 3–4 leaf stage wheat (or other) seedlings using needleless syringe.

2. Sample leaf blades in and around the infiltration area as ~1 cm leaf segments at 0 h (for the treatment with Cya reporter strain only), 8 h and 18 or 24 h in triplicates (each with two leaf segments), snap-freeze in liquid nitrogen and store at –80 °C until use.

3. Set up an Styrofoam freezer box with liquid N_2, place a cardboard freezer box and fill with liquid N_2 up to the top of the tube divider for sample processing.

4. Pre-cool Eppi-grinder tip in liquid N_2 freezer box.

5. Process one sample at a time quickly by pulverizing sample to a compact pellet and remove the grinder and return the tube to liquid N_2 freezer box.

6. After pulverizing all samples add (one at a time) 325 μL of 1.1 M $HClO_4$ (1.1 M) and grind at room temperature until liquid and then placed on ice.

7. Process the rest of the samples one at a time.

8. Centrifuge samples (10 min, max speed at room temperature) and collect supernatant (~300 μl) in a 2.0 mL tube.

9. Save the pellets for protein quantification by Bradford assay (**steps 12–17**) and store at –20 °C until use.

10. Neutralize the supernatant (300 μl) with 40 μL K_2CO_3 (6 M) (adding slowly, this generates lots of foam) then mix well (*see* **Note 19**).

11. Centrifuge samples for 8 min at max speed and transfer 200 μL of supernatant into a 1.5 mL tube.

12. Use 10 μL of this supernatant for Cayman Cya assay and store rest at –80 °C for further use if required (*see* **Note 20**).

13. Measure the cAMP levels in the extracted samples using a cAMP enzyme immunoassay kit (Cayman Chemical Company) according to manufacturer's instructions (*see* **Note 21**).

14. Resuspend the pellet from **step 7** in 500 μL of 1.5 M Tris–HCl, pH 8.8 by mild sonication.

15. Transfer 5 μL in duplicates to a flat bottom micro-titer plate.

16. Also prepare BSA standards of 0, 10, 20, 40, 80, 160, 320, 640 and 1,280 μg/mL from a 10.0 mg/mL stock and transfer 5 μL in duplicates to the micro-titer plate.

17. Add 150 μL of Quick Start Bradford Dye Reagent to each sample and standards.

18. Read OD_{750} after 15 min using a spectrophotometer (plate reader).

19. Calculate protein concentration from the standard curve (*see* **Note 22**).

20. Express cAMP in pmol/mg protein.

4   Notes

1. Fertilize once a week with Aquasol solution (0.2 %).

2. Prepare a 2× stock of minimal media and store at 4 °C. Add fructose and antibiotic just before use while making the 1× minimal media.

3. A high-fidelity DNA polymerase such as Phusion (Thermo Scientific) should be used for initial PCR amplification from cDNA of the pathogen of interest (we have used RNA isolated from either purified haustoria of wheat stem rust, or from infected wheat tissue).

4. The candidate effector entry clones should include the coding sequence of the predicted mature protein without the signal peptide region.

5. The destination vector pNR526 (or pEDV-AvrRpm1-Des1N) contains the promoter (189 nt) plus first 267 nt of the coding region (aa 1–89) of the *Pseudomonas* syringae *AvrRpm1* gene and an HA tag flanking the Reading frame A *att*R1-*ccdB*-cm^R-*att*R2 cassette (Invitrogen) (Fig. 1a). The backbone of this construct is the broad host-range vector pBBR 1MCS-5 [12, 23]. This vector is compatible with Topo entry clones produced using pENTR-D-Topo (Invitrogen). Although this vector includes a 3′ region encoding an HA tag for antibody detection of the encoded protein, we have opted to generate clones with their native stop codon in case the HA tag could interfere with function.

6. We do the diagnostic PCR using the gene-specific forward primer (the ones used in the initial amplification step of effector candidates for Topo entry clone production) and a Destination vector sequence-specific primer binding (in antisense) to the region downstream of the HA tag (5′-TCACGACGTTGTAAAACGACGGCCAG-3′). We do a double RE digestion of the isolated plasmid DNA with *Hin*dIII and *Asc*1 (*see* Fig. 1) for further confirmation.

7. We have also generated a derivative of pNR526 (pNR526-G2AC3A) with the myristoylation and palmitoylation sites at the N-terminal of AvrRpm1 [24] mutated (Gly at position 2 to Ala and Cys at position 3 to Ala) to abolish plasma membrane targeting of the delivered protein, which could interfere with function or recognition of pathogen effectors. Although this version showed reduced Cya activity in wheat infiltration assays, this is likely because the calmodulins required for Cya activity are plasma-membrane associated. Delivery of the flax rust AvrM protein into transgenic tobacco with the corresponding M resistance protein using this vector system resulted in cell death [20]. Furthermore, the candidate *Pgt* effector PGTAUSPE10-1 triggered a strong cell death only when this protein was expressed using this vector system in wheat [20].

8. If using tri-parental mating, the preferred media for plating the mating mix (for transconjugant selection) is peptone (or tryptone) sucrose agar (PSA or TSA). In PSA or TSA EtHAn exhibits a characteristic colony color (yellowish) and it glows under UV light because of siderophore and levan productions, thus can easily be distinguished from *E. coli*. Cultures need to be grown at 33 °C to avoid interference from the restriction modification system of EtHAn.

9. Transformation efficiency is often very low and so we tend to use at least 500 ng of the expression vector for EtHAn transformation.

10. *Pf* EtHAn cultures can be prepared after growth in Luria Broth. The bacterial T3SS system is inactive during growth in rich media cultures such as LB and is only induced after infiltration into plants [25].

11. Viability of glycerol stocks (LB with 25 % glycerol) of EtHAn is relatively poor. Take out a good amount of the frozen culture for plating on an LB (with selection) plate just taken out from the 4 °C store room and let it thaw slowly.

12. For Cya assays we have used $OD_{600} = 0.4$, but for screening effector candidates we use $OD_{600} = 2.0$.

13. The *Pf* EtHAn TT3S can be pre-induced by growth in minimal media (osmolarity below 50 mM and pH 5.5–6.00) with certain carbon sources such as fructose [25, 26], which may mimic conditions in the plant apoplast. Although we found that *EtHAn* could effectively deliver the Cya reporter protein into wheat and barley cells after growth in LB (Fig. 2), we found that pre-induction increased delivered Cya activity in wheat by about twofold and also induced a faster cell death response in the tobacco AvrM/M assay [20]. Furthermore, we observed HR induction with a *Pgt* effector candidate on wheat line W3534 only after infiltration of a pre-induced culture (Fig. 3).

14. It is better not to add the second antibiotic chloramphenicol at this stage but just add gentamicin. We have found a reduction in cell density with the addition of chloramphenicol.

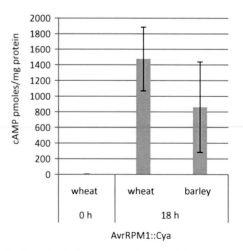

Fig. 2 cAMP production after infiltration of wheat (cv BW56) and barley (cv Golden Promise) seedling leaves with the engineered *Pseudomonas fluorescens* Pf0-1 strain EtHAn carrying the Cya expression vector pNR527 (pEDV-AvrRpm1::Cya)

Candidate effector PGTAUSPE-10-1

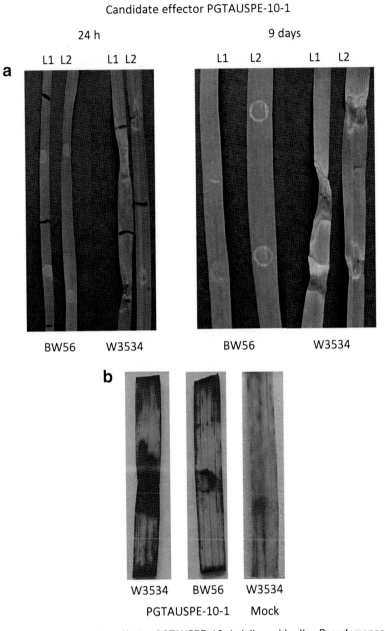

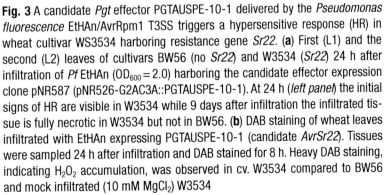

Fig. 3 A candidate *Pgt* effector PGTAUSPE-10-1 delivered by the *Pseudomonas fluorescence* EtHAn/AvrRpm1 T3SS triggers a hypersensitive response (HR) in wheat cultivar WS3534 harboring resistance gene *Sr22*. (**a**) First (L1) and the second (L2) leaves of cultivars BW56 (no *Sr22*) and W3534 (*Sr22*) 24 h after infiltration of *Pf* EtHAn (OD$_{600}$ = 2.0) harboring the candidate effector expression clone pNR587 (pNR526-G2AC3A::PGTAUSPE-10-1). At 24 h (*left panel*) the initial signs of HR are visible in W3534 while 9 days after infiltration the infiltrated tissue is fully necrotic in W3534 but not in BW56. (**b**) DAB staining of wheat leaves infiltrated with EtHAn expressing PGTAUSPE-10-1 (candidate *AvrSr22*). Tissues were sampled 24 h after infiltration and DAB stained for 8 h. Heavy DAB staining, indicating H$_2$O$_2$ accumulation, was observed in cv. W3534 compared to BW56 and mock infiltrated (10 mM MgCl$_2$) W3534

15. We are now routinely using the optimized EtHAn T3SS system (pEDV-AvrRpm1_G2AC3A derived expression vectors (Fig. 1), induced cultures with OD_{600} 2.00 and 3 leaf stage seedlings) for obtaining most consistent results.

16. Different wheat cultivars vary a great deal in leaf blade lengths and one has to adjust the number of spots of infiltration per leaf based on the length of the leaf blade. Normally any HR is much more robust in the first leaf but usually the first leaf is very brittle and prone to damage during infiltration and so care should be taken not to damage the leaves. Cultivars do vary with efficacy with which infiltration can be achieved. Make sure that the culture spreads at least one cm on either side of point of contact of the syringe head.

17. Initial sign is collapse and darkening of the tissue around infiltration zone and can be observed in 24–48 h. After 5–9 days necrotic tissue is dried out.

18. Diaminobenzidine tetrahydrochloride hydrate is water soluble and pH of the 0.1 % solution is ~2. To increase the sensitivity of staining raise the pH to 5–6 with NaOH. Please note that increasing the pH reduces the specificity of staining. Therefore determine the optimum pH for the specific plant material beforehand.

19. K_2CO_3 needs to be added very slowly as it forms lots of froth.

20. For samples expected to have high cAMP concentration a 1 in 10 dilution is required for accuracy of measurement.

21. Employ "no acetylation" method and prepare standard curves accordingly (range 0.3–750 pmol/mL) as described in manufacturer's (cAMP EIA Kit) booklet. Calculate the $\%B/B_0$ value for each sample using the equation obtained from the standard curve plot. Remember to account for any concentration or dilution of the sample (dilution factor) prior to the addition to the well. Samples with $\%B/B_0$ values >80 % or less than 20 % need to be re-assayed for accuracy. Please note: GENSTAT and the MS Excel plug-in XLSTAT both have the modules for fitting the data to a four-parameter logistic equation.

22. Determine the protein concentration from the fitted linear regression ($y = ax + b$).

Acknowledgements

The authors wish to thank Robyn East, Dhara Bhat and Lina Ma for excellent technical assistance. Authors also wish to thank Dr. Brian J. Staskawicz, University of California, Berkeley and Dr. David Joly, Agriculture and Agri-Food Canada for sharing some of their laboratory protocols. Funding from this work comes from 2 Blades Foundation.

References

1. Rafiqi M, Ellis JG, Ludowici VA et al (2012) Challenges and progress towards understanding the role of effectors in plant-fungal interactions. Curr Opin Plant Biol 15:477–482

2. Bozkurt TO, Schornack S, Banfield MJ et al (2012) Oomycetes, effectors, and all that jazz. Curr Opin Plant Biol 15:483–492

3. Fabro G, Steinbrenner J, Coates M et al (2011) Multiple candidate effectors from the oomycete pathogen *Hyaloperonospora arabidopsidis* suppress host plant immunity. PLoS Pathog 7:e1002348

4. Yoshida K, Saitoh H, Fujisawa S et al (2009) Association genetics reveals three novel avirulence genes from the rice blast fungal pathogen *Magnaporthe oryzae*. Plant Cell 21:1573–1591

5. Duplessis S, Cuomo CA, Lin YC et al (2011) Obligate biotrophy features unraveled by the genomic analysis of rust fungi. Proc Natl Acad Sci USA 108(22):9166–9171

6. Catanzariti AM, Dodds PN, Lawrence GJ et al (2006) Haustorially expressed secreted proteins from flax rust are highly enriched for avirulence elicitors. Plant Cell 18:243–256

7. Dodds PN, Lawrence GJ, Catanzariti AM et al (2004) The *Melampsora lini AvrL567* avirulence genes are expressed in haustoria and their products are recognized inside plant cells. Plant Cell 16:755–768

8. Andrieu A, Breitler JC, Siré C et al (2012) An *in planta*, *Agrobacterium*-mediated transient gene expression method for inducing gene silencing in rice (*Oryza sativa* L.) leaves. Rice (N Y) 5:23

9. Leister RT, Ausubel FM, Katagiri F (1996) Molecular recognition of pathogen attack occurs inside of plant cells in plant disease resistance specified by the *Arabidopsis* genes *RPS2* and *RPM1*. Proc Natl Acad Sci USA 93:15497–15502

10. Jia Y, McAdams SA, Bryan GT et al (2000) Direct interaction of resistance gene and avirulence gene products confers rice blast resistance. EMBO J 19:4004–4014

11. Bai S, Liu J, Chang C et al (2012) Structure-function analysis of barley NLR immune receptor MLA10 reveals its cell compartment specific activity in cell death and disease resistance. PLoS Pathog 8:e1002752

12. Sohn KH, Lei R, Nemri A et al (2007) The downy mildew effector proteins ATR1 and ATR13 promote disease susceptibility in *Arabidopsis thaliana*. Plant Cell 19:4077–4090

13. Rentel MC, Leonelli L, Dahlbeck D et al (2008) Recognition of the *Hyaloperonospora parasitica* effector ATR13 triggers resistance against oomycete, bacterial, and viral pathogens. Proc Natl Acad Sci USA 105:1091–1096

14. Whisson SC, Boevink PC, Moleleki L et al (2007) A translocation signal for delivery of oomycete effector proteins into host plant cells. Nature 450:115–118

15. Sharma S, Sharma S, Hirabuchi A et al (2013) Deployment of the *Burkholderia glumae* type III secretion system as an efficient tool for translocating pathogen effectors to monocot cells. Plant J. doi:10.1111/tpj.12148

16. Casper-Lindley C, Dahlbeck D, Clark ET et al (2002) Direct biochemical evidence for type III secretion-dependent translocation of the AvrBs2 effector protein into plant cells. Proc Natl Acad Sci USA 99:8336–8341

17. Schechter LM, Roberts KA, Jamir Y et al (2004) *Pseudomonas syringae* type III secretion system targeting signals and novel effectors studied with a Cya translocation reporter. J Bacteriol 186:543–555

18. Triplett LR, Melotto M, Sundin GW (2009) Functional analysis of the N terminus of the *Erwinia amylovora* secreted effector DspA/E reveals features required for secretion, translocation, and binding to the chaperone DspB/F. Mol Plant-Microbe Interact 22:1282–1292

19. Thomas WJ, Thireault CA, Kimbrel JA et al (2009) Recombineering and stable integration of the Pseudomonas syringae pv. syringae 61 hrp/hrc cluster into the genome of the soil bacterium Pseudomonas fluorescens p f0-1. Plant J 60:919–928

20. Upadhyaya NM, Mago R, Staskawicz BJ et al (2014) A bacterial type III secretion assay for delivery of fungal effector proteins into wheat. Mol Plant-Microbe Interact. http://dx.doi.org/10.1094/MPMI-07-13-0187-FI

21. Sambrook J, Fritsch EF, Maniatis T (1989) Molecular cloning: a laboratory manual, 2nd edn. Cold Spring Harbor Laboratory Press, New York

22. Thordal-Christensen H, Zhang Z, Wei Y et al (1997) Subcellular localization of H_2O_2 in plants: H_2O_2 accumulation in papillae and hypersensitive response during the barley-powdery mildew interaction. Plant J 11:1187–1194

23. Kovach ME, Elzer PH, Hill DS et al (1995) Four new derivatives of the broad-host-range cloning vector pBBR1MCS, carrying different antibiotic-resistance cassettes. Gene 166:175–176

24. Nimchuk Z, Marois E, Kjemtrup S et al (2000) Eukaryotic fatty acylation drives plasma membrane targeting and enhances function of several type III effector proteins from *Pseudomonas syringae*. Cell 101: 353–363

25. Huynh TV, Dahlbeck D, Staskawicz BJ (1989) Bacterial blight of soybean: regulation of a pathogen gene determining host cultivar specificity. Science 245:1374–1377

26. Stauber JL, Loginicheva E, Schechter LM (2012) Carbon source and cell density-dependent regulation of type III secretion system gene expression in *Pseudomonas syringae* pathovar tomato DC3000. Res Microbiol 163:531–539

Chapter 22

Genomic DNA Library Preparation for Resistance Gene Enrichment and Sequencing (RenSeq) in Plants

Florian Jupe, Xinwei Chen, Walter Verweij, Kamel Witek, Jonathan D.G. Jones, and Ingo Hein

Abstract

Enrichment of genomic DNA for genes of interest prior to high-throughput sequencing offers an efficient and cost-effective approach to reduce genome complexity. Target enrichment typically yields higher read-depth for selected genes and is therefore suitable for determination of sequence polymorphisms and enables multiplexing of samples. Target enrichment also provides a means to annotate specific gene families within the sequenced organisms without the requirements for gene models. Here we describe enrichment procedures for NB-LRR-type plant resistance genes that can, for example, be used to establish the NB-LRR gene complements of individual plants and to map resistances more rapidly using a bulked segregant analysis.

Key words Agilent SureSelect, Target enrichment, High-throughput sequencing, NB-LRR genes

1 Introduction

Recent advances in genome sequencing technologies have led to a dramatic reduction in the associated costs, facilitating rapid analysis of entire crop genomes. These genomes provide a blueprint to identify new genes for important traits, including disease resistance. In all plants, disease resistance can be mediated by resistance (R) proteins that contain nucleotide binding (NB) and leucine-rich repeat (LRR) domains and are collectively known as NB-LRRs [1]. Their presence, absence or allelic identity is decisive for host disease resistance and, in addition, they are thought to be important contributors to non-host resistance [2].

A recent study in potato identified 438 NB-LRR genes in the sequenced *Solanum tuberosum* group Phureja clone DM1-3 516 R44 (DM) and described their phylogenetic relationship as well as their physical locations on the 12 potato chromosomes [3, 4].

Florian Jupe, Xinwei Chen, and Walter Verweij contributed equally to this manuscript.

Paul Birch et al. (eds.), *Plant-Pathogen Interactions: Methods and Protocols*, Methods in Molecular Biology, vol. 1127, DOI 10.1007/978-1-62703-986-4_22, © Springer Science+Business Media New York 2014

This study formed the basis of a novel *R* gene enrichment and sequencing platform (RenSeq) [5]. The sequences of the re-annotated potato and tomato NB-LRRs [6], alongside functionally validated pepper resistance genes, were used to design over 48,500 NB-LRR gene-specific biotinylated RNA probes, which were subsequently ordered as a customized Agilent SureSelect® target enrichment kit [7]. The enrichment for NB-LRR genes was used to significantly reduce the genome complexity of wild *Solanum* species and within the segregating populations prior to Illumina GAII 76 bp paired-end sequencing. When applied to the sequenced clone DM, over 300 additional NB-LRR containing regions were found [5]. This approach might thus be used to re-annotate the NB-LRR gene complements from at least partially available genomes without reference gene models. A comparison between the newly identified NB-LRRs and the corresponding baits with the highest sequence similarity demonstrated that approximately 80 % identity is sufficient for enrichment. In the same study, RenSeq was also successfully utilized to determine the genetic map position for novel resistances using bulked resistant (BR) and bulked susceptible (BS) samples from segregating populations.

As high-throughput sequencing technologies produce longer and more accurate reads, the assembly of entire NB-LRR genes, and thus a way to discriminate between highly similar allelic and paralogous sequences, is becoming feasible. RenSeq can be used to establish the smallest number of high-quality contigs, based on long sequencing reads, that best describe the NB-LRRs in plants with contrasting disease resistance phenotypes. This can then be followed by an NB-LRR-specific bulked segregant analysis (BSA) to identify *R* gene candidates that are specific to the resistant parent and bulk, and not present in the susceptible parent and corresponding bulk.

Here we describe a protocol that is largely based on the Agilent SureSelect procedure and is suitable for the generation of target-enriched genomic DNA libraries with an average insert length of 400 bp. The method can be adapted to enable enrichment of fragments that are larger than 1 kb and which are, for example, appropriate for PacBio sequencing technology. PacBio in combination with other platforms such as paired-end MiSeq sequencing can yield error corrected, long and accurate sequencing reads for downstream analysis [8].

2 Materials

2.1 Reagents

1. Customized RNA bait library (*see* Subheading 3.1 for bait design; Agilent SureSelect Technologies).
2. AMPure® XP Beads (Beckman Coulter, Inc. #A63881).
3. NEBNext Ultra DNA library Prep Kit (New England BioLabs, E7370S).

4. NEBNext Singleplex or Multiplex Oligos for Illumina (E7350, E7335 or E7500).

5. Nuclease-free water (Ambion Cat #AM9930).

6. PCE—plant capture enhancer (http://www.nimblegen.com/products/lit/06703526001_SeqCapEZ_SoybeanTechNote_0512.pdf).

7. Dynabeads magnetic beads (Invitrogen Cat #11205D).

8. Quant-iT PicoGreen Reagent (Invitrogen Cat # P11496).

2.2 Equipment Required

1. Equipment for DNA shearing: Covaris.

2. Equipment for electrophoresis and gel viewing.

3. PCR and qPCR machine.

4. Vacuum concentration device.

5. Magnetic tube rack.

6. Thermomixer.

7. Fluorometer.

3 Methods

3.1 Design of a Customized Bait Library for NB-LRR Gene Enrichment

Target enrichment-specific, biotinylated RNA-based probes ("baits") with a length of 120 nt that correspond to previously established NB-LRRs [3–6] can be designed using the freely available software eArray (https://earray.chem.agilent.com) from Agilent Technologies. Overlapping baits are typically designed over the sequence of interest (e.g. exons or a whole genomic region) allowing for a 60 nt overlap to neighboring baits to facilitate 2× coverage. Regions with ambiguous nucleotides will be discarded by eArray, alongside repeat-rich regions. As a precaution we routinely conduct a Blast search [9] of the designed baits against the entire genome and remove baits that yield an aberrantly high number of potential target sites.

3.2 DNA Extraction and Preparation

We typically extract >3 μg of high quality genomic DNA (gDNA), which is sufficient to yield 250–500 ng of size selected, PCR amplified library material for the enrichment procedure (Subheading 3.5). Any protocol resulting in high quality DNA can be used for the DNA extraction. We routinely use the Qiagen DNeasy Plant Mini Kit for extraction of DNA from 100 mg of young and newly emerging leaf tissue. Extracted DNA is eluted from the columns in not more than 100 μl of elution buffer AE. The DNA quantity and quality is checked on a Nanodrop and by running 2 μl of the eluted sample on a 1 % agarose gel. High quality gDNA samples should have an A_{260}/A_{280} ratio of 1.8–2.0 indicating the absence of contaminating proteins, and an A_{260}/A_{230} ratio of >2.0, indicating the absence of other organic compounds. High molecular weight genomic DNA viewed on a gel should appear as a single band.

3.3 DNA Fragmentation

We have successfully used a Covaris sonicator for fragmentation and sizing of genomic DNA. The protocol is detailed below:

1. Prepare the Covaris sonicator as detailed in the corresponding user manual. Degas and cool the water to a temperature of 4 °C.

2. Open the SonoLab Single Software and use the recommended settings based on the desired target size range for shearing the DNA.

3. Adjust the volume of the >3 µg of extracted gDNA to a total volume of 100 µl with water.

4. Mix and transfer the 100 µl of the gDNA to an opened Covaris microTube.

5. Secure the microTube in the tube holder.

6. Hit "Start" to shear the DNA with the settings detailed in Table 1 to obtain an average fragment size of approximately 400 bp.

7. While keeping the snap-cap on, insert a pipette tip through the pre-split septa, and then slowly transfer the sheared DNA into a fresh 1.5 ml LoBind tube. Transfer the tube onto ice and restart the cycle for the next sample.

8. Take a 5 µl aliquot of the sheared gDNA and separate on a 1 % agarose gel to check the size range.

9. If the appropriate size range has not been achieved, repeat steps **1–8**.

3.4 Library Preparation

The generation of gDNA libraries prior to enrichment comprises purification of sheared gDNA, end repair, and adapter ligation. The adapters are specific to the chosen sequencing platform and can be used to incorporate barcodes if multiplexing is desired. A subsequent second purification step allows for further size selection. During a PCR-step the Illumina y-shaped adapters are brought into the right conformation. Finally, the library can be further size selected on an agarose gel, followed by purification and adjustment of the concentration.

Table 1
Covaris shearing parameters for 400 bp insert libraries

Fragment size	400 bp
Duty cycle	20
Intensity	5
Cycle/burst	200
Time (s)	30
Temperature of bath	4 °C

<table>
<tr><td>3.4.1 Purification of Sheared gDNA Using AMPure® XP Beads</td></tr>
</table>

3.4.1 Purification of Sheared gDNA Using AMPure® XP Beads

1. Bring the AMPure® XP beads to room temperature and homogenize by vortexing.

2. Add an equal volume (95 µl) of AMPure® XP beads (1×) to the 95 µl of sheared gDNA (Subheading 3.3), mix by vortexing and incubate at room temperature for 5 min.

   Briefly centrifuge the tube to collect solution from the sides of the tube. Be careful not to pellet the magnetic beads. Place the tube in a magnetic stand to separate the beads from the supernatant. Wait for 5 min for the solution to become clear. Carefully transfer the supernatant to a fresh tube without disturbing the beads (*see* **Note 1**).

3. Add 200 µl of freshly prepared 80 % ethanol to the tube with the beads whilst positioned in the magnetic stand. Incubate at room temperature for 30 s to let the beads settle down and then carefully remove and discard the supernatant. Repeat the wash with freshly prepared 80 % ethanol a further two times.

4. Carefully remove the residual ethanol using a small-volume (10 µl) pipette after the third wash. Air dry the beads for 10 min while the tube remains in the magnetic stand with the lid open.

5. Elute the DNA from the beads by adding 60 µl 0.1× TE. Mix by vortexing, briefly centrifuge the tube to collect the solution from the sides and lid of the tube. Place tube in the magnetic stand for 5 min to collect the beads and wait until the solution becomes clear.

6. Transfer 56 µl of the supernatant which contains the eluted DNA to a fresh 1.5 ml tube.

3.4.2 End Repair of the Sheared gDNA

1. Set up the end repair reaction using the NEBNext Ultra DNA library Prep Kit (New England BioLabs, E7370S), by mixing the following components in a sterile, nuclease-free PCR tube:

End Prep Enzyme Mix	3.0 µl
10× End Repair Reaction Buffer	6.5 µl
Sheared DNA	55.5 µl
Total volume	65 µl

2. Mix the components by pipetting, followed by a brief centrifugation step to collect all liquid from the sides of the tube.

3. Transfer the tube to a thermocycler, with the "heated lid" option engaged and run the following program:

   20 °C for 30 min.

   60 °C for 30 min.

   Hold at 4 °C.

3.4.3   Adapter Ligation

1. Add the following components from the NEBNext Ultra DNA library Prep Kit (New England BioLabs, E7370S) directly to the 65 μl of the end repair reaction and mix well by pipetting. For adapter sequence, *see* **Note 2**:

Blunt/TA ligase Master Mix	15 μl
NEBNext Adapter for Illumina	2.5 μl
Ligation Enhancer	1.0 μl
Total volume	83.5 μl

2. Briefly centrifuge the sample to collect liquid from the sides of the tube before incubating the reaction mixture at 20 °C for 15 min in a thermal cycler with the "heated lid" option enabled.

3. Add 3 μl of USER™ enzyme to the ligation mixture, mix well by pipetting followed by a brief centrifugation to collect liquid from the sides of the tube.

4. Place the reaction in a thermal cycler and incubate at 37 °C for 15 min, with the "heated lid" option enabled.

3.4.4   Purification of Adapter-Ligated gDNA Using AMPure® XP Beads with Size Selection

1. Transfer the 86.5 μl ligation reaction into a fresh 1.5 ml tube and adjust the volume to 100 μl dH$_2$O.

2. The recovered, size selected and adapter-ligated products are purified using 1× (100 μl) AMPure® XP beads as described in Subheading 3.4.1. DNA should be eluted in 28 μl of 0.1× TE.

3. Transfer 22 μl to a fresh PCR tube for amplification without disturbing the beads.

3.4.5   PCR Amplification of the Purified Adapter-Ligated DNA to get the Adapters into the Right Conformation

For primer sequences, *see* **Note 2**:

1. Mix the following components of the NEBNext Singleplex or Multiplex Oligos for Illumina (E7350, E7335 or E7500) in the PCR tube:

Adapter-ligated DNA fragments	22 μl
NEBNext high Fidelity 2× PCR Master Mix	25 μl
Index primer	1 μl
Universal PCR primer	2 μl
Total volume	50 μl

2. Run the PCR using the cycling conditions detailed in Table 2 (*see* **Note 3**).

3. Run 4 μl of the PCR product on a 1.5 % agarose gel to ensure that sufficient DNA has been amplified.

Table 2
PCR amplification of the purified adapter-ligated DNA
to get the adapters into the right conformation

Conditions according to amplicon size		400 bp
98 °C		30 s
98 °C		10 s
65 °C	6–15 cycles	30 s
72 °C		30 s
72 °C		5 min
4 °C		∞

3.4.6 Size Selection of Fragments on a 1.5 % Agarose Gel

1. Prepare a 1.5 % agarose gel with 1× TAE and with wells large enough to enable loading of the remaining PCR mix.

2. Load a standard molecular weight DNA ladder and the entire PCR samples leaving at least a gap of one empty well between samples to avoid cross contamination. Optimize electrophoresis parameters according to the size of the gel. Run the gel for long enough to sufficiently separate the ladder (usually 30 min at 100 V).

3. Visualize the DNA under UV light and identify the region of the gel that contains the DNA fragments of the required size (initial gDNA fragment size + the adapters). Use a clean scalpel or razor blade to excise the appropriate region for DNA recovery.

4. Recover DNA from the excised gel using a Qiagen Gel Extraction kit and elute DNA in 50 μl elution buffer.

3.4.7 Purification and Concentration of PCR Amplified and Size-Selected Libraries Using AMPure® XP Beads

The recovered, size selected PCR products are purified using 1× (50 μl) AMPure® XP beads as described in Subheading 3.4.1. DNA is eluted in 10 μl of 0.1× TE.

3.4.8 Measurement and Adjustment of the Sample Concentrations

Concentrations of the purified and size selected libraries are established using a NanoDrop. Typically, a yield of 250–300 ng (25–30 ng/μl) per sample is obtained from 1 μg of gDNA (Subheading 3.2). If required, independently created libraries from identical plants or, for pooling strategies, distinct libraries with individual indexes (barcodes) can be pooled in equimolar amounts. For full target enrichment reactions (Subheading 3.5), 500 ng of the prepared DNA library is required whereas 250 ng is sufficient for half reactions. At this stage, the volume of the sample is not critical (*see* **Note 4**).

3.5 Target Enrichment Using the Agilent SureSelect Kit

The enrichment of target DNA fragments is achieved through hybridisation of the PCR amplified genomic libraries generated in Subheading 3.4 with complementary RNA (cRNA) baits designed in Subheading 3.1. These reactions can be carried out in a "full" reaction volume as suggested by the manufacturer for 500 ng library DNA, or in a "half" reaction volume for 250 ng of DNA (*see* **Note 5**). The following protocol describes "half" reaction volumes. All reagents, unless stated otherwise, are part of the Agilent SureSelect kit.

3.5.1 Set Up a SureSelect Blocking Mix

Add the following components in a 1.5 ml tube (*see* **Note 6**):

Prepped library (Subheading 3.4)	250 ng
PCE (Roche; **Note 7**)	5 μl
Hyb-Block #3	0.6 μl

3.5.2 Adjust the Volume for the Hybridisation Reaction

Mix by vortexing and then place in any vacuum concentrator device, set to 40 °C to dry the mixture to a final volume of 4.5 μl.

3.5.3 Prepare Hybridization Buffer

Add the following components at room temperature in a 1.5 ml tube. The following volumes are required for each enrichment:

SureSelect Hyb # 1	25 μl
SureSelect Hyb # 2	1 μl
SureSelect Hyb # 3	10 μl
SureSelect Hyb # 4	13 μl

Mix the components by vortexing the samples followed by a quick centrifugation step to collect the liquid at the bottom of the tube.

Do NOT place on ice.

3.5.4 Prepare the SureSelect RNA Bait Library Mix for Target Enrichment

1. Prepare a dilution of the RNase Block by adding an equal volume of nuclease-free water (0.25 μl RNase Block + 0.25 μl water per reaction).

2. Add 0.5 μl of diluted RNase block to 2.5 μl of the SureSelect bait library and keep on ice.

3.5.5 Prepare the Hybridization Mix

1. Incubate the size selected, blocked gDNA library in a thermocycler at 95 °C for 5 min then hold at 65 °C, with the "heated lid" option enabled.

2. Incubate both the RNase blocked SureSelect bait library (Subheading 3.5.4) at 65 °C for 2 min and the hybridization buffers prepared in Subheading 3.5.3 at 65 °C for 5 min.

3. Whilst keeping the samples at 65 °C, add 6.5 μl hybridization buffer to the blocked SureSelect bait library and mix by pipetting carefully up and down.

4. Add the 9.5 μl of this mix to the 4.5 μl of prepped genomic DNA library (Subheading 3.5.2) to obtain a final volume of 14 μl. Do NOT remove the mixture from the thermocycler.

5. Incubate the hybridisation mixture at 62–65 °C for 24–36 h (*see* **Note 8**).

3.5.6 Selection for RNA Baits-Target Genomic DNA Hybrids

1. Pre-warm SureSelect Wash Buffer #2 to 65 °C.

2. Vigorously re-suspend Dynabeads (Invitrogen Cat #11205D) and bring to room temperature.

3. For each hybridization add 50 μl Dynabeads to a 1.5 ml tube, and wash the beads in total three times following steps (a–d):

   (a) Add 200 μl SureSelect Binding Buffer.

   (b) Mix the beads on a vortex mixer for 5 s.

   (c) Separate beads on a magnetic rack.

   (d) Remove and discard supernatant
       Re-suspend the beads in 200 μl SureSelect Binding Buffer.

4. Whilst keeping the hybridisation mix at 62–65 °C, add 14 μl Hybridization Buffer from Subheading 3.5.3 (also incubated at 62–65 °C in Subheading 3.5.5) to the hybridization mixture.

5. Add the hybridization mixture from the thermocycler directly to the 200 μl Dynabeads suspension. Invert tubes 3–5 times.

6. Incubate for 30 min at room temperature (21 °C) on a thermomixer, moving vigorously.

7. Separate the Dynabeads and supernatants for 5 min in a magnetic rack. Remove and discard the supernatant. The magnetic beads contain the captured library.

8. Re-suspend magnetic beads with 500 μl SureSelect Wash Buffer #1 and mix by vortexing. Incubate at room temperature (21 °C) for 15 min in a shaking thermomixer or vortex at regular intervals.

9. Separate the beads from the buffer in a magnetic rack. Remove and discard the supernatant.

10. Re-suspend magnetic beads in 500 μl of the pre-warmed SureSelect Wash Buffer #2 and wash the beads a total of three times following steps (a–c):

    (a) Mix the beads in 500 μl of the pre-warmed SureSelect Wash Buffer #2 by vortexing.

    (b) Incubate at 65 °C for 10 min on a shaker block.

    (c) Separate beads and buffer in a magnetic rack. Remove and discard the supernatant.

11. Elute the captured and washed library by adding 50 μl of SureSelect elution buffer. Mix by vortexing.

12. Incubate at room temperature for 10 min.

13. Briefly centrifuge the sample to collect liquid from the sides of the tube and then place in a magnetic stand for 5 min to separate the beads and buffer. The buffer now contains the captured and eluted DNA library.

14. Carefully move the supernatant to a fresh 1.5 ml tube without disturbing the beads.

15. Add 50 μl of SureSelect Neutralization Buffer and mix by vortexing.

3.5.7 Desalt the Captured Solution Using AMPure® XP Beads

1. Desalting the capture solution is performed by adding 1× AMPure® XP beads (100 μl) to the captured library followed by standard AMPure® XP beads purification as described in Subheading 3.4.1. The desalted and enriched library is eluted in 30–50 μl nuclease-free water.

2. Concentrate the eluted library to a volume of 10–20 μl using the previously applied vacuum protocol (Subheading 3.5.2).

3.6 Post-capture Amplification

To run the sample in one lane of an Illumina GAII flowcell, a total of 6 pM NaOH denatured library is required as input. Amplifying the captured library to a total of 20–50 ng therefore provides sufficient input material not only for the sequencing reaction but also for a qPCR-based quality control (Subheading 3.8).

1. Set up a test PCR amplification using 5 μl of enriched library by mixing the following components:

5× Q5 buffer	5 μl
10 mM dNTP	0.5 μl
Corresponding primers (*see* **Note 2**)	2.5 μl
Q5 high-fidelity DNA polymerase	0.25 μl
Enriched library DNA	5 μl
H₂O	11.75 μl
Total volume	25 μl

2. Run a PCR program described earlier in Subheading 3.4 using different numbers of PCR cycles* (*see* **Note 3**).

3. Amplify the remaining library using the optimal number of PCR cycles established in **step 2** above.

4. Purification of the amplified enriched library is conducted by using 1× AMPure® XP Beads as described earlier in Subheading 3.4.1.

3.7 Quantification of Enriched Libraries

To precisely measure the concentration of the captured and amplified DNA libraries, we use the Quant-iT PicoGreen Reagent from Invitrogen (Cat. No. P11496), but other fluorometric techniques are also suitable. Measurements should be performed in duplicate to ensure precise results, using black 96-well plates.

1. Dilute concentrated (200×) PicoGreen in 1× TE to a 2× concentration

2. Prepare a standard curve in 100 μl of 1× TE using DNA of known concentration provided with the PicoGreen Reagents. The standard curve should encompass 0, 1, 2.5, 5, 10, 25, 50 and 100 ng of DNA standard.

3. Prepare your samples by mixing 2 μl of amplified library from Subheading 3.6 and 98 μl of 1× TE

4. Add 100 μl of diluted (2×) PicoGreen Reagent to reach the final concentration of 1× in 200 μl.

5. Measure in Fluorescence Microplate Reader with the following program:

   Shake 20 s, 600 rpm, diameter 1 mm.

   Pause 3 min.

   Fluorescence reading: Exc. 485 nm; Emit. 535 nm.

6. Draw a standard curve of the dilution series and calculate the concentration of your samples based on this reference.

3.8 Enrichment Estimate Using qPCR

To assess the target enrichment efficiency, in our case captured NB-LRR gene sequences, we perform a qPCR based on the amplification of conserved NB-LRR domains such as the NB-ARC domains. In this example, three conserved NB-LRR gene fragments were amplified (*see* Table 3 for sequences). We normalize results to the amount of input DNA, rather than 18S, as during the enrichment the relative amount of 18S changes. We usually use

Table 3
Primers used for qPCR

Target	Sequence 5′–3′
NB-ARC forward	ACGAATTCGTTGTTGGTAGAGACAAAGATG
NB-ARC reverse	ACGGATCCGCTCTTAGTTTCTGACATTTCAGG
R3a NB-ARC forward	ACGAATTCAGAGCAGTCTTGAAGGTTGGAGC
R3a NB-ARC reverse	ACGGATCCATCTCCTTTCCGATTGCCACAAGG
R2 NB-ARC forward	ACGAATTCCAGCAGAGTCATTATTACCACG
R2 NB-ARC reverse	ACGGATCCAAGTAGTCCGCTCAATACAACAATTGC

Table 4
qPCR cycling conditions

95 °C		30 s
95 °C		10 s
55 °C	40 cycles	30 s
72 °C		30 s
72 °C		5 min
Melting curve 65–95 °C, increment 0.5 °C		5 s

SYBR Green JumpStart Taq ReadyMix from Sigma (Cat. No. S4438-500RXN) to setup reactions.

1. Use 1 ng of DNA before and after enrichment as input for the qPCR and perform three replicates per sample to ensure accurate quantification.

2. Set up a qPCR amplification by mixing the following components:

2× SYBR buffer	10 µl
5 µM primer forward	1 µl
5 µM primer reverse	1 µl
1 ng of DNA	1 µl
H_2O	7 µl
Total volume	20 µl

3. Run the program detailed in Table 4 and calculate ΔCt between enriched and non-enriched samples. Typically, we observe a difference between 7 and 10 ΔCt (128–1,024 fold enrichment for NB-LRR sequences).

4   Notes

1. Store the supernatant on ice until successful DNA recovery has been confirmed.

2. Sequences for Adapters and primers used with the NEBNext Ultra DNA Library Prep Kit and the NEBNext Multiplex Oligos for Illumina can be found under NEB catalog number E7335S.

3. It is recommended to start with a lower amplification cycle number, to assess the quantity on an agarose gel and, if

required, conduct additional cycles. The enriched library should not be over-amplified at this step.

4. The volume will be adjusted after the blocking agents have been added by using a vacuum concentrator.

5. We routinely use half reactions for enrichment procedures.

6. When handling small volumes for multiple samples, prepare a master mix to minimize pipetting errors.

7. The Roche proprietary plant capture enhancer (PCE) reagent increases enrichment specificity during the hybridization process and is used as a substitute to species specific Cot-1 DNA, which represents the repetitive fraction of a genome [10, 11]. The PCE material is equivalent to the Roche SeqCap EZ Developer Reagent.

8. We have used the lower temperature for hybridisations between baits and prepped gDNA from more distantly related plants.

References

1. Meyers BC, Dickerman AW, Michelmore RW, Sivaramakrishnan S, Sobral BW, Young ND (1999) Plant disease resistance genes encode members of an ancient and diverse protein family within the nucleotide-binding superfamily. Plant J 20:317–332

2. Schulze-Lefert P, Panstruga R (2011) A molecular evolutionary concept connecting nonhost resistance, pathogen host range, and pathogen speciation. Trends Plant Sci 16: 117–125

3. Potato Genome Sequencing Consortium (2011) Genome sequence and analysis of the tuber crop potato. Nature 475:189–195

4. Jupe F, Pritchard L, Etherington GJ, Mackenzie K, Cock PJ, Wright F, Sharma SK, Bolser D, Bryan GJ, Jones JD, Hein I (2012) Identification and localisation of the NB-LRR gene family within the potato genome. BMC Genomics 13:75

5. Jupe F, Witek K, Verweij W, Sliwka J, Pritchard L, Etherington GJ, Maclean D, Cock PJ, Leggett RM, Bryan GJ, Cardle L, Hein I, Jones JD (2013) Resistance gene enrichment sequencing (RenSeq) enables reannotation of the NB-LRR gene family from sequenced plant genomes and rapid mapping of resistance loci in segregating populations. Plant J 76:530–544

6. Tomato Genome Consortium (2012) The tomato genome sequence provides insights into fleshy fruit evolution. Nature 485:635–641

7. Gnirke A, Melnikov A, Maguire J, Rogov P, LeProust EM, Brockman W, Fennell T, Giannoukos G, Fisher S, Russ C, Gabriel S, Jaffe DB, Lander ES, Nusbaum C (2009) Solution hybrid selection with ultra-long oligonucleotides for massively parallel targeted sequencing. Nat Biotechnol 27:182–189

8. Koren S, Schatz MC, Walenz BP, Martin J, Howard JT, Ganapathy G, Wang Z, Rasko DA, McCombie WR, Jarvis ED, Adam MP (2012) Hybrid error correction and de novo assembly of single-molecule sequencing reads. Nat Biotechnol 30:693–700

9. Altschul SF, Gish W, Miller W, Myers EW, Lipman DJ (1990) Basic local alignment search tool. J Mol Biol 215:403–410

10. Haun WJ, Hyten DL, Xu WW, Gerhardt DJ, Albert TJ, Richmond T, Jeddeloh JA, Jia G, Springer NM, Vance CP, Stupar RM (2011) The composition and origins of genomic variation among individuals of the soybean Reference Cultivar Williams 82. Plant Physiol 155:645–655

11. Bolon YT, Haunm WJ, Xum WW, Grant D, Stacey MG, Nelson RT, Gerhardt DJ, Jeddeloh JA, Stacey G, Muehlbauer GJ, Orf JH, Naeve SL, Stupar RM, Vance CP (2011) Phenotypic and genomic analyses of a fast neutron mutant population resource in soybean. Plant Physiol 156:240–253

INDEX